THE GULAG ARCHIPELAGO

SELECTED WORKS BY
ALEKSANDR I. SOLZHENITSYN

August 1914
Cancer Ward
Candle in the Wind
The First Circle
The Gulag Archipelago 1918–1956
Invisible Allies
Lenin in Zurich
Letter to the Soviet Leaders
The Love-Girl and the Innocent
The Mortal Danger
Nobel Lecture
November 1916
The Oak and the Calf
One Day in the Life of Ivan Denisovich
Prussian Nights
Rebuilding Russia
"The Russian Question" at the End of the Twentieth Century
Stories and Prose Poems
Warning to the West
A World Split Apart

Aleksandr I. Solzhenitsyn

THE GULAG ARCHIPELAGO

1918–1956

An Experiment in Literary Investigation

*Translated from the Russian by Thomas P. Whitney
(Parts I–IV) and Harry Willets (Parts V–VII)*

Abridged by Edward E. Ericson, Jr.

Foreword by Anne Applebaum

HARPER**PERENNIAL** ● MODERN**CLASSICS**

NEW YORK ● LONDON ● TORONTO ● SYDNEY

HARPER**PERENNIAL** ● MODERN**CLASSICS**

Designed by Sidney Feinberg

The Library of Congress has catalogued the previous edition as follows:

Solzhenitsyn, Aleksandr Isayevich.
 [Arkhipelag Gulag, 1918–1956. English]
 The Gulag Archipelago 1918–1956 / Aleksandr I. Solzhenitsyn ; translated from the Russian by Thomas P. Whitney and Harry Willetts ; abridged by Edward E. Ericson, Jr.—Abridged ed.
  p. cm.—(Perennial classics)
 Previously published: New York : Harper & Row, 1985.
 ISBN 0-06-000776-1 (pbk.)
 1. Prisons—Soviet Union. 2. Political prisoners—Soviet Union. 3. Concentration camps—Soviet Union. I. Title. II. Perennial classic.
HV9713.S6413 2002
365'.45'0947—dc21   2001046504

ISBN: 978-0-06-125380-5 (reissue)
ISBN-10: 0-06-125380-4 (reissue)

24 25 26 27 28 LBC 38 37 36 35 34

I dedicate this
to all those who did not live
to tell it.
And may they please forgive me
for not having seen it all
nor remembered it all,
for not having divined all of it.

Contents

Author's Note

In this book there are no fictitious persons, nor fictitious events. People and places are named with their own names. If they are identified by initials instead of names, it is for personal considerations. If they are not named at all, it is only because human memory has failed to preserve their names. But it all took place just as it is here described.

This book could never have been created by one person alone. In addition to what I myself was able to take away from the Archipelago —on the skin of my back, and with my eyes and ears—material for this book was given me in reports, memoirs, and letters by 227 witnesses, whose names were to have been listed here.

What I here express to them is not personal gratitude, because this is our common, collective monument to all those who were tortured and murdered.

From among them I would like to single out in particular those who worked hard to help me obtain supporting bibliographical material from books to be found in contemporary libraries or from books long since removed from libraries and destroyed; great persistence was often required to find even one copy which had been preserved. Even more would I like to pay tribute to those who helped me keep this manuscript concealed in difficult periods and then to have it copied.

But the time has not yet come when I dare name them.

The old Solovetsky Islands prisoner Dmitri Petrovich Vitkovsky was to have been editor of this book. But his half a lifetime spent *there* —indeed, his own camp memoirs are entitled "Half a Lifetime"—

resulted in untimely paralysis, and it was not until after he had already been deprived of the gift of speech that he was able to read several completed chapters only and see for himself that everything *will be told.*

And if freedom still does not dawn on my country for a long time to come, then the very reading and handing on of this book will be very dangerous, so that I am bound to salute future readers as well—on behalf of *those* who have perished.

When I began to write this book in 1958, I knew of no memoirs nor works of literature dealing with the camps. During my years of work before 1967 I gradually became acquainted with the *Kolyma Stories* of Varlam Shalamov and the memoirs of Dmitri Vitkovsky, Y. Ginzburg, and O. Adamova-Sliozberg, to which I refer in the course of my narrative as literary facts known to all (as indeed they someday shall be).

Despite their intent and against their will, certain persons provided invaluable material for this book and helped preserve many important facts and statistics as well as the very air they breathed: M. I. Sudrabs-Latsis, N. V. Krylenko, the Chief State Prosecutor for many years, his heir A. Y. Vyshinsky, and those jurists who were his accomplices, among whom one must single out in particular I. L. Averbakh.

Material for this book was also provided by *thirty-six* Soviet writers, headed by *Maxim Gorky,* authors of the disgraceful book on the White Sea Canal, which was the first in Russian literature to glorify slave labor.

Foreword

Although more than three decades have now passed since the winter of 1974, when unbound, hand-typed, samizdat manuscripts of Aleksandr Solzhenitsyn's *Gulag Archipelago* first began circulating around what was then the Soviet Union, the emotions the book stirred have left marks which remain today. Usually, readers were given only twenty-four hours to finish the lengthy manuscript before it had to be passed on to the next person. That meant spending an entire day and a whole night absorbed in Solzhenitsyn's prose—not an experience anyone was likely to forget. Members of that first generation of readers remember who gave the book to them, who else knew about it, whom they passed it on to next. They remember what the book felt like—the blurry, mimeographed text, the dog-eared paper, the dim glow of the lamp switched on late at night—and with whom they later discussed it.

In part, Russians responded so strongly because *The Gulag Archipelago*'s author was, at that time, simultaneously very famous and strictly taboo. Twelve year earlier, in 1962, Solzhenitsyn had attained an unusual distinction, becoming both the first authentic Gulag author to be published in the official press, as well as the last. In that year—the height of the post-Stalinist "Thaw"—the Soviet general secretary Nikita Khrushchev had personally permitted the publication of Solzhenitsyn's short novel *One Day in the Life of Ivan Denisovich*. The book, based on Solzhenitsyn's own camp experiences—like Ivan Denisovich, he too had been a camp bricklayer—described a single, ordinary day in the life of a Gulag prisoner.

Reading it now, it can be hard for contemporary readers to understand why Solzhenitsyn's only published work had created such a furor in the Soviet literary world. But in 1962, *Ivan Denisovich* came as a revelation. Instead of speaking vaguely about 'repressions,' as some other books did at the time, *Ivan Denisovich* was blunt and specific. The sufferings of its heroes were pointless. The work they did was boring and exhausting, and they tried to avoid it. They spoke using camp slang and were rude to one another. The Party did not triumph at the end of the story, and communism did not win out in the end. This honesty, unusual in an era of morality tales and social realism, won Solzhenitsyn admirers, particularly among camp survivors, who wrote him long letters of praise. Each new printing of the novel sold out instantly, and copies were eagerly shared among groups of friends.

Solzhenitsyn's honesty also quickly won him detractors. Within a month of publication, the novel had already been denounced at a meeting of the Soviet Writers' Union. Critics wrote that it was too bleak, too "amoral." Within a few more months, Solzhenitsyn himself was under personal attack, falsely accused of having surrendered to the Germans during the war, and of having been convicted on criminal charges. He fought back, but to no avail: thanks to the furor caused by this first published novel, none of his work would ever be officially published in the Soviet Union again.

Yet his name and his novels remained in circulation thanks to the world of underground publishing in Russia, which at that time was growing rapidly. In fact, in the years between his first burst of "official" fame and the appearance of *The Gulag*

Archipelago in samizdat form, Solzhenitsyn became if anything more notorious, and more celebrated, despite the official ban. The KGB began to follow him closely, and at one point stole his entire personal archive. His wife lost her job. Recently released archival documents show that his every move was closely analyzed at the highest levels of the Soviet security apparatus, and sometimes even by the Politburo itself. At the same time, his occasional lectures were wildly popular: six hundred people showed up for one of his first public readings in 1966. His books began to appear in foreign translations, to great acclaim, and were copied and re-copied in secret.

Then, in 1970, Solzhenitsyn won the Nobel Prize. Fearing he would be barred from returning to Russia, he decided not to travel to Stockholm to accept the award. But he issued a statement to be read out at the Nobel banquet, among other things noting the "remarkable fact that the day of the Nobel Prize presentation coincides with Human Rights day," and calling on all Nobel Prize winners to remember that fact: "Let none at this festive table forget that political prisoners are on hunger-strike this very day in defence of rights that have been curtailed or trampled underfoot."

The Swedish government was unnerved, and the Nobel Committee failed to read out that part of the statement. The Soviet authorities were furious, and boycotted the ceremony. The Soviet Writers' Union denounced Solzhenitsyn as the darling of "reactionary circles in the West," and reviewers described him as "a run-of-the-mill writer with an exaggerated idea of his own importance" whose literary gifts were "inferior to many of his Soviet contemporaries—writers the West chooses to ignore because it finds the impact of truth in their writing unbearable."

Still, millions of Russians learned of the prize through Western radio, as well as through the underground press (which circulated the statement that the Swedes had feared to read), and celebrated the award to their countryman. Thus when news that Solzhenitsyn had written a history of the Soviet Gulag began to filter out too, there was an enormous reading public—and a listening public, for excerpts were immediately read out on Radio Liberty—already waiting to receive it.

Yet the impression which *The Gulag Archipelago* made on its first Russian readers was not solely due to the author's notoriety, or to his Nobel Prize, or to the denunciations of him in the Soviet press. More importantly, the book's appearance also marked the first time that anyone inside Russia had ever tried to write a complete history of the Soviet concentration camps, using what information was then available, mostly the "reports, memoirs and letters by 227 witnesses," whom Solzhenitsyn cites in his introduction. Many knew fragments of the story, from the cousin who had been there or the neighbor's nephew who worked in the police. No one, however, had attempted to put it all together, to tell, in effect, an alternative history of the Soviet Union.

And the result was unique. Solzhenitsyn called *The Gulag Archipelago* an "experiment in literary investigation," and that remains the best description of a work which is otherwise impossible to categorize. The book is not quite a straight history— obviously, Solzhenitsyn did not have access to archives or historical records—and large sections are autobiographical. Solzhenitsyn describes in great detail his own arrest and interrogation, his first prison cell, and, courageously, his flirtation with camp police who asked him to serve as an informer. Other parts of the book rely heavily on the words and experiences of others, including some of Solzhenitsyn's camp friends, as well as many people he did not know but who wrote to him after the publication of *Ivan Denisovich*. Still other sections are based on Solzhenitsyn's research into what sources were available: legal tomes, official histories, and the Soviet press.

But all of the material was then filtered through Solzhenitsyn's unique sensibility, and retold in a style which was simultaneously angry, prophetic, ironic—and always opinionated. Thus *The Gulag Archipelago* is a history, but it is also an interpretation of history, and one which many at first found shocking. Up until the publication of *Gulag*, many in Russia and elsewhere were content to blame Stalin for Soviet terror, concentration camps, and mass arrests. Solzhenitsyn argued, and with real evidence, that Lenin, not Stalin, was responsible for creating the Gulag, and that the first Soviet concentration camps for political prisoners were built in the 1920s, not the 1930s. He also showed that the famous "great purge" of the 1930s, during which many leaders of the Bolshevik Revolution were put on public trial and then executed, was no aberration. In reality, it was only one of the many "waves which strained the murky, stinking pipes of our prison sewers to bursting," and not even the largest at that: far more people were killed during the era of mass collectivization, and the Gulag population actually reached its zenith a decade later, at the end of the 1940s and in the early 1950s.

Most importantly, Solzhenitsyn aimed to show that, contrary to what many believed, the Gulag was not an incidental phenomenon, something which the Soviet Union could eventually eliminate or outgrow. Rather, the prison system had been an essential part of the Soviet economic and political system from the very beginning. "We never did have empty prisons," he wrote, "merely prisons which were full or prisons which were very, very overcrowded." In fact, *The Gulag Archipelago* was intended to serve as a condemnation not just of the Soviet camp system, but of the Soviet Union itself. It succeeded—so much so that Soviet authorities decided they could no longer tolerate Solzhenitsyn's presence at all. As a result of its publication, not just in Russia but in multiple foreign countries, Solzhenitsyn was expelled from the country. He would not return until after the collapse of the Soviet Union in 1991.

With his expulsion, Solzhenitsyn became a true international celebrity, and the influence of *The Gulag Archipelago* began to spread rapidly outside Russia. The first German translation was received rapturously and in exactly the spirit its author had intended. One left-wing German newspaper wrote that *The Gulag Archipelago* constituted a "burning question mark over fifty years of Soviet power, over the whole Soviet experiment from 1918 on." The French and English translations appeared somewhat later, thanks to some misunderstandings over ownership of publication rights, but were equally influential. In the United States, where Solzhenitsyn ultimately chose to reside after his expulsion, the paperback edition of *The Gulag Archipelago*'s first volume sold more than two million copies. In France, it is no exaggeration to say that the book effectively ended the long-standing French intellectual flirtation with Soviet communism. So threatening was the book to the French status quo that Jean-Paul Sartre himself described Solzhenitsyn as a "dangerous element."

The West had heard of the Soviet camp system before, of course: credible witnesses had begun reporting on the growth of the Gulag as early as the 1920s. But what Solzhenitsyn produced was simply more thorough, more monumental, and more detailed than anything that had been produced previously. It could not be ignored, or dismissed as a single man's experience. No one who dealt with the Soviet Union, diplomatically or intellectually, could ignore it. Among other things, its horrific portrait of Soviet terror certainly contributed to the development, first under President Jimmy Carter and then under President Ronald Reagan, of an American foreign policy which recognized "human rights" as a legitimate element of international debate.

Since then, the stature of *The Gulag Archipelago*—now published in hundreds of editions, and in dozens of languages—has continued to grow. True, as open debate about the book and its subject has become possible in recent years, some legitimate criticisms of the work have been aired. Some camp survivors felt their memoirs, entrusted to Solzhenitsyn, were used in ways they didn't like, or to illustrate points they hadn't been making. Others objected to his almost fanatical insistence that any form of cooperation with the Gulag authorities had amounted to collaboration. The writer Lev Razgon, another Gulag memoirist, argued that for himself, as for many others, choosing to take an indoor accounting job was a matter of survival, not moral weakness: it was not immoral, Razgon wrote, to choose life.

It is also true that in the fifteen years that have passed since the Russian archives have opened, some errors have been found in Solzhenitsyn's work. His statistics are often wrong, and he sometimes garbles names and dates. Some of the stories he tells are impossible to verify. Some of the information he presents is partial or incomplete.

Nevertheless, what is most extraordinary about re-reading *The Gulag Archipelago* more than fifteen years after the collapse of the Soviet Union is how much it does get right. Although he did not have access to archival documents or government records, Solzhenitsyn's general outline of the history of the Gulag— from its origins in the Solovetsky Islands in the 1920s, through its expansion at the time of collectivization in the early 1930s, through the death of Stalin and the subsequent camp rebellions— has been proven correct. His description of the moral issues faced by the prisoners has never been disputed. His sociology of camp life, though presented in literary form, is unquestionably accurate. Among other things, the general reliability of the history presented in *The Gulag Archipelago* proves that "prisoners' gossip," so often dismissed by scholars as inaccurate, was often right. Indeed, part of the book's impact at the time of publication derived from the fact that both former victims and former perpetrators recognized Solzhenitsyn's descriptions and chronology as accurate, reflecting their own experiences.

This truthfulness continues to give the book a freshness and an importance which will never be challenged. For a contemporary reader, the book brilliantly evokes a mentality which no longer exists, and which is increasingly difficult to describe or explain: the atmosphere of constant fear; the constant temptation of betrayal; the ubiquity of secret police; the reversal of "normal" values; the generalized cruelty that permeated the culture of the Gulag, and of the Soviet Union itself.

And yet—no twenty-first century reader who picks up Solzhenitsyn's masterpiece for the first time should imagine that he or she is about to read a straightforward historical account. His book not only describes history: it is itself history. Thanks to Solzhenitsyn's obsessive attention to detail and his literary and polemical gifts, *The Gulag Archipelago* helped create the world that we live in today—a world in which Soviet communism is no longer held up as anybody's political ideal.

Anne Applebaum

Preface

In 1949 some friends and I came upon a noteworthy news item in *Nature,* a magazine of the Academy of Sciences. It reported in tiny type that in the course of excavations on the Kolyma River a subterranean ice lens had been discovered which was actually a frozen stream—and in it were found frozen specimens of prehistoric fauna some tens of thousands of years old. Whether fish or salamander, these were preserved in so fresh a state, the scientific correspondent reported, that those present immediately broke open the ice encasing the specimens and devoured them *with relish* on the spot.

The magazine no doubt astonished its small audience with the news of how successfully the flesh of fish could be kept fresh in a frozen state. But few, indeed, among its readers were able to decipher the genuine and heroic meaning of this incautious report.

As for us, however—we understood instantly. We could picture the entire scene right down to the smallest details: how those present broke up the ice in frenzied haste; how, flouting the higher claims of ichthyology and elbowing each other to be first, they tore off chunks of the prehistoric flesh and hauled them over to the bonfire to thaw them out and bolt them down.

We understood because we ourselves were the same kind of people as *those present* at that event. We, too, were from that powerful tribe of *zeks,* unique on the face of the earth, the only people who could devour prehistoric salamander *with relish.*

And the Kolyma was the greatest and most famous island, the pole of ferocity of that amazing country of *Gulag* which, though scattered in an Archipelago geographically, was, in the psychological sense, fused into a continent—an almost invisible, almost imperceptible country inhabited by the zek people.

And this Archipelago crisscrossed and patterned that other country within which it was located, like a gigantic patchwork, cutting into its cities, hovering over its streets. Yet there were many who did not even guess at its presence and many, many others who had heard something vague. And only those who had been there knew the whole truth.

But, as though stricken dumb on the islands of the Archipelago, they kept their silence.

By an unexpected turn of our history, a bit of the truth, an insignificant part of the whole, was allowed out in the open. But those same hands which once screwed tight our handcuffs now hold out their palms in reconciliation: "No, don't! Don't dig up the past! Dwell on the past and you'll lose an eye."

But the proverb goes on to say: "Forget the past and you'll lose both eyes."

Decades go by, and the scars and sores of the past are healing over for good. In the course of this period some of the islands of the Archipelago have shuddered and dissolved and the polar sea of oblivion rolls over them. And someday in the future, this Archipelago, its air, and the bones of its inhabitants, frozen in a lens of ice, will be discovered by our descendants like some improbable salamander.

I would not be so bold as to try to write the history of the Archipelago. I have never had the chance to read the documents. And, in fact, will anyone ever have the chance to read them? Those who do not wish to *recall* have already had enough time—and will have more —to destroy all the documents, down to the very last one.

I have absorbed into myself my own eleven years there not as something shameful nor as a nightmare to be cursed: I have come almost to love that monstrous world, and now, by a happy turn of events, I have also been entrusted with many recent reports and letters. So perhaps I shall be able to give some account of the bones and flesh of that salamander—which, incidentally, is still alive.

Introduction to the Abridgment

If it were possible for any nation to fathom another people's bitter experience through a book, how much easier its future fate would become and how many calamities and mistakes it could avoid. But it is very difficult. There always is this fallacious belief: "It would not be the same here; here such things are impossible."

Alas, all the evil of the twentieth century is possible everywhere on earth.

Yet I have not given up all hope that human beings and nations may be able, in spite of all, to learn from the experience of other people without having to live through it personally. Therefore, I gratefully accepted Professor Ericson's suggestion to create a one-volume abridgment of my three-volume work, *The Gulag Archipelago,* in order to facilitate its reading for those who do not have much time in this hectic century of ours. I thank Professor Ericson for his generous initiative as well as for the tactfulness, the literary taste, and the understanding of Western readers which he displayed during the work on the abridgment.

ALEKSANDR I. SOLZHENITSYN

Cavendish, Vermont
December, 1983

THE GULAG ARCHIPELAGO

THE DESTRUCTIVE-LABOR CAMPS

NOVAYA ZEMLYA

Murmansk

Monchegorsk

KOLA

Amderma

Naryan-Mar

Kem

Solovetsky
Islands

Archangel

Ukhta

Knyazh-Pogost

Ust-Vym

Svir

Kargopol

Kotlas

Vychegda

Kronstadt

Totma

Solikamsk

GULF OF FINLAND

Kady

Berezniki

LENINGRAD

Novgorod

Staraya Russa

Molotovsk Perm

RIGA

Dvina

Rybinsk

Jaroslavl

(Nizhni Novgorod)

Volokolamsko

Ivanovo

GORKY

KAZAN

MOSCOW

Moscow

Smolensk

Ryazan

Oka

KUIBYSHEV

Katyn

Kaluga

Penza

Syzran

MINSK

Tambov

Kuznetsk

Voronezh

Saratov

Volga

KIEV

Dnieper

KHARKOV

Kamyshino

Don

Dniester

Donets

STALINGRAD

Shakhty

Odessa

Rostov

Don

Volga

Krasnodar

Astrakhan

Sevastopol

Kuban

BLACK SEA

TBILISI

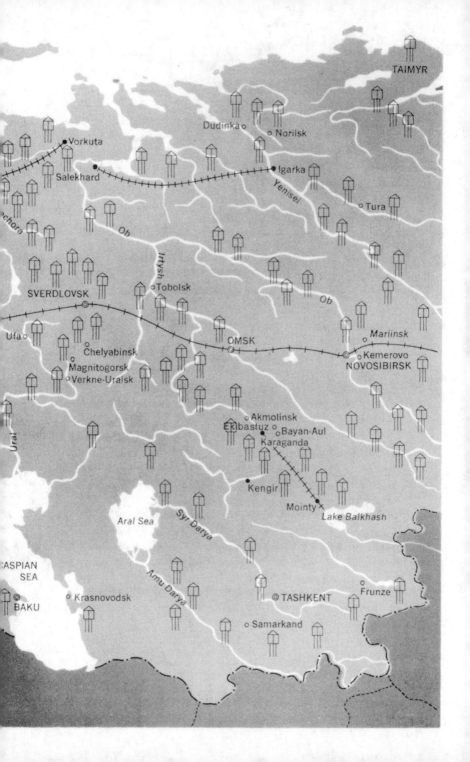

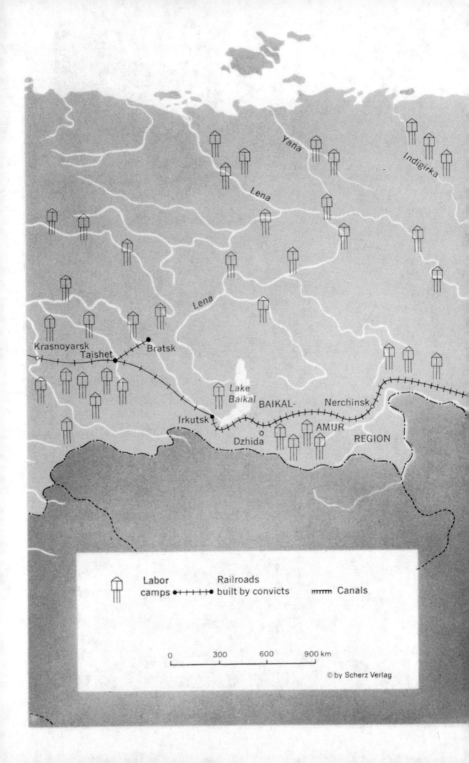

PART I

The Prison Industry

■

Aleksandr Isayevich Solzhenitsyn—in the army

. . . in detention  . . . after his release from camp

Chapter 1

■

Arrest

How do people get to this clandestine Archipelago? Hour by hour planes fly there, ships steer their course there, and trains thunder off to it—but all with nary a mark on them to tell of their destination. And at ticket windows or at travel bureaus for Soviet or foreign tourists the employees would be astounded if you were to ask for a ticket to go there. They know nothing and they've never heard of the Archipelago as a whole or of any one of its innumerable islands.

Those who go to the Archipelago to administer it get there via the training schools of the Ministry of Internal Affairs.

Those who go there to be guards are conscripted via the military conscription centers.

And those who, like you and me, dear reader, go there to die, must get there solely and compulsorily via arrest.

Arrest! Need it be said that it is a breaking point in your life, a bolt of lightning which has scored a direct hit on you? That it is an unassimilable spiritual earthquake not every person can cope with, as a result of which people often slip into insanity?

The Universe has as many different centers as there are living beings in it. Each of us is a center of the Universe, and that Universe is shattered when they hiss at you: *"You are under arrest."*

If *you* are arrested, can anything else remain unshattered by this cataclysm?

But the darkened mind is incapable of embracing these displacements in our universe, and both the most sophisticated and the veriest simpleton among us, drawing on all life's experience, can gasp out only: "Me? What for?"

And this is a question which, though repeated millions and millions of times before, has yet to receive an answer.

Arrest is an instantaneous, shattering thrust, expulsion, somersault from one state into another.

We have been happily borne—or perhaps have unhappily dragged our weary way—down the long and crooked streets of our lives, past all kinds of walls and fences made of rotting wood, rammed earth, brick, concrete, iron railings. We have never given a thought to what lies behind them. We have never tried to penetrate them with our vision or our understanding. But there is where the *Gulag* country begins, right next to us, two yards away from us. In addition, we have failed to notice an enormous number of closely fitted, well-disguised doors and gates in these fences. All those gates were prepared for us, every last one! And all of a sudden the fateful gate swings quickly open, and four white male hands, unaccustomed to physical labor but nonetheless strong and tenacious, grab us by the leg, arm, collar, cap, ear, and drag us in like a sack, and the gate behind us, the gate to our past life, is slammed shut once and for all.

That's all there is to it! You are arrested!

And you'll find nothing better to respond with than a lamblike bleat: "Me? What for?"

That's what arrest is: it's a blinding flash and a blow which shifts the present instantly into the past and the impossible into omnipotent actuality.

That's all. And neither for the first hour nor for the first day will you be able to grasp anything else.

Except that in your desperation the fake circus moon will blink at you: "It's a mistake! They'll set things right!"

And everything which is by now comprised in the traditional, even literary, image of an arrest will pile up and take shape, not in your own disordered memory, but in what your family and your neighbors in your apartment remember: The sharp nighttime ring or the rude knock at the door. The insolent entrance of the unwiped jackboots of the unsleeping State Security operatives. The frightened and cowed civilian witness at their backs. (And what function does this civilian witness serve? The victim doesn't even dare think about it and the operatives don't remember, but that's what the regulations call for, and so he has to sit there all night long and sign in the morning. For the witness, jerked from his bed, it is torture too—to go out night after night to help arrest his own neighbors and acquaintances.)

The traditional image of arrest is also trembling hands packing for the victim—a change of underwear, a piece of soap, something to eat; and no one knows what is needed, what is permitted, what clothes are best to wear; and the Security agents keep interrupting and hurrying you:

"You don't need anything. They'll feed you there. It's warm there." (It's all lies. They keep hurrying you to frighten you.)

The traditional image of arrest is also what happens afterward, when the poor victim has been taken away. It is an alien, brutal, and crushing force totally dominating the apartment for hours on end, a breaking, ripping open, pulling from the walls, emptying things from wardrobes and desks onto the floor, shaking, dumping out, and ripping apart—piling up mountains of litter on the floor—and the crunch of things being trampled beneath jackboots. And nothing is sacred in a search! During the arrest of the locomotive engineer Inoshin, a tiny coffin stood in his room containing the body of his newly dead child. The *"jurists"* dumped the child's body out of the coffin and searched it. They shake sick people out of their sickbeds, and they unwind bandages to search beneath them.

For those left behind after the arrest there is the long tail end of a wrecked and devastated life. And the attempts to go and deliver food parcels. But from all the windows the answer comes in barking voices: "Nobody here by that name!" "Never heard of him!" Yes, and in the worst days in Leningrad it took five days of standing in crowded lines just to get to that window. And it may be only after half a year or a year that the arrested person responds at all. Or else the answer is tossed out: "Deprived of the right to correspond." And that means once and for all. "No right to correspondence"—and that almost for certain means: "Has been shot."

That's how we picture arrest to ourselves.

The kind of night arrest described is, in fact, a favorite, because it has important advantages. Everyone living in the apartment is thrown into a state of terror by the first knock at the door. The arrested person is torn from the warmth of his bed. He is in a daze, half-asleep, helpless, and his judgment is befogged. In a night arrest the State Security men have a superiority in numbers; there are many of them, armed, against one person who hasn't even finished buttoning his trousers. During the arrest and search it is highly improbable that a crowd of potential supporters will gather at the entrance. The unhurried, step-by-step visits, first to one apartment, then to another, tomorrow to a third and a fourth, provide an opportunity for the Security operations personnel to be deployed with the maximum efficiency and to imprison many more citizens of a given town than the police force itself numbers.

In addition, there's an advantage to night arrests in that neither the people in neighboring apartment houses nor those on the city streets can see how many have been taken away. Arrests which frighten the closest neighbors are no event at all to those farther away. It's as if they had not taken place. Along that same asphalt ribbon on which the Black Marias scurry at night, a tribe of youngsters strides by day with banners, flowers, and gay, untroubled songs.

But those who *take,* whose work consists solely of arrests, for whom the horror is boringly repetitive, have a much broader understanding of how arrests operate. They operate according to a large body of theory, and innocence must not lead one to ignore this. The science of arrest is an important segment of the course on general penology and has been propped up with a substantial body of social theory. Arrests are classified according to various criteria: nighttime and daytime; at home, at work, during a journey; first-time arrests and repeats; individual and group arrests. Arrests are distinguished by the degree of surprise required, the amount of resistance expected (even though in tens of millions of cases no resistance was expected and in fact there was none). Arrests are also differentiated by the thoroughness of the required search; by instructions either to make out or not to make out an inventory of confiscated property or seal a room or apartment; to arrest the wife after the husband and send the children to an orphanage, or to send the rest of the family into exile, or to send the old folks to a labor camp too.

No, no: arrests vary widely in form. In 1926 Irma Mendel, a Hungarian, obtained through the Comintern two front-row tickets to the Bolshoi Theatre. Interrogator Klegel was courting her at the time and she invited him to go with her. They sat through the show very affectionately, and when it was over he took her—straight to the Lubyanka. And if on a flowering June day in 1927 on Kuznetsky Most, the plump-cheeked, redheaded beauty Anna Skripnikova, who had just bought some navy-blue material for a dress, climbed into a hansom cab with a young man-about-town, you can be sure it wasn't a lovers' tryst at all, as the cabman understood very well and showed by his frown (he knew the *Organs* don't pay). It was an arrest. In just a moment they would turn on the Lubyanka and enter the black maw of the gates. No, one certainly cannot say that daylight arrest, arrest during a journey, or arrest in the middle of a crowd has ever been neglected in our country. However, it has always been clean-cut—and, most surprising of all, the victims, in cooperation with the Security men, have conducted themselves in the noblest conceivable manner, so as to spare the living from witnessing the death of the condemned.

Not everyone can be arrested at home, with a preliminary knock at the door (and if there is a knock, then it has to be the house manager or else the postman). And not everyone can be arrested at work either. If the person to be arrested is vicious, then it's better to seize him *outside* his ordinary milieu—away from his family and colleagues, from those who share his views, from any hiding places. It is essential that he have no chance to destroy, hide, or pass on anything to anyone. VIP's in the military or the Party were sometimes first given new assignments, ensconced in a private railway car, and then arrested en route. Some obscure, ordinary mortal, scared to death by epidemic arrests all around him and already depressed for a week by sinister glances from his chief, is suddenly summoned to the local Party committee, where he is beamingly presented with a vacation ticket to a Sochi sanatorium. The rabbit is overwhelmed and immediately concludes that his fears were groundless. After expressing his gratitude, he hurries home, triumphant, to pack his suitcase. It is only two hours till train time, and he scolds his wife for being too slow. He arrives at the station with time to spare. And there in the waiting room or at the bar he is hailed by an extraordinarily pleasant young man: "Don't you remember me, Pyotr Ivanich?" Pyotr Ivanich has difficulty remembering: "Well, not exactly, you see, although . . ." The young man, however, is overflowing with friendly concern: "Come now, how can that be? I'll have to remind you. . . ." And he bows respectfully to Pyotr Ivanich's wife: "You must forgive us. I'll keep him only *one minute.*" The wife accedes, and trustingly the husband lets himself be led away by the arm—forever or for ten years!

The station is thronged—and no one notices anything. . . . Oh, you citizens who love to travel! Do not forget that in every station there are a GPU Branch and several prison cells.

This importunity of alleged acquaintances is so abrupt that only a person who has not had the wolfish preparation of camp life is likely to pull back from it. Do not suppose, for example, that if you are an employee of the American Embassy by the name of Alexander Dolgun you cannot be arrested in broad daylight on Gorky Street, right by the Central Telegraph Office. Your unfamiliar friend dashes through the press of the crowd, and opens his plundering arms to embrace you: "Saaasha!" He simply shouts at you, with no effort to be inconspicuous. "Hey, pal! Long time no see! Come on over, let's get out of the way."

At that moment a Pobeda sedan draws up to the curb. . . . And several days later TASS will issue an angry statement to all the papers alleging that informed circles of the Soviet government have no information on the disappearance of Alexander Dolgun. But what's so unusual about that? Our boys have carried out such arrests in Brussels—which was where Zhora Blednov was seized—not just in Moscow.

One has to give the *Organs* their due: in an age when public speeches, the plays in our theaters, and women's fashions all seem to have come off assembly lines, arrests can be of the most varied kind. They take you aside in a factory corridor after you have had your pass checked—and you're arrested. They take you from a military hospital with a temperature of 102, as they did with Ans Bernshtein, and the doctor will not raise a peep about your arrest—just let him try! They'll take you right off the operating table—as they took N. M. Vorobyev, a school inspector, in 1936, in the middle of an operation for stomach ulcer—and drag you off to a cell, as they did him, half-alive and all bloody (as Karpunich recollects). In the Gastronome—the fancy food store—you are invited to the special-order department and arrested there. You are arrested by a religious pilgrim whom you have put up for the night "for the sake of Christ." You are arrested by a meterman who has come to read your electric meter. You are arrested by a bicyclist who has run into you on the street, by a railway conductor, a taxi driver, a savings bank teller, the manager of a movie theater. Any one of them can arrest you, and you notice the concealed maroon-colored identification card only when it is too late.

Sometimes arrests even seem to be a game—there is so much superfluous imagination, so much well-fed energy, invested in them. After all, the victim would not resist anyway. Is it that the Security agents want to justify their employment and their numbers? After all, it would seem enough to send notices to all the rabbits marked for arrest, and they would show up obediently at the designated hour and minute at the iron gates of State Security with a bundle in their hands —ready to occupy a piece of floor in the cell for which they were intended. And, in fact, that's the way collective farmers are arrested. Who wants to go all the way to a hut at night, with no roads to travel on? They are summoned to the village soviet—and arrested there. Manual workers are called into the office.

Of course, every machine has a point at which it is overloaded, beyond which it cannot function. In the strained and overloaded years of 1945 and 1946, when trainload after trainload poured in from Europe, to be swallowed up immediately and sent off to *Gulag,* all that excessive theatricality went out the window, and the whole theory suffered greatly. All the fuss and feathers of ritual went flying in every direction, and the arrest of tens of thousands took on the appearance of a squalid roll call: they stood there with lists, read off the names of those on one train, loaded them onto another, and that was the whole arrest.

For several decades political arrests were distinguished in our country precisely by the fact that people were arrested who were guilty of nothing and were therefore unprepared to put up any resistance whatsoever. There was a general feeling of being destined for destruction, a sense of having nowhere to escape from the GPU-NKVD (which, incidentally, given our internal passport system, was quite accurate). And even in the fever of epidemic arrests, when people leaving for work said farewell to their families every day, because they could not be certain they would return at night, even then almost no one tried to run away and only in rare cases did people commit suicide. And that was exactly what was required. A submissive sheep is a find for a wolf.

This submissiveness was also due to ignorance of the mechanics of epidemic arrests. By and large, the *Organs* had no profound reasons for their choice of whom to arrest and whom not to arrest. They merely had over-all assignments, quotas for a specific number of arrests. These quotas might be filled on an orderly basis or wholly arbitrarily. In 1937 a woman came to the reception room of the Novocherkassk NKVD to ask what she should do about the unfed unweaned infant of a neighbor who had been arrested. They said: "Sit down, we'll find out." She sat there for two hours—whereupon they took her and tossed her into a cell. They had a total plan which had to be fulfilled in a hurry, and there was no one available to send out into the city—and here was this woman already in their hands!

Universal innocence also gave rise to the universal failure to act. Maybe they *won't take* you? Maybe it will all blow over? A. I. Lady-zhensky was the chief teacher in a school in remote Kologriv. In 1937 a peasant approached him in an open market and passed him a message from a third person: "Aleksandr Ivanich, get out of town, *you are on the list!*" But he stayed: After all, the whole school rests on my shoulders, and *their own* children are pupils here. How can they arrest me? (Several days later he was arrested.) Not everyone was so fortunate as to understand at the age of fourteen, as did Vanya Levitsky: "Every honest man is sure to go to prison. Right now my papa is serving time, and when I grow up they'll put me in too." (They put him in when he was twenty-three years old.) The majority sit quietly and dare to hope. Since you aren't guilty, then how can they arrest you? *It's a mistake!* They are already dragging you along by the collar, and you still keep on exclaiming to yourself: "It's a mistake! *They'll set things straight and let me out!*" Others are being arrested en masse, and that's a bothersome fact, but in those other cases there is always some dark area: "Maybe *he* was guilty . . . ?" But as for you, you are obviously innocent! You still believe that the *Organs* are humanly logical institutions: they will set things straight and let you out.

Why, then, should you run away? And how can you resist right then? After all, you'll only make your situation worse; you'll make it more difficult for them to sort out the mistake. And it isn't just that you don't put up any resistance; you even walk down the stairs on tiptoe, as you are ordered to do, so your neighbors won't hear.

At what exact point, then, should one resist? When one's belt is taken away? When one is ordered to face into a corner? When one crosses the threshold of one's home? An arrest consists of a series of incidental irrelevancies, of a multitude of things that do not matter, and there seems no point in arguing about any one of them individually— especially at a time when the thoughts of the person arrested are wrapped tightly about the big question: "What for?"—and yet all these incidental irrelevancies taken together implacably constitute the arrest.

Almost anything can occupy the thoughts of a person who has just been arrested! This alone would fill volumes. There can be feelings which we never suspected. When nineteen-year-old Yevgeniya Doyarenko was arrested in 1921 and three young Chekists were poking about her bed and through the underwear in her chest of drawers, she was not disturbed. There was nothing there, and they would find nothing. But all of a sudden they touched her personal diary, which she would not have shown even to her own mother. And these hostile young strangers reading the words she had written was more devastating to her than the whole Lubyanka with its bars and its cellars. It is true of many that the outrage inflicted by arrest on their personal feelings and attachments can be far, far stronger than their political beliefs or their fear of prison. A person who is not inwardly prepared for the use of violence against him is always weaker than the person committing the violence.

There are a few bright and daring individuals who understand instantly. Grigoryev, the Director of the Geological Institute of the Academy of Sciences, barricaded himself inside and spent two hours burning up his papers when they came to arrest him in 1948.

Sometimes the principal emotion of the person arrested is relief and even *happiness!* This is another aspect of human nature. It happened before the Revolution too: the Yekaterinodar schoolteacher Serdyukova, involved in the case of Aleksandr Ulyanov, felt only relief when she was arrested. But this feeling was a thousand times stronger during epidemics of arrests when all around you they were hauling in people like yourself and still had not come for you; for some reason they were taking their time. After all, that kind of exhaustion, that kind of suffering, is worse than any kind of arrest, and not only for a person of limited courage. Vasily Vlasov, a fearless Communist, whom we shall recall more than once later on, renounced the idea of escape proposed by his non-Party assistants, and pined away because the entire leadership of the Kady District was arrested in 1937, and they kept delaying and delaying his own arrest. He could only endure the blow head on. He did endure it, and then he relaxed, and during the first days after his arrest he felt marvelous. In 1934 the priest Father Irakly went to Alma-Ata to visit some believers in exile there. During his absence they came three times to his Moscow apartment to arrest him. When he returned, members of his flock met him at the station and refused to let him go home, and for eight years hid him in one apartment after another. The priest suffered so painfully from this harried life that when he was finally arrested in 1942 he sang hymns of praise to God.

"Resistance! Why didn't you resist?" Today those who have continued to live on in comfort scold those who suffered.

Yes, resistance should have begun right there, at the moment of the arrest itself.

But it did not begin.

And so they are *leading* you. During a daylight arrest there is always that brief and unique moment when they are *leading* you, either inconspicuously, on the basis of a cowardly deal you have made, or else quite openly, their pistols unholstered, through a crowd of hundreds of just such doomed innocents as yourself. You aren't gagged. You really can and you really ought to *cry out*—to *cry out* that you are being arrested! That villains in disguise are trapping people! That arrests are being made on the strength of false denunciations! That millions are being subjected to silent reprisals! If many such outcries had been heard all over the city in the course of a day, would not our fellow citizens perhaps have begun to bristle? And would arrests perhaps no longer have been so easy?

In 1927, when submissiveness had not yet softened our brains to such a degree, two Chekists tried to arrest a woman on Serpukhov Square during the day. She grabbed hold of the stanchion of a street-lamp and began to scream, refusing to submit. A crowd gathered. (There had to have been that kind of woman; there had to have been that kind of crowd too! Passers-by didn't all just close their eyes and hurry by!) The quick young men immediately became flustered. They can't *work* in the public eye. They got into their car and fled. (Right then and there she should have gone to a railroad station and left! But she went home to spend the night. And during the night they took her off to the Lubyanka.)

Instead, not one sound comes from *your* parched lips, and that passing crowd naïvely believes that you and your executioners are friends out for a stroll.

I myself often had the chance to *cry out.*

On the eleventh day after my arrest, three SMERSH bums, more burdened by four suitcases full of war booty than by me (they had come to rely on me in the course of the long trip), brought me to the Byelorussian Station in Moscow. They were called a *Special Convoy*—in other words, a special escort guard—but in actual fact their automatic pistols only interfered with their dragging along the four terribly heavy bags of loot they and their chiefs in SMERSH counterintelligence on the Second Byelorussian Front had plundered in Germany and were now bringing to their families in the Fatherland under the pretext of convoying me. I myself lugged a fifth suitcase with no great joy since it contained my diaries and literary works, which were being used as evidence against me.

Not one of the three knew the city, and it was up to me to pick the shortest route to the prison. I had personally to conduct them to the Lubyanka, where they had never been before (and which, in fact, I confused with the Ministry of Foreign Affairs).

I had spent one day in the counterintelligence prison at army headquarters and three days in the counterintelligence prison at the headquarters of the front, where my cellmates had educated me in the deceptions practiced by the interrogators, their threats and beatings; in the fact that once a person was arrested he was never released; and in the inevitability of a *tenner,* a ten-year sentence; and then by a miracle I had suddenly burst out of there and for four days had traveled like a *free* person among *free* people, even though my flanks had already lain on rotten straw beside the latrine bucket, my eyes had already beheld beaten-up and sleepless men, my ears had heard the truth, and my mouth had tasted prison gruel. So why did I keep silent? Why, in my last minute out in the open, did I not attempt to enlighten the hoodwinked crowd?

I kept silent, too, in the Polish city of Brodnica—but maybe they didn't understand Russian there. I didn't call out one word on the streets of Bialystok—but maybe it wasn't a matter that concerned the Poles. I didn't utter a sound at the Volkovysk Station—but there were very few people there. I walked along the Minsk Station platform beside those same bandits as if nothing at all were amiss—but the station was still a ruin. And now I was leading the SMERSH men through the circular upper concourse of the Byelorussian-Radial subway station on the Moscow circle line, with its white-ceilinged dome and brilliant electric lights, and opposite us two parallel escalators, thickly packed with Muscovites, rising from below. It seemed as though they were all looking at me! They kept coming in an endless ribbon from down there, from the depths of ignorance—on and on beneath the gleaming dome, reaching toward me for at least one word of truth—so why did I keep silent?

Every man always has handy a dozen glib little reasons why he is right not to sacrifice himself.

Some still have hopes of a favorable outcome to their case and are afraid to ruin their chances by an outcry. (For, after all, we get no news from that other world, and we do not realize that from the very moment of arrest our fate has almost certainly been decided in the worst possible sense and that we cannot make it any worse.) Others have not yet attained the mature concepts on which a shout of protest to the crowd must be based. Indeed, only a revolutionary has slogans on his lips that are crying to be uttered aloud; and where would the uninvolved, peaceable average man come by such slogans? He simply *does not know what* to shout. And then, last of all, there is the person whose heart is too full of emotion, whose eyes have seen too much, for that whole ocean to pour forth in a few disconnected cries.

As for me, I kept silent for one further reason: because those Muscovites thronging the steps of the escalators were too few for me, *too few!* Here my cry would be heard by 200 or twice 200, but what about the 200 million? Vaguely, unclearly, I had a vision that someday I would cry out to the 200 million.

But for the time being I did not open my mouth, and the escalator dragged me implacably down into the nether world.

And when I got to Okhotny Ryad, I continued to keep silent.

Nor did I utter a cry at the Metropole Hotel.

Nor wave my arms on the Golgotha of Lubyanka Square.

■

Mine was, probably, the easiest imaginable kind of arrest. It did not tear me from the embrace of kith and kin, nor wrench me from a deeply cherished home life. One pallid European February it took me from our narrow salient on the Baltic Sea, where, depending on one's point of view, either we had surrounded the Germans or they had surrounded us, and it deprived me only of my familiar artillery battery and the scenes of the last three months of the war.

The brigade commander called me to his headquarters and asked me for my pistol; I turned it over without suspecting any evil intent, when suddenly, from a tense, immobile suite of staff officers in the corner, two counterintelligence officers stepped forward hurriedly, crossed the room in a few quick bounds, their four hands grabbed simultaneously at the star on my cap, my shoulder boards, my officer's belt, my map case, and they shouted theatrically:

"You are under arrest!"

Burning and prickling from head to toe, all I could exclaim was: "Me? What for?"

And even though there is usually no answer to this question, surprisingly I received one! This is worth recalling, because it is so contrary to our usual custom. Across the sheer gap separating me from those left behind, the gap created by the heavy-falling word "arrest," across that quarantine line not even a sound dared penetrate, came the unthinkable, magic words of the brigade commander:

"Solzhenitsyn. Come back here."

With a sharp turn I broke away from the hands of the SMERSH men and stepped back to the brigade commander. I had never known him very well. He had never condescended to run-of-the-mill conversations with me. To me his face had always conveyed an order, a command, wrath. But right now it was illuminated in a thoughtful way. Was it from shame for his own involuntary part in this dirty business? Was it from an impulse to rise above the pitiful subordination of a whole lifetime? Ten days before, I had led my own reconnaissance battery almost intact out of the *fire pocket* in which the twelve heavy guns of his artillery battalion had been left, and now he had to renounce me because of a piece of paper with a seal on it?

"You have . . ." he asked weightily, "a friend on the First Ukrainian Front?"

"It's forbidden! You have no right!" the captain and the major of counterintelligence shouted at the colonel. But I had already understood: I knew instantly I had been arrested because of my correspondence with a school friend, and understood from what direction to expect danger.

Zakhar Georgiyevich Travkin could have stopped right there! But no! Continuing his attempt to expunge his part in this and to stand erect before his own conscience, he rose from behind his desk—he had never stood up in my presence in my former life—and reached across the quarantine line that separated us and gave me his hand, although he would never have reached out his hand to me had I remained a free man. And pressing my hand, while his whole suite stood there in mute horror, showing that warmth that may appear in an habitually severe face, he said fearlessly and precisely:

"I wish you happiness, Captain!"

Not only was I no longer a captain, but I had been exposed as an enemy of the people (for among us every person is totally exposed from the moment of arrest). And he had wished happiness—to an enemy?

This is not going to be a volume of memoirs about my own life. Therefore I am not going to recount the truly amusing details of my arrest, which was like no other. That night the SMERSH officers gave up their last hope of being able to make out where we were on the map —they never had been able to read maps anyway. So they politely handed the map to me and asked me to tell the driver how to proceed to counterintelligence at army headquarters. I, therefore, led them and myself to that prison, and in gratitude they immediately put me not in an ordinary cell but in a punishment cell. And I really must describe that closet in a German peasant house which served as a temporary punishment cell.

It was the length of one human body and wide enough for three to lie packed tightly, four at a pinch. As it happened, I was the fourth, shoved in after midnight. The three lying there blinked sleepily at me in the light of the smoky kerosene lantern and moved over, giving me enough space to lie on my side, half between them, half on top of them, until gradually, by sheer weight, I could wedge my way in. And so four overcoats lay on the crushed-straw-covered floor, with eight boots pointing at the door. They slept and I burned. The more self-assured I had been as a captain half a day before, the more painful it was to crowd onto the floor of that closet. Once or twice the other fellows woke up numb on one side, and we all turned over at the same time.

Toward morning they awoke, yawned, grunted, pulled up their legs, moved into various corners, and our acquaintance began.

"What are you in for?"

But a troubled little breeze of caution had already breathed on me beneath the poisoned roof of SMERSH and I pretended to be surprised: "No idea. Do the bastards tell you?"

However, my cellmates—tankmen in soft black helmets—hid nothing. They were three honest, openhearted soldiers—people of a kind I had become attached to during the war years because I myself was more complex and worse. All three had been officers. Their shoulder boards also had been viciously torn off, and in some places the cotton batting stuck out. On their stained field shirts light patches indicated where decorations had been removed, and there were dark and red scars on their faces and arms, the results of wounds and burns. Their tank unit had, unfortunately, arrived for repairs in the village where the SMERSH counterintelligence headquarters of the Forty-eighth Army was located. Still damp from the battle of the day before, yesterday they had gotten drunk, and on the outskirts of the village broke into a bath where they had noticed two raunchy broads going to bathe. The girls, half-dressed, managed to get away all right from the soldiers' staggering, drunken legs. But one of them, it turned out, was the property of the army Chief of Counterintelligence, no less.

Yes! For three weeks the war had been going on inside Germany, and all of us knew very well that if the girls were German they could be raped and then shot. This was almost a combat distinction. Had they been Polish girls or our own displaced Russian girls, they could have been chased naked around the garden and slapped on the behind—an amusement, no more. But just because this one was the "campaign wife" of the Chief of Counterintelligence, right off some deep-in-the-rear sergeant had viciously torn from three front-line officers the shoulder boards awarded them by the front headquarters and had taken off the decorations conferred upon them by the Presidium of the Supreme Soviet. And now these warriors, who had gone through the whole war and who had no doubt crushed more than one line of enemy trenches, were waiting for a court-martial, whose members, had it not been for their tank, could have come nowhere near the village.

We put out the kerosene lamp, which had already used up all the air there was to breathe. A *Judas hole* the size of a postage stamp had been cut in the door and through it came indirect light from the corridor. Then, as if afraid that with the coming of daylight we would have too much room in the punishment cell, they *tossed in* a fifth person. He stepped in wearing a newish Red Army tunic and a cap that was also new, and when he stopped opposite the peephole we could see a fresh face with a turned-up nose and red cheeks.

"Where are you from, brother? Who are you?"

"From the *other* side," he answered briskly. "A shhpy."

"You're kidding!" We were astounded. (To be a spy and to admit it—Sheinin and the brothers Tur had never written that kind of spy story!)

"What is there to kid about in wartime?" the young fellow sighed reasonably. "And just how else can you get back home from being a POW? Well, you tell me!"

He had barely begun to tell us how, some days back, the Germans had led him through the front lines so that he could play the spy and blow up bridges, whereupon he had gone immediately to the nearest battalion headquarters to turn himself in; but the weary, sleep-starved battalion commander hadn't believed his story about being a spy and had sent him off to the nurse to get a pill. And at that moment new impressions burst upon us:

"Out for toilet call! Hands behind your backs!" hollered a master sergeant *hardhead* as the door sprang open; he was just built for swinging the tail of a 122-millimeter cannon.

A circle of machine gunners had been strung around the peasant courtyard, guarding the path which was pointed out to us and which went behind the barn. I was bursting with indignation that some ignoramus of a master sergeant dared to give orders to us officers: "Hands behind your backs!" But the tank officers put their hands behind them and I followed suit.

Back of the barn was a small square area in which the snow had been all trampled down but had not yet melted. It was soiled all over with human feces, so densely scattered over the whole square that it was difficult to find a spot to place one's two feet and squat. However, we spread ourselves about and the five of us did squat down. Two machine gunners grimly pointed their machine pistols at us as we squatted, and before a minute had passed the master sergeant brusquely urged us on:

"Come on, hurry it up! With us they do it quickly!"

Not far from me squatted one of the tankmen, a native of Rostov, a tall, melancholy senior lieutenant. His face was blackened by a thin film of metallic dust or smoke, but the big red scar stretching across his cheek stood out nonetheless.

"What do you mean, *with us?*" he asked quietly, indicating no intention of hurrying back to the punishment cell that still stank of kerosene.

"In SMERSH counterintelligence!" the master sergeant shot back proudly and more resonantly than was called for. (The counterintelligence men used to love that tastelessly concocted word "SMERSH," manufactured from the initial syllables of the words for "death to spies." They felt it intimidated people.)

"And *with us* we do it slowly," replied the senior lieutenant thoughtfully. His helmet was pulled back, uncovering his still untrimmed hair. His oaken, battle-hardened rear end was lifted toward the pleasant coolish breeze.

"Where do you mean, *with us?*" the master sergeant barked at him more loudly than he needed to.

"In the Red Army," the senior lieutenant replied very quietly from his heels, measuring with his look the cannon-tailer that never was.

Such were my first gulps of prison air.

Chapter 2

■

The History of Our Sewage Disposal System

When people today decry *the abuses of the cult,* they keep getting hung up on those years which are stuck in our throats, '37 and '38. And memory begins to make it seem as though arrests were never made *before* or *after,* but only in those two years.

Although I have no statistics at hand, I am not afraid of erring when I say that the *wave* of 1937 and 1938 was neither the only one nor even the main one, but only one, perhaps, of the three biggest waves which strained the murky, stinking pipes of our prison sewers to bursting.

Before it came the wave of 1929 and 1930, the size of a good River Ob, which drove a mere fifteen million peasants, maybe even more, out into the taiga and the tundra. But peasants are a silent people, without a literary voice, nor do they write complaints or memoirs. No interrogators sweated out the night with them, nor did they bother to draw up formal indictments—it was enough to have a decree from the village soviet. This wave poured forth, sank down into the permafrost, and even our most active minds recall hardly a thing about it. It is as if it had not even scarred the Russian conscience. And yet Stalin (and you and I as well) committed no crime more heinous than this.

And *after* it there was the wave of 1944 to 1946, the size of a good Yenisei, when they dumped whole *nations* down the sewer pipes, not to mention millions and millions of others who (because of us!) had been prisoners of war, or carried off to Germany and subsequently repatriated. (This was Stalin's method of cauterizing the wounds so that scar tissue would form more quickly, and thus the body politic as a whole would not have to rest up, catch its breath, regain its strength.) But in this wave, too, the people were of the simpler kind, and they wrote no memoirs.

But the wave of 1937 swept up and carried off to the Archipelago people of position, people with a Party past, yes, educated people, around whom were many who had been wounded and remained in the cities . . . and what a lot of them had pen in hand! And today they are all writing, speaking, remembering: "Nineteen thirty-seven!" A whole Volga of the people's grief!

But just say "Nineteen thirty-seven" to a Crimean Tatar, a Kalmyk, a Chechen, and he'll shrug his shoulders. And what's 1937 to Leningrad when 1935 had come before it? And for the *second-termers* (i.e., *repeaters*), or people from the Baltic countries—weren't 1948 and 1949 harder on them? And if sticklers for style and geography should accuse me of having omitted some Russian rivers, and of not yet having named some of the waves, then just give me enough paper! There were enough waves to use up the names of all the rivers of Russia!

It is well known that any *organ* withers away if it is not used. Therefore, if we know that the Soviet Security organs, or *Organs* (and they christened themselves with this vile word), praised and exalted above all living things, have not died off even to the extent of one single tentacle, but, instead, have grown new ones and strengthened their muscles—it is easy to deduce that they have had *constant* exercise.

Through the sewer pipes the flow pulsed. Sometimes the pressure was higher than had been projected, sometimes lower. But the prison sewers were never empty. The blood, the sweat, and the urine into which we were pulped pulsed through them continuously. The history of this sewage system is the history of an endless swallow and flow; flood alternating with ebb and ebb again with flood; waves pouring in, some big, some small; brooks and rivulets flowing in from all sides; trickles oozing in through gutters; and then just plain individually scooped-up droplets.

The chronological list which follows, in which waves made up of millions of arrested persons are given equal attention with ordinary streamlets of unremarkable handfuls, is quite incomplete, meager, miserly, and limited by my own capacity to penetrate the past. What is really needed is a great deal of additional work by survivors familiar with the material.

■

In considering now the period from 1918 to 1920, we are in difficulties: Should we classify among the prison waves all those who were done in before they even got to prison cells? And in what classification should we put those whom the Committees of the Poor took behind the wing of the village soviet or to the rear of the courtyard, and *finished off* right there? Did the participants in the clusters of plots uncovered in every province at least succeed in setting foot on the land of the Archipelago, or did they not—and are they therefore not related to the subject of our investigations? Bypassing the repression of the now famous rebellions (Yaroslavl, Murom, Rybinsk, Arzamas), we know of certain events only by their names—for instance, the Kolpino executions of June, 1918. What were they? Who were they? And where should they be classified?

There is also no little difficulty in deciding whether we should classify among the prison waves or on the balance sheets of the Civil War those tens of thousands of *hostages,* i.e., people not personally accused of anything, those peaceful citizens not even listed by name, who were taken off and destroyed simply to terrorize or wreak vengeance on a military enemy or a rebellious population.

This action was, in fact, explained openly (Latsis, in the newspaper *Red Terror,* November 1, 1918): "We are not fighting against single individuals. We are exterminating the bourgeoisie as a class. It is not necessary during the interrogation to look for evidence proving that the accused opposed the Soviets by word or action. The first question which you should ask him is what class does he belong to, what is his origin, his education and his profession. These are the questions which will determine the fate of the accused. Such is the sense and the essence of red terror." A decree of the Defense Council on February 15, 1919 (the meeting was evidently presided over by Lenin), suggests that the Cheka and the NKVD take hostages among the peasants of those regions where "the cleaning of snow from the railroads does not proceed quite satisfactorily" and that "these hostages be executed if the cleaning is not completed."

But even restricting ourselves to ordinary arrests, we can note that by the spring of 1918 a torrent of socialist traitors had already begun that was to continue without slackening for many years.

In 1919, suspicion of our Russians returning from abroad was already having its effect (Why? What was their alleged assignment?)—thus the officers of the Russian expeditionary force in France were imprisoned on their homecoming.

In 1919, too, what with the big hauls in connection with such actual and pseudo plots as the "National Center" and the "Military Plot," executions were carried out in Moscow, Petrograd, and other cities *on the basis of lists*—in other words, free people were simply arrested and executed immediately. . . .

From January, 1919, on, food requisitioning was organized and food-collecting detachments were set up. They encountered resistance everywhere in the rural areas, sometimes stubborn and passive, sometimes violent. The suppression of this opposition gave rise to an abundant flood of arrests during the course of the next two years, not counting those who were shot on the spot.

In May, 1920, came the well-known decree of the Central Committee "on Subversive Activity in the Rear." We know from experience that every such decree is a call for a new wave of widespread arrests; it is the outward sign of such a wave.

It was in 1920 that we knew (or failed to know) of the trial of the "Siberian Peasants' Union." And at the end of 1920 the repression of the Tambov peasants' rebellion began. There was no trial for them.

But the main drive to uproot people from the Tambov villages took place mostly in June, 1921. Throughout the province concentration camps were set up for the families of peasants who had taken part in the revolts.

Even earlier, in March, 1921, the rebellious Kronstadt sailors, minus those who had been shot, were sent to the islands of the Archipelago. . . .

In that same year the practice of arresting *students* began. . . . Also in 1921 the arrests of members of all non-Bolshevik parties were expanded and systematized. In fact, all Russia's political parties had been buried, except the victorious one.

In the spring of 1922 the Extraordinary Commission for Struggle Against Counterrevolution, Sabotage, and Speculation, the Cheka, recently renamed the GPU, decided to intervene in church affairs. It was called on to carry out a "church revolution"—to remove the existing leadership and replace it with one which would have only one ear turned to heaven and the other to the Lubyanka. The so-called "Living Church" people seemed to go along with this plan, but without outside help they could not gain control of the church apparatus. For this reason, the Patriarch Tikhon was arrested and two resounding trials were held, followed by the execution in Moscow of those who had publicized the Patriarch's appeal and, in Petrograd, of the Metropolitan Veniamin, who had attempted to hinder the transfer of ecclesiastical power to the "Living Church" group. Here and there in the provincial centers and even further down in the administrative districts, met-ropolitans and bishops were arrested, and, as always, in the wake of the big fish, followed shoals of smaller fry: archpriests, monks, and dea-cons. These arrests were not even reported in the press. They also arrested those who refused to swear to support the "Living Church" "renewal" movement.

Men of religion were an inevitable part of every annual "catch," and their silver locks gleamed in every cell and in every prisoner trans-port en route to the Solovetsky Islands.

From the early twenties on, arrests were also made among groups of theosophists, mystics, spiritualists. (Count Palen's group used to keep official transcripts of its communications with the spirit world.) Also, religious societies and philosophers of the Berdyayev circle. The so-called "Eastern Catholics"—followers of Vladimir Solovyev—were arrested and destroyed in passing, as was the group of A. I. Abrikosova. And, of course, ordinary Roman Catholics—Polish Catholic priests, etc.—were arrested, too, as part of the normal course of events.

However, the root destruction of religion in the country, which throughout the twenties and thirties was one of the most important goals of the GPU-NKVD, could be realized only by mass arrests of Orthodox believers. Monks and nuns, whose black habits had been a distinctive feature of Old Russian life, were intensively rounded up on every hand, placed under arrest, and sent into exile. They arrested and sentenced active laymen. The circles kept getting bigger, as they raked in ordinary believers as well, old people, and particularly women, who were the most stubborn believers of all and who, for many long years to come, would be called "nuns" in transit prisons and in camps.

True, they were supposedly being arrested and tried not for their actual faith but for openly declaring their convictions and for bringing up their children in the same spirit. As Tanya Khodkevich wrote:

> You can pray *freely*
> But just so God alone can hear.

(She received a ten-year sentence for these verses.) A person convinced that he possessed spiritual truth was required to conceal it from his own children! In the twenties the religious education of children was clas-sified as a political crime under Article 58-10 of the Code—in other words, counterrevolutionary propaganda! True, one was still permitted to renounce one's religion at one's trial: it didn't often happen but it nonetheless did happen that the father would renounce his religion and remain at home to raise the children while the mother went to the

Solovetsky Islands. (Throughout all those years women manifested great firmness in their faith.) All persons convicted of religious activity received *tenners,* the longest term then given.

(In those years, particularly in 1927, in purging the big cities for the pure society that was coming into being, they sent prostitutes to the Solovetsky Islands along with the "nuns." Those lovers of a sinful earthly life were given *three-*year sentences under a more lenient article of the Code. The conditions in prisoner transports, in transit prisons, and on the Solovetsky Islands were not of a sort to hinder them from plying their merry trade among the administrators and the convoy guards. And three years later they would return with laden suitcases to the places they had come from. Religious prisoners, however, were prohibited from ever returning to their children and their home areas.)

As early as the early twenties, waves appeared that were purely national in character. . . .

The waves flowed underground through the pipes; they provided sewage disposal for the life flowering on the surface.

In 1931, following the trial of the Promparty, a grandiose trial of the Working Peasants Party was being prepared—on the grounds that they existed (never, in actual fact!) as an enormous organized underground force among the rural intelligentsia, including leaders of consumer and agricultural cooperatives and the more advanced upper layer of the peasantry, and supposedly were preparing to overthrow the dictatorship of the proletariat. At the trial of the Promparty this Working Peasants Party—the TKP—was referred to as if it were already well known and under detention. . . .

Then all of a sudden, one lovely night, Stalin *reconsidered.* Why? Maybe we will never know. Did he perhaps wish to save his soul? Too soon for that, it would seem. Did his sense of humor come to the fore —was it all so deadly, monotonous, so bitter-tasting? But no one would ever dare accuse Stalin of having a sense of humor! Likeliest of all, Stalin simply figured out that the whole countryside, not just 200,000 people, would soon die of famine anyway, so why go to the trouble? And instantly the whole TKP trial was called off. All those who had "confessed" were told they could *repudiate* their confessions (one can picture their happiness!).

Paragraph piles on paragraph, year on year—and yet there is no way we can describe in sequence everything that took place (but the GPU did its job effectively! The GPU never let anything get by!). But we must always remember that:

Religious believers, of course, were being arrested uninterruptedly. (There were, nonetheless, certain special dates and peak periods. There was a "night of struggle against religion" in Leningrad on Christmas Eve, 1929, when they arrested a large part of the religious intelligentsia and held them—not just until morning either. And that was certainly no "Christmas tale." Then in February, 1932, again in Leningrad, many churches were closed simultaneously, while, at the same time, large-scale arrests were made among the clergy. And there are still more dates and places, but they haven't been reported to us by anyone.)

Non-Orthodox *sects* were also under constant attack. . . .

The Big Solitaire game played with the socialists went on and on uninterruptedly—of course.

In 1929, also, those historians who had not been sent abroad in time were arrested. . . .

From one end of the country to the other, nationalities kept pouring in. . . .

From 1928 on, it was time to call to a reckoning those late stragglers after the bourgeoisie—the NEPmen. The usual practice was to impose on them ever-increasing and finally totally intolerable taxes. At a certain point they could no longer pay; they were immediately arrested for bankruptcy, and their property was confiscated. The state needed property and gold. The famous *gold fever* began at the end of 1929.

Who was arrested in the "gold" wave? All those who, at one time or another, fifteen years before, had had a private "business," had been involved in retail trade, had earned wages at a craft, and *could have,* according to the GPU's deductions, hoarded gold. But it so happened that they often had no gold. They had put their money into real estate or securities, which had melted away or been taken away in the Revolution, and nothing remained. They had high hopes, of course, in arresting dental technicians, jewelers, and watch repairmen. All were arrested, all were crammed into GPU cells in numbers no one had considered possible up to then—but that was all to the good: they would *cough it up* all the sooner! It even reached a point of such confusion that men and women were imprisoned in the same cells and used the latrine bucket in each other's presence—who cared about those niceties? Give up your gold, vipers! The interrogators had one universal method: feed the prisoners nothing but salty food and give them no water. Whoever coughed up gold got water! One gold piece for a cup of fresh water!

People perish for cold metal.

The crudest detective stories and operas about brigands were played out in real life on a vast national scale.

And so the waves foamed and rolled. But over them all, in 1929–1930, billowed and gushed the multimillion wave of *dispossessed kulaks.* It was immeasurably large and it could certainly not have been housed in even the highly developed network of Soviet interrogation prisons (which in any case were packed full by the "gold" wave). Instead, it bypassed the prisons, going directly to the transit prisons and camps, onto prisoner transports, into the Gulag country. In sheer size this nonrecurring tidal wave (it was an ocean) swelled beyond the bounds of anything the penal system of even an immense state can permit itself. There was nothing to be compared with it in all Russian history. It was the forced resettlement of a whole people, an ethnic catastrophe.

This wave was also distinct from all those which preceded it because no one fussed about with taking the head of the family first and then working out what to do with the rest of the family. On the contrary, in this wave they burned out whole nests, whole families, from the start; and they watched jealously to be sure that none of the children —fourteen, ten, even six years old—got away: to the last scrapings, all had to go down the same road, to the same common destruction. (This was the *first* such experiment—at least in modern history. It was subsequently repeated by Hitler with the Jews, and again by Stalin with nationalities which were disloyal to him or suspected by him.)

Like raging beasts, abandoning every concept of "humanity," abandoning all humane principles which had evolved through the millennia, the authorities began to round up the very best farmers and their families, and to drive them, stripped of their possessions, naked, into the northern wastes, into the tundra and the taiga.

But new waves rolled from the collectivized villages: one of them was a wave of agricultural *wreckers.* Everywhere they began to discover *wrecker* agronomists.

There was even a wave for *snipping ears,* the nighttime snipping of individual ears of grain in the field—a totally new type of agricultural activity, a new type of harvesting! The wave of those caught doing this was not small—it included many tens of thousands of peasants, many of them not even adults but boys, girls, and small children whose elders had sent them out at night to *snip,* because they had no hope of receiving anything from the collective farm for their daytime labor. For this bitter and not very productive occupation (an extreme of poverty to which the peasants had not been driven even in serfdom) the courts handed out a full measure: *ten* years for what ranked as an especially dangerous theft of socialist property.

■

Paradoxically enough, every act of the all-penetrating, eternally wakeful *Organs,* over a span of many years, was based solely on *one* article of the 140 articles of the nongeneral division of the Criminal Code of 1926. One can find more epithets in praise of this article than Turgenev once assembled to praise the Russian language, or Nekrasov to praise Mother Russia: great, powerful, abundant, highly ramified, multiform, wide-sweeping 58, which summed up the world not so much through the exact terms of its sections as in their extended dialectical interpretation.

Who among us has not experienced its all-encompassing embrace? In all truth, there is no step, thought, action, or lack of action under the heavens which could not be punished by the heavy hand of Article 58. . . .

There was no section in Article 58 which was interpreted as broadly and with so ardent a revolutionary conscience as Section 10. Its definition was: "Propaganda or agitation, containing an appeal for the overthrow, subverting, or weakening of the Soviet power . . . and, equally, the dissemination or preparation or possession of literary materials of similar content." For this section in *peacetime* a minimum penalty only was set (not any less! not too light!); *no upper limit* was set for the maximum penalty.

Here is one vignette from those years as it actually occurred. A district Party conference was under way in Moscow Province. It was presided over by a new secretary of the District Party Committee, replacing one recently *arrested.* At the conclusion of the conference, a tribute to Comrade Stalin was called for. Of course, everyone stood up (just as everyone had leaped to his feet during the conference at every mention of his name). The small hall echoed with "stormy applause, rising to an ovation." For three minutes, four minutes, five minutes, the "stormy applause, rising to an ovation," continued. But palms were getting sore and raised arms were already aching. And the older people were panting from exhaustion. It was becoming insufferably silly even to those who really adored Stalin. However, who would dare be the *first* to stop? The secretary of the District Party Committee could have done it. He was standing on the platform, and it was he who had just called for the ovation. But he was a newcomer. He had taken the place of a man who'd been arrested. He was afraid! After all, NKVD men were standing in the hall applauding and watching to see *who* quit first! And in that obscure, small hall, unknown to the Leader, the applause went on—six, seven, eight minutes! They were done for! Their goose was cooked! They couldn't stop now till they collapsed with heart attacks! At the rear of the hall, which was crowded, they could of course cheat a bit, clap less frequently, less vigorously, not so eagerly—but up there with the presidium where everyone could see them? The director of the local paper factory, an independent and strong-minded man, stood with the presidium. Aware of all the falsity and all the impossibility of the situation, he still kept on applauding! Nine minutes! Ten! In anguish he watched the secretary of the District Party Committee, but the latter dared not stop. Insanity! To the last man! With make-believe enthusiasm on their faces, looking at each other with faint hope, the district leaders were just going to go on and on applauding till they fell where they stood, till they were carried out of the hall on stretchers! And even then those who were left would not falter. . . . Then, after eleven minutes, the director of the paper factory assumed a businesslike expression and sat down in his seat. And, oh, a miracle took place! Where had the universal, uninhibited, indescribable enthusiasm gone? To a man, everyone else stopped dead and sat down. They had been saved! The squirrel had been smart enough to jump off his revolving wheel.

That, however, was how they discovered who the independent people were. And that was how they went about eliminating them. That same night the factory director was arrested. They easily pasted ten years on him on the pretext of something quite different. But after he had signed Form 206, the final document of the interrogation, his interrogator reminded him:

"Don't ever be the first to stop applauding!"

(And just what are we supposed to do? How are we supposed to stop?)

Now that's what Darwin's natural selection is. And that's also how to grind people down with stupidity.

But today a new myth is being created. Every story of 1937 that is printed, every reminiscence that is published, relates without exception the tragedy of the Communist leaders. They have kept on assuring us, and we have unwittingly fallen for it, that the history of 1937 and 1938 consisted chiefly of the arrests of the big Communists—and virtually no one else. But out of the *millions* arrested at that time, important

Party and state officials could not possibly have represented more than 10 percent. Most of the relatives standing in line with food parcels outside the Leningrad prisons were lower-class women, the sort who sold milk.

The real law underlying the arrests of those years was *the assignment of quotas,* the norms set, the planned allocations. Every city, every district, every military unit was assigned a specific quota of arrests to be carried out by a stipulated time. From then on everything else depended on the ingenuity of the Security operations personnel.

The former Chekist Aleksandr Kalganov recalls that a telegram arrived in Tashkent: "Send 200!" They had just finished one clean-out, and it seemed as if there was "no one else" to take. Well, true, they had just brought in about fifty more from the districts. And then they had an idea! They would reclassify as 58's all the nonpolitical offenders being held by the police. No sooner said than done. But despite that, they had still not filled the quota. At that precise moment the police reported that a gypsy band had impudently encamped on one of the city squares and asked what to do with them. Someone had another bright idea! They surrounded the encampment and raked in all the gypsy men from seventeen to sixty as 58's! They had fulfilled the plan!

Just as the intelligentsia had never been overlooked in previous waves, it was not neglected in this one. A student's denunciation that a certain lecturer in a higher educational institution kept citing Lenin and Marx frequently but Stalin not at all was all that was needed for the lecturer not to show up for lectures any more. And what if he *cited no one?*

Arrests rolled through the streets and apartment houses like an epidemic. Just as people transmit an epidemic infection from one to another without knowing it, by such innocent means as a handshake, a breath, handing someone something, so, too, they passed on the infection of inevitable arrest by a handshake, by a breath, by a chance meeting on the street. For if you are destined to confess tomorrow that you organized an underground group to poison the city's water supply, and if today I shake hands with you on the street, that means I, too, am doomed.

■

The *reverse wave* of 1939 was an unheard-of incident in the history of the *Organs,* a blot on their record! But, in fact, this reverse wave was not large; it included about 1 to 2 percent of those who had been arrested but not yet convicted, who had not yet been sent away to far-off places and had not yet perished. It was not large, but it was put to effective use. It was like giving back one kopeck change from a ruble, but it was necessary in order to heap all the blame on that dirty Yezhov, to strengthen the newcomer, Beria, and to cause the Leader himself to shine more brightly. With this kopeck they skillfully drove the ruble right into the ground. After all, if "they had sorted things out and freed some people" (and even the newspapers wrote intrepidly about *individual* cases of persons who had been slandered), it meant that the rest of those arrested were indeed scoundrels! And those who returned kept silent. They had signed pledges not to speak out. They were mute with terror. And there were very few who knew even a little about the secrets of the Archipelago.

But for that matter they soon took that kopeck back—during those same years and via those same sections of the boundless Article 58. Well, who in 1940 noticed the wave of wives arrested for *failure to renounce* their husbands? And who in Tambov remembers that during that year of peace they arrested an entire jazz orchestra playing at the "Modern" Cinema Theatre because they all turned out to be enemies of the people? And who noticed the thirty thousand Czechs who in 1939 fled from occupied Czechoslovakia to their Slavic kinfolk in the U.S.S.R.? It was impossible to guarantee that a single one of them was not a spy. They sent them all off to northern camps. And was it not, indeed, in 1939 that we reached out our helping hands to the West Ukrainians and the West Byelorussians, and, in 1940, to the Baltic states and to the Moldavians? It turned out that our brothers badly needed to be purged, and from them, too, flowed waves of *social prophylaxis.* They took those who were too independent, too influential, along with those who were too well-to-do, too intelligent, too noteworthy; they took, particularly, many Poles from former Polish provinces. They arrested officers everywhere. Thus the population was shaken up, forced into silence, and left without any possible leaders of resistance. Thus it was that wisdom was instilled, that former ties and former friendships were cut off.

Finland ceded its isthmus to us with zero population. Nevertheless, the removal and resettlement of all persons with Finnish blood took place throughout Soviet Karelia and in Leningrad in 1940. We didn't notice that wavelet: we have no Finnish blood.

In the Finnish War we undertook our first experiment in convicting our war prisoners as traitors to the Motherland. The first such experiment in human history; and would you believe it?—we didn't notice!

That was the rehearsal—just at that moment the war burst upon us. And with it a massive retreat. It was essential to evacuate swiftly everyone who could be got out of the western republics that were being abandoned to the enemy. In the rush, entire military units—regiments, antiaircraft and artillery batteries—were left behind intact in Lithuania. But they still managed to get out several thousand families of unreliable Lithuanians. From June 23 on, in Latvia and Estonia, they speeded up the arrests. But the ground was burning under them, and they were forced to leave even faster. They forgot to take whole fortresses with them, like the one at Brest, but they did not forget to shoot down political prisoners in the cells and courtyards of Lvov, Rovno, Tallinn, and many other Western prisons. In the Tartu Prison they shot 192 prisoners and threw their corpses down a well.

How can one visualize it? You know nothing. The door of your cell opens, and they shoot you. You cry out in your death agony, and there is no one to hear your cries or tell of them except the prison stones. They say, however, that there were some who weren't successfully finished off, and we may someday read a book about that too.

In 1941 the Germans went round Taganrog, cutting it off so swiftly that prisoners were left in freight wagons at the railway station where they had been brought to be evacuated. What should one do with them? Certainly not set them free nor leave them to the Germans. Oil tank trucks were rushed to the station, and the wagons were drenched with oil and set on fire. All the prisoners were burned alive.

In the rear, the first wartime wave was for *those spreading rumors and panic.* That was the language of a special decree, outside the Code, issued in the first days of the war.

Then there was a wave of those who *failed to turn in radio receivers* or radio parts. For one radio tube found (as a result of denunciation) they gave ten years.

Then there was the wave of *Germans*—Germans living on the Volga, colonists in the Ukraine and the North Caucasus, and all Germans in general who lived anywhere in the Soviet Union. The determining factor here was *blood,* and even heroes of the Civil War and old members of the Party who were German were sent off into exile.

By the end of the summer of 1941, becoming bigger in the autumn, the wave of *the encircled* was surging in. These were the defenders of their native land, the very same warriors whom the cities had seen off to the front with bouquets and bands a few months before, who had then sustained the heaviest tank assaults of the Germans, and in the general chaos, and through no fault of their own, had spent a certain time as isolated units not in enemy imprisonment, not at all, but in temporary encirclement, and later had broken out. And instead of being given a brotherly embrace on their return, such as every other army in the world would have given them, instead of being given a chance to rest up, to visit their families, and then return to their units—they were held on suspicion, disarmed, deprived of all rights, and taken away in groups to identification points and screening centers where officers of the Special Branches started interrogating them, distrusting not only their every word but their very identity.

The victory outside Moscow gave rise to a new wave: guilty Muscovites. Looking at things after the event, it turned out that those Muscovites who had not run away and who had not been evacuated but had fearlessly remained in the threatened capital, which had been abandoned by the authorities, were by that very token under suspicion either of subverting governmental authority (58-10); or of staying on to await the Germans. . . .

From 1943 on, when the war turned in our favor, there began the multimillion wave from the occupied territories and from Europe, which got larger every year up to 1946.

And dismiss the thought that honorable participation in an underground anti-German organization would surely protect one from being arrested in this wave. More than one case proved this.

Those who were in Europe got the stiffest punishments of all, even though they went there as conscripted German slaves. That was because they had seen something of European life and could talk about it.

That also was the reason why they sentenced the majority of *war prisoners* (it was not simply because they had allowed themselves to be captured), particularly those POW's who had seen a little more of the West than a German death camp. This was obvious from the fact that *interned persons* were sentenced as severely as POW's. For example, during the first days of the war one of our destroyers went aground on Swedish territory. Its crew proceeded to live freely in Sweden during all the rest of the war. After the war Sweden returned them to us along with the destroyer. Their treason to the Motherland was indubitable—but somehow the case didn't get off the ground. They let them go their different ways and then pasted them with Anti-Soviet Agitation for their lovely stories in praise of freedom and good eating in capitalist Sweden. (This was the Kadenko group.)

What happened to this group later makes an anecdote. In camp they kept their mouths shut about Sweden, fearing they'd get a second term. But people in Sweden somehow found out about their fate and published slanderous reports in the press. By that time the boys were scattered far and near among various camps. Suddenly, on the strength of special orders, they were all yanked out and taken to the Kresty Prison in Leningrad. There they were fed for two months as though for slaughter and allowed to let their hair grow. Then they were dressed with modest elegance, rehearsed on what to say and to whom, and warned that any bastard who dared to squeak out of turn would get a bullet in his skull—and they were led off to a press conference for selected foreign journalists and some others who had known the entire crew in Sweden. The former internees bore themselves cheerfully, described where they were living, studying, and working, and expressed their indignation at the bourgeois slander they had *read* about not long before in the Western press (after all, Western papers are sold in the Soviet Union at every corner newsstand!). And so they had written to one another and decided to gather in Leningrad. (Their travel expenses didn't bother them in the least.) Their fresh, shiny appearance completely gave the lie to the newspaper canard. The discredited journalists went off to write their apologies. It was wholly inconceivable to the Western imagination that there could be any other explanation. And the men who had been the subjects of the interview were taken off to a bath, had their hair cut off again, were dressed in their former rags, and sent back to the same camps. But because they had conducted themselves properly, none of them was given a second term.

During the last years of the war, of course, there was a wave of German *war criminals* who were selected from the POW camps and transferred by court verdict to the jurisdiction of Gulag.

In 1945, even though the war with Japan didn't last three weeks, great numbers of Japanese war prisoners were raked in for urgent construction projects in Siberia and Central Asia, and the same process of selecting *war criminals* for Gulag was carried out among them.

At the end of 1944, when our army entered the Balkans, and especially in 1945, when it reached into Central Europe, a wave of Russian émigrés flowed through the channels of Gulag. Most were old men, who had left at the time of the Revolution, but there were also young people, who had grown up outside Russia. They usually dragged off the menfolk and left the women and children where they were. It is true that they did not take everyone, but they took all those who, in the course of twenty-five years, had expressed even the mildest political views, or who had expressed them earlier, during the Revolution. They did not touch those who had lived a purely vegetable existence. The main waves came from Bulgaria, Yugoslavia, and Czechoslovakia; there were fewer from Austria and Germany. In the other countries of

Eastern Europe, there were hardly any Russians.

As if in response to 1945, a wave of émigrés poured from Manchuria too. (Some of them were not arrested immediately. Entire families were encouraged to return to the homeland as free persons, but once back in Russia they were separated and sent into exile or taken to prison.)

All during 1945 and 1946 a big wave of genuine, at-long-last, enemies of the Soviet government flowed into the Archipelago. (These were the Vlasov men, the Krasnov Cossacks, and Moslems from the national units created under Hitler.) Some of them had acted out of conviction; others had been merely involuntary participants.

Along with them were seized *not less than one million fugitives from the Soviet government*— civilians of all ages and of both sexes who had been fortunate enough to find shelter on Allied territory, but who in 1946–1947 were perfidiously returned by Allied authorities into Soviet hands.

It is surprising that in the West, where political secrets cannot be kept long, since they inevitably come out in print or are disclosed, the secret of *this* particular act of betrayal has been very well and carefully kept by the British and American governments. This is truly the last secret, or one of the last, of the Second World War. Having often encountered these people in camps, I was unable to believe for a whole quarter-century that the public in the West knew *nothing* of this action of the Western governments, this massive handing over of ordinary Russian people to retribution and death. Not until 1973—in the *Sunday Oklahoman* of January 21—was an article by Julius Epstein published. And I am here going to be so bold as to express gratitude on behalf of the mass of those who perished and those few left alive. One random little document was published from the many volumes of the hitherto concealed case history of forced repatriation to the Soviet Union. "After having remained unmolested in British hands for two years, they had allowed themselves to be lulled into a false sense of security and they were therefore taken completely by surprise. . . . They did not realize they were being repatriated. . . . They were mainly simple peasants with bitter personal grievances against the Bolsheviks." The English authorities gave them the treatment "reserved in the case of every other nation for war criminals alone: that of being handed over against their will to captors who, incidentally, were not expected to give them a fair trial." They were all sent to destruction on the Archipelago. The American authorities did the same: in Bavaria as well as on the U.S. territory, they delivered tens of thousands of Soviet citizens to a cruel fate, turning them over to the Soviets against their will.

A certain number of *Poles,* members of the Home Army, followers of Mikolajczyk, arrived in Gulag in 1945 via our prisons.

There were a certain number of *Rumanians* and *Hungarians.*

At war's end and for many years after, there flowed uninterruptedly an abundant wave of Ukrainian nationalists.

■

We have to remind our readers once again that this chapter does not attempt by any means to list *all* the waves which fertilized Gulag—but only those which had a political coloration. And just as, in a course in physiology, after a detailed description of the circulation of the blood, one can begin over again and describe in detail the lymphatic system, one could begin again and describe the waves of *nonpolitical offenders* and *habitual criminals* from 1918 to 1953. And this description, too, would run long. It would bring to light many famous decrees, now in part forgotten (even though they have never been repealed), which supplied abundant human material for the insatiable Archipelago.

We are not going to go into a lengthy and lavish examination of the waves of nonpolitical offenders and common criminals. But, having reached 1947, we cannot remain silent about one of the most grandiose of Stalin's decrees. We have already mentioned the famous law of "Seven-Eight" or "Seven-eighths," on the basis of which they arrested people right and left—for taking a stalk of grain, a cucumber, two small potatoes, a chip of wood, a spool of thread—all of whom got ten years.

But the requirements of the times, as Stalin understood them, had changed, and the *tenner,* which had seemed adequate on the eve of a terrible war, seemed now, in the wake of a world-wide historical victory, inadequate. And so again, in complete disregard of the Code, and totally overlooking the fact that many different articles and decrees on the subject of thefts and robberies already existed, on June 4, 1947, a decree was issued which outdid them all. It was instantly christened "Four-sixths" by the undismayed prisoners.

The advantages of the new decree lay first of all in its newness. From the very moment it appeared, a torrent of the crimes it specified would be bound to burst forth, thereby providing an abundant wave of newly sentenced prisoners. But it offered an even greater advantage in prison terms. If a young girl sent into the fields to get a few ears of grain took along two friends for company ("an organized gang") or some twelve-year-old youngsters went after cucumbers or apples, they were liable to get *twenty* years in camp. In factories, the maximum sentence was raised to *twenty-five years.* (This sentence, called the *quarter,* had been introduced a few days earlier to replace the death penalty, which had been abolished as a humane act.)

And then, at long last, an ancient shortcoming of the law was corrected. Previously the only failure to make a denunciation which qualified as a crime against the state had been in connection with political offenses. But now simple failure to report the theft of state or collective farm property earned three years of camp or seven years of exile.

Stalin's new line, suggesting that it was necessary, in the wake of the victory over fascism, *to jail* more people more energetically and for longer terms than ever before, had immediate repercussions, of course, on political prisoners.

The year 1948–1949, notable throughout Soviet public life for intensified persecution and vigilance, was marked by one tragicomedy hitherto unheard of even in Stalinist antijustice—that of the *repeaters.*

That is what, in the language of Gulag, they called those still undestroyed unfortunates of 1937 vintage, who had succeeded in surviving ten impossible, unendurable years, and who in 1947–1948, had timidly stepped forth onto the land of *freedom* . . . worn out, broken in health, but hoping to live out in peace what little of their lives remained. But some sort of savage fantasy (or stubborn malice, or unsated vengeance) pushed the Victorious Generalissimo into issuing the order to arrest all those cripples over again, without any new charges! It was even disadvantageous, both economically and politically, to clog the meat grinder with its own refuse. But Stalin issued the order anyway. Here was a case in which a historical personality simply behaved capriciously toward historical necessity.

At this point the Autocrat decided it wasn't enough to arrest just those who had survived since 1937! What about the *children* of his sworn enemies? They, too, must be imprisoned! They were growing up, and they might have notions of vengeance.

By 1948, after the great European displacement, Stalin had succeeded once again in tightly barricading himself in and pulling the ceiling down closer to him: in this reduced space he had recreated the tension of 1937.

And so in 1948, 1949, and 1950 there flowed past:

- Alleged spies (ten years earlier they had been German and Japanese, now they were Anglo-American).
- Believers (this wave non-Orthodox for the most part).
- Those geneticists and plant breeders, disciples of the late Vavilov and of Mendel, who had not previously been arrested.

- Just plain ordinary thinking people (and students, with particular severity) who had not been sufficiently scared away from the West. It was fashionable to charge them with:
- VAT—Praise of American Technology;
- VAD—Praise of American Democracy; and
- PZ—Toadyism Toward the West.

These waves were not unlike those of 1937, but the *sentences* were different. The standard sentence was no longer the patriarchal *ten-ruble bill,* but the new Stalinist *twenty-five.* By now the *tenner* was for *juveniles.*

By this time resistance in Lithuania and Estonia had already come to an end. But in 1949 new waves of new "social prophylaxis" to assure collectivization kept coming. They took whole trainloads of city dwellers and peasants from the three Baltic republics into Siberian exile. (The historical rhythm was disrupted in these republics: they were forced to recapitulate in brief, limited periods the more extended experience of the rest of the country.)

In 1948 one more nationalist wave went into exile—that of the *Greeks* who inhabited the areas around the Sea of Azov, the Kuban, and Sukhumi. They had done nothing to offend the Father during the war, but now he avenged himself on them for his failure in Greece, or so it seemed. This wave, too, was evidently the fruit of his personal insanity.

During the last years of Stalin's life, a wave of *Jews* became noticeable. (From 1950 on they were hauled in little by little as *cosmopolites.* And that was why the *doctors'* case was cooked up. It would appear that Stalin intended to arrange a great massacre of the Jews.)

But this became the first plan of his life to fail. God told him— apparently with the help of human hands—to depart from his rib cage.

The preceding exposition should have made it clear, one would think, that in the removal of millions and in the populating of Gulag, consistent, cold-blooded planning and never-weakening persistence were at work.

That we never did have any *empty* prisons, merely prisons which were full or prisons which were very, very overcrowded.

And that while you occupied yourself to your heart's content studying the safe secrets of the atomic nucleus, researching the influence of Heidegger on Sartre, or collecting Picasso reproductions; while you rode off in your railroad sleeping compartment to vacation resorts, or finished building your country house near Moscow—the Black Marias rolled incessantly through the streets and the *gaybisty*—the State Security men—knocked at doors and rang doorbells.

And I think this exposition proves that the *Organs* always earned their pay.

Chapter 3

■

The Interrogation

If the intellectuals in the plays of Chekhov who spent all their time guessing what would happen in twenty, thirty, or forty years had been told that in forty years interrogation by torture would be practiced in Russia; that prisoners would have their skulls squeezed within iron rings, that a human being would be lowered into an acid bath; that they would be trussed up naked to be bitten by ants and bedbugs; that a ramrod heated over a primus stove would be thrust up their anal canal (the "secret brand"); that a man's genitals would be slowly crushed beneath the toe of a jackboot; and that, in the luckiest possible circumstances, prisoners would be tortured by being kept from sleeping for a week, by thirst, and by being beaten to a bloody pulp, not one of Chekhov's plays would have gotten to its end because all the heroes would have gone off to insane asylums.

Yes, not only Chekhov's heroes, but what normal Russian at the beginning of the century, including any member of the Russian Social Democratic Workers' Party, could have believed, would have tolerated, such a slander against the bright future? What had been acceptable under Tsar Aleksei Mikhailovich in the seventeenth century, what had already been regarded as barbarism under Peter the Great, what might have been used against ten or twenty people in all during the time of Biron in the mid-eighteenth century, what had already become totally impossible under Catherine the Great, was all being practiced during the flowering of the glorious twentieth century—in a society based on socialist principles, and at a time when airplanes were flying and the radio and talking films had already appeared—not by one scoundrel alone in one secret place only, but by tens of thousands of specially trained human beasts standing over millions of defenseless victims.

Was it only that explosion of atavism which is now evasively called "the cult of personality" that was so horrible? Or was it even more horrible that during those same years, in 1937 itself, we celebrated Pushkin's centennial? And that we shamelessly continued to stage those self-same Chekhov plays, even though the answers to them had already come in? Is it not still more dreadful that we are now being told, thirty years later, "Don't talk about it!"? If we start to recall the sufferings of millions, we are told it will distort the historical perspective! If we doggedly seek out the essence of our morality, we are told it will darken our material progress! Let's think rather about the blast furnaces, the rolling mills that were built, the canals that were dug . . . no, better not talk about the canals. . . . Then maybe about the gold of the Kolyma? No, maybe we ought not to talk about that either. . . . Well, we can talk about anything, so long as we do it adroitly, so long as we glorify it. . . .

It is really hard to see why we condemn the Inquisition. Wasn't it true that beside the autos-da-fé, magnificent services were offered the Almighty? It is hard to see why we are so down on serfdom. After all, no one forbade the peasants to work every day. And they could sing carols at Christmas too. And for Trinity Day the girls wove wreaths. . . .

■

In his *Dictionary of Definitions* Dal makes the following distinction: "An *inquiry* is distinguished from an *investigation* by the fact that it is carried out to determine whether there is a basis for proceeding to an *investigation.*"

On, sacred simplicity! The *Organs* have never heard of such a thing as an *inquiry!* Lists of names prepared up above, or an initial suspicion, or a denunciation by an informer, or any anonymous denunciation, were all that was needed to bring about the arrest of the suspect, followed by the inevitable formal charge. The time allotted for investigation was not used to unravel the crime but, in ninety-five cases out of a hundred, to exhaust, wear down, weaken, and render helpless the defendant, so that he would want it to end at any cost.

As long ago as 1919 the chief method used by the interrogator was *a revolver on the desk.* That was how they investigated not only political but also ordinary misdemeanors and violations. At the trial of the Main Fuels Committee (1921), the accused Makhrovskaya complained that at her interrogation she had been drugged with cocaine. The prosecutor replied: "If she had declared that she had been treated rudely, that they had *threatened to shoot her, this might be just barely believable.*" The frightening revolver lies there and sometimes it is aimed at you, and the interrogator doesn't tire himself out thinking up what you are guilty of, but shouts: "Come on, talk! You know what about!" That was what the interrogator Khaikin demanded of Skripnikova in 1927. That was what they demanded of Vitkovsky in 1929. And twenty-five years later nothing had changed. In 1952 Anna Skripnikova was undergoing her *fifth* imprisonment, and Sivakov, Chief of the Investigative Department of the Ordzhonikidze State Security Administration, said to her: "The prison doctor reports you have a blood pressure of 240/120. That's too low, you bitch! We're going to drive it up to 340 so you'll kick the bucket, you viper, and with no black and blue marks; no beatings; no broken bones. We'll just not let you sleep." She was in her fifties at the time. And if, back in her cell, after a night spent in interrogation, she closed her eyes during the day, the jailer broke in and shouted: "Open your eyes or I'll haul you off that cot by the legs and tie you to the wall standing up."

As early as 1921 interrogations usually took place at night. At that time, too, they shone automobile lights in the prisoner's face. And at the Lubyanka in 1926 they made use of the hot-air heating system to fill the cell first with icy-cold and then with stinking hot air. And there was an airtight cork-lined cell in which there was no ventilation and they cooked the prisoners. A participant in the Yaroslavl uprising of 1918, Vasily Aleksandrovich Kasyanov, described how the heat in such a cell was turned up until your blood began to ooze through your pores. When they saw this happening through the peephole, they would put the prisoner on a stretcher and take him off to sign his confession. The "hot" and "salty" methods of the "gold" period are well known. And in Georgia in 1926 they used lighted cigarettes to burn the hands of prisoners under interrogation. In Metekhi Prison they pushed prisoners into a cesspool in the dark.

There is a very simple connection here. Once it was established that charges had to be brought at any cost and despite everything, threats, violence, tortures became inevitable. And the more fantastic the charges were, the more ferocious the interrogation had to be in order to force the required confession. Given the fact that the cases were always fabricated, violence and torture had to accompany them. This was not peculiar to 1937 alone. It was a chronic, general practice. And that is why it seems strange today to read in the recollections of former zeks that "torture was permitted from the spring of 1938 on."

There were never any spiritual or moral barriers which could have held the *Organs* back from torture. In the early postwar years, in the *Cheka Weekly, The Red Sword,* and *Red Terror,* the admissibility of torture from a Marxist point of view was openly debated. Judging by the subsequent course of events, the answer deduced was positive, though not universally so.

It is more accurate to say that if before 1938 some kind of formal documentation was required as a preliminary to torture, as well as specific permission for each case under investigation (even though such permission was easy to obtain), then in the years 1937–1938, in view of the extraordinary situation prevailing (the specified millions of admissions to the Archipelago had to be ground through the apparatus of individual interrogation in specified, limited periods, something which had simply not happened in the mass waves of kulaks and nationalities), interrogators were allowed to use violence and torture on an unlimited basis, at their own discretion, and in accordance with the demands of their work quotas and the amount of time they were given. The types of torture used were not regulated and every kind of ingenuity was permitted, no matter what.

In 1939 such indiscriminate authorization was withdrawn, and once again written permission was required for torture, and perhaps it may not have been so easily granted. (Of course, simple threats, blackmail, deception, exhaustion through enforced sleeplessness, and punishment cells were never prohibited.) Then, from the end of the war and throughout the postwar years, certain *categories* of prisoners were established by decree for whom a broad range of torture was automatically permitted. Among these were nationalists, particularly the Ukrainians and the Lithuanians, especially in those cases where an underground organization existed (or was suspected) that had to be completely uncovered, which meant obtaining the names of everyone involved from those already arrested.

It would also be incorrect to ascribe to 1937 the "discovery" that the personal confession of an accused person was more important than any other kind of proof or facts. This concept had already been formulated in the twenties. And 1937 was just the year when the brilliant teaching of Vyshinsky came into its own. Incidentally, even at that time, his teaching was transmitted only to interrogators and prosecutors—for the sake of their morale and steadfastness. The rest of us only learned about it twenty years later—when it had already come into disfavor—through subordinate clauses and minor paragraphs of newspaper articles, which treated the subject as if it had long been widely known to all.

It turns out that in that terrible year Andrei Yanuaryevich (one longs to blurt out, "Jaguaryevich") Vyshinsky, availing himself of the most flexible dialectics (of a sort nowadays not permitted either Soviet citizens or electronic calculators, since to them *yes* is *yes* and *no* is *no*), pointed out in a report which became famous in certain circles that it is never possible for mortal men to establish absolute truth, but relative truth only. He then proceeded to a further step, which jurists of the last two thousand years had not been willing to take: that the truth established by interrogation and trial could not be absolute, but only, so to speak, relative. Therefore, when we sign a sentence ordering someone to be shot we can never be *absolutely* certain, but only approximately, in view of certain hypotheses, and in a certain sense, that we are punishing a *guilty person.* Thence arose the most practical conclusion: that it was useless to seek absolute evidence—for evidence is always relative—or unchallengeable witnesses—for they can say different things at different times. The proofs of guilt were *relative,* approximate, and the interrogator could find them, even when there was no evidence and no witness, without leaving his office, "basing his conclusions not only on his own intellect but also on his Party sensitivity, his *moral forces*" (in other words, the superiority of someone who has slept well, has been well fed, and has not been beaten up) "and on his *character*" (i.e., his willingness to apply cruelty!).

In only one respect did Vyshinsky fail to be consistent and retreat from dialectical logic: for some reason, the executioner's *bullet* which he allowed was not relative but *absolute.* . . .

Thus it was that the conclusions of advanced Soviet jurisprudence, proceeding in a spiral, returned to barbaric or medieval standards. Like medieval torturers, our interrogators, prosecutors, and judges agreed to accept the confession of the accused as the chief proof of guilt.

However, the simple-minded Middle Ages used dramatic and picturesque methods to squeeze out the desired confessions: the rack, the wheel, the bed of nails, impalement, hot coals, etc. In the twentieth century, taking advantage of our more highly developed medical knowledge and extensive prison experience (and someone seriously defended a doctoral dissertation on this theme), people came to realize that the accumulation of such impressive apparatus was superfluous and that, on a mass scale, it was also cumbersome. And in addition . . .

In addition, there was evidently one other circumstance. As always, Stalin did not pronounce that final word, and his subordinates had to guess what he wanted. Thus, like a jackal, he left himself an escape hole, so that he could, if he wanted, beat a retreat and write about "dizziness from success." After all, for the first time in human history the calculated torture of millions was being undertaken, and, even with all his strength and power, Stalin could not be absolutely sure of success. In dealing with such an enormous mass of material, the effects of the experiment might differ from those obtained from a smaller sample. An unforeseen explosion might take place, a slippage in a geological fault, or even world-wide disclosure. In any case, Stalin had to remain innocent, his sacred vestments angelically pure.

We are therefore forced to conclude that no list of tortures and torments existed in printed form for the guidance of interrogators! Instead, all that was required was for every Interrogation Department to supply the tribunal within a specified period with a stipulated number of rabbits who had confessed everything. *And it was simply stated,* orally but often, that any measures and means employed were good, since they were being used for a lofty purpose; that no interrogator would be made to answer for the death of an accused; and that the prison doctor should interfere as little as possible with the course of the investigation. In all probability, they exchanged experiences in comradely fashion; "they learned from the most successful workers." Then, too, "material rewards" were offered—higher pay for night work, bonus pay for fast work—and there were also definite warnings that interrogators who could not cope with their tasks . . . Even the chief of some provincial NKVD administration, if some sort of mess developed, could show Stalin his hands were clean: he had issued no direct instructions to use torture! But at the same time he had ensured that torture would be used!

Let us try to list some of the simplest methods which break the will and the character of the prisoner without leaving marks on his body.

Let us begin with *psychological* methods. These methods have enormous and even annihilating impact on rabbits who have never been prepared for prison suffering. And it isn't easy even for a person who holds strong convictions.

1. First of all: *night.* Why is it that all the main work of breaking down human souls went on at *night?* Why, from their very earliest years, did the *Organs* select the *night?* Because at night, the prisoner torn from sleep, even though he has not yet been tortured by sleepless-

ness, lacks his normal daytime equanimity and common sense. He is more vulnerable.

2. *Persuasion* in a sincere tone is the very simplest method. Why play at cat and mouse, so to speak? After all, having spent some time among others undergoing interrogation, the prisoner has come to see what the situation is. And so the interrogator says to him in a lazily friendly way: "Look, you're going to get a prison term whatever happens. But if you resist, *you'll croak* right here in prison, you'll lose your health. But if you go to camp, you'll have fresh air and sunlight. . . . So why not sign right now?" Very logical. And those who agree and sign are smart, if . . . if the matter concerns only themselves! But that's rarely so. A struggle is inevitable.

Another variant of persuasion is particularly appropriate to the Party member. "If there are shortages and even famine in the country, then you as a Bolshevik have to make up your mind: can you admit that the whole Party is to blame? Or the whole Soviet government?" "No, of course not!" the director of the flax depot hastened to reply. "Then be brave, and shoulder the blame yourself!" And he did!

3. *Foul language* is not a clever method, but it can have a powerful impact on people who are well brought up, refined, delicate. I know of two cases involving priests, who capitulated to foul language alone. One of them, in the Butyrki in 1944, was being interrogated by a woman. At first when he'd come back to our cell he couldn't say often enough how polite she was. But once he came back very despondent, and for a long time he refused to tell us how, with her legs crossed high, she had begun to *curse.* (I regret that I cannot cite one of her little phrases here.)

4. *Psychological contrast* was sometimes effective: sudden reversals of tone, for example. For a whole or part of the interrogation period, the interrogator would be extremely friendly, addressing the prisoner formally by first name and patronymic, and promising everything. Suddenly he would brandish a paperweight and shout: "Foo, you rat! I'll put nine grams of lead in your skull!" And he would advance on the accused, clutching hands outstretched as if to grab him by the hair, fingernails like needles. (This worked very, very well with women prisoners.)

Or as a variation on this: two interrogators would take turns. One would shout and bully. The other would be friendly, almost gentle. Each time the accused entered the office he would tremble—which would it be? He wanted to do everything to please the gentle one because of his different manner, even to the point of signing and confessing to things that had never happened.

5. Preliminary *humiliation* was another approach. In the famous cellars of the Rostov-on-the-Don GPU (House 33), which were lit by lenslike insets of thick glass in the sidewalk above the former storage basement, prisoners awaiting interrogation were made to lie face down for several hours in the main corridor and forbidden to raise their heads or make a sound. They lay this way, like Moslems at prayer, until the guard touched a shoulder and took them off to interrogation. Another case: At the Lubyanka, Aleksandra O———va refused to give the testimony demanded of her. She was transferred to Lefortovo. In the admitting office, a woman jailer ordered her to undress, allegedly for a medical examination, took away her clothes, and locked her in a "box" naked. At that point the men jailers began to peer through the peephole and to appraise her female attributes with loud laughs. If one were systematically to question former prisoners, many more such examples would certainly emerge. They all had but a single purpose: to dishearten and humiliate.

6. Any method of inducing extreme *confusion* in the accused might be employed. Here is how F.I.V. from Krasnogorsk, Moscow Province, was interrogated. (This was reported by I. A. P———ev.) During the interrogation, the interrogator, a woman, undressed in front of him by stages (a striptease!), all the time continuing the interrogation as if nothing were going on. She walked about the room and came close to him and tried to get him to give in. Perhaps this satisfied some personal quirk in her, but it may also have been cold-blooded calculation, an attempt to get the accused so muddled that he would sign. And she was in no danger. She had her pistol, and she had her alarm bell.

7. *Intimidation* was very widely used and very varied. It was often accompanied by *enticement* and by *promises* which were, of course, false. In 1924: "If you don't confess, you'll go to the Solovetsky Islands. Anybody who confesses is turned loose." In 1944: "Which camp you'll be sent to depends on us. Camps are different. We've got hard-labor camps now. If you confess, you'll go to an easy camp. If you're stubborn, you'll get twenty-five years in handcuffs in the mines!" Another form of intimidation was threatening a prisoner with a prison worse than the one he was in. "If you keep on being stubborn, we'll send you to Lefortovo" (if you are in the Lubyanka), "to Sukhanovka" (if you are at Lefortovo). "They'll find another way to talk to you there." You have already gotten used to things where you are; the regimen seems

to be *not so bad;* and what kind of torments await you *elsewhere?* Yes, and you also have to be transported there. . . . Should you give in?

Intimidation worked beautifully on those who had not yet been arrested but had simply received an official summons to the Bolshoi Dom—the Big House. He (or she) still had a lot to lose. He (or she) was frightened of everything—that they wouldn't let him (or her) out today, that they would confiscate his (or her) belongings or apartment. He would be ready to give all kinds of testimony and make all kinds of concessions in order to avoid these dangers. She, of course, would be ignorant of the Criminal Code, and, at the very least, at the start of the questioning they would push a sheet of paper in front of her with a fake citation from the Code: "I have been warned that for giving false testimony . . . five years of imprisonment." (In actual fact, under Article 95, it is two years.) "For refusal to give testimony—five years . . ." (In actual fact, under Article 92, it is up to three months.) Here, then, one more of the interrogator's basic methods has entered the picture and will continue to re-enter it.

8. *The lie.* We lambs were forbidden to lie, but the interrogator could tell all the lies he felt like. Those articles of the law did not apply to him. We had even lost the yardstick with which to gauge: what does he get for lying? He could confront us with as many documents as he chose, bearing the forged signatures of our kinfolk and friends—and it would be just a skillful interrogation technique.

Intimidation through enticement and lies was the fundamental method for bringing pressure on the *relatives* of the arrested person when they were called in to give testimony. "If you don't tell us such and such" (whatever was being asked), "it's going to be the worse for *him.* . . . You'll be destroying him completely." (How hard for a mother to hear that!) "Signing this paper" (pushed in front of the relatives) "is the only way you can save him" (destroy him).

Under the "harsh laws" of the Tsarist Empire, close relatives could refuse to testify. And even if they gave testimony at a preliminary investigation, they could choose to repudiate it and refuse to permit it to be used in court. And, curiously enough, kinship or acquaintance with a criminal was never in itself considered evidence.

9. *Playing on one's affection* for those one loved was a game that worked beautifully on the accused as well. It was the most effective of all methods of intimidation. One could break even a totally fearless

person through his concern for those he loved. (Oh, how foresighted was the saying: "A man's family are his enemies.") Remember the Tatar who bore his sufferings—his own and those of his wife—but could not endure his daughter's! In 1930, Rimalis, a woman interrogator, used to threaten: "We'll arrest your daughter and lock her in a cell with syphilitics!" And that was a woman!

They would threaten to arrest everyone you loved. Sometimes this would be done with sound effects: Your wife has already been arrested, but her further fate depends on you. They are questioning her in the next room—just listen! And through the wall you can actually hear a woman weeping and screaming. (After all, they all sound alike; you're hearing it through a wall; and you're under terrific strain and not in a state to play the expert on voice identification. Sometimes they simply play a recording of the voice of a "typical wife"—soprano or contralto —a labor-saving device suggested by some inventive genius.) And then, without fakery, they actually show her to you through a glass door, as she walks along in silence, her head bent in grief. Yes! Your own wife in the corridors of State Security! You have destroyed her by your stubbornness! She has already been arrested! (In actual fact, she has simply been summoned in connection with some insignificant procedural question and sent into the corridor at just the right moment, after being told: "Don't raise your head, or you'll be kept here!") Or they give you a letter to read, and the handwriting is exactly like hers: "I renounce you! After the filth they have told me about you, I don't need you any more!" (And since such wives do exist in our country, and such letters as well, you are left to ponder in your heart: Is that the kind of wife she really is?)

Just as there is no classification in nature with rigid boundaries, it is impossible rigidly to separate psychological methods from *physical* ones. Where, for example, should we classify the following amusement?

10. *Sound effects:* The accused is made to stand twenty to twenty-five feet away and is then forced to speak more and more loudly and to repeat everything. This is not easy for someone already weakened to the point of exhaustion. Or two megaphones are constructed of rolled-up cardboard, and two interrogators, coming close to the prisoner, bellow in both ears: "Confess, you rat!" The prisoner is deafened; sometimes he actually loses his sense of hearing. But this method is uneconomical. The fact is that the interrogators like some diversion in their monotonous work, and so they vie in thinking up new ideas.

11. *Tickling:* This is also a diversion. The prisoner's arms and legs are bound or held down, and then the inside of his nose is tickled with

a feather. The prisoner writhes; it feels as though someone were drilling into his brain.

12. *A cigarette is put out* on the accused's skin (already mentioned above).

13. *Light effects* involve the use of an extremely bright electric light in the small, white-walled cell or "box" in which the accused is being held—a light which is never extinguished. (The electricity saved by the economies of schoolchildren and housewives!) Your eyelids become inflamed, which is very painful. And then in the interrogation room searchlights are again directed into your eyes.

14. Here is another imaginative trick: On the eve of May 1, 1933, in the Khabarovsk GPU, for *twelve* hours—all night—Chebotaryev was not interrogated, no, but was simply kept in a continual state of being *led to* interrogation. "Hey, you—hands behind your back!" They led him out of the cell, up the stairs quickly, into the interrogator's office. The guard left. But the interrogator, without asking one single question, and sometimes without even allowing Chebotaryev to sit down, would pick up the telephone: "Take away the prisoner from 107!" And so they came to get him and took him back to his cell. No sooner had he lain down on his board bunk than the lock rattled: "Chebotaryev! To interrogation. Hands behind your back!" And when he got there: "Take away the prisoner from 107!"

For that matter, the methods of bringing pressure to bear can begin a long time before the interrogator's office.

15. Prison begins with the *box*, in other words, what amounts to a closet or packing case. The human being who has just been taken from freedom, still in a state of inner turmoil, ready to explain, to argue, to struggle, is, when he first sets foot in prison, clapped into a "box," which sometimes has a lamp and a place where he can sit down, but which sometimes is dark and constructed in such a way that he can only stand up and even then is squeezed against the door. And he is held there for several hours, or for half a day, or a day. During those hours he knows absolutely nothing! Will he perhaps be confined there all his life? He has never in his life encountered anything like this, and he cannot guess at the outcome. Those first hours are passing when everything inside him is still ablaze from the unstilled storm in his heart. Some become despondent—and that's the time to subject them to their first interrogation. Others become angry—and that, too, is all to the good, for they may insult the interrogator right at the start or make a slip, and it will be all the easier to cook up their case.

16. When boxes were in short supply, they used to have another

method. In the Novocherkassk NKVD, Yelena Strutinskaya was forced to remain seated on a stool in the corridor for six days in such a way that she did not lean against anything, did not sleep, did not fall off, and did not get up from it. Six days! Just try to sit that way for six hours!

Then again, as a variation, the prisoner can be forced to sit on a tall chair, of the kind used in laboratories, so that his feet do not reach the floor. They become very numb in this position. He is left sitting that way from eight to ten hours.

Or else, during the interrogation itself, when the prisoner is out in plain view, he can be forced to sit in this way: as far forward as possible on the front edge ("Move further forward! Further still!") of the chair so that he is under painful pressure during the entire interrogation. He is not allowed to stir for several hours. Is that all? Yes, that's all. Just try it yourself!

17. Depending on local conditions, a *divisional pit* can be substituted for the box, as was done in the Gorokhovets army camps during World War II. The prisoner was pushed into such a pit, ten feet in depth, six and a half feet in diameter; and beneath the open sky, rain or shine, this pit was for several days both his cell and his latrine. And ten and a half ounces of bread, and water, were lowered to him on a cord. Imagine yourself in this situation just after you've been arrested, when you're all in a boil.

Either identical orders to all Special Branches of the Red Army or else the similarities of their situations in the field led to broad use of this method. Thus, in the 36th Motorized Infantry Division, a unit which took part in the battle of Khalkhin-Gol, and which was encamped in the Mongolian desert in 1941, a newly arrested prisoner was, without explanation, given a spade by Chief of the Special Branch Samulyev and ordered to dig a pit the exact dimensions of a *grave.* (Here is a hybridization of physical and psychological methods.) When the prisoner had dug deeper than his own waist, they ordered him to stop and sit down on the bottom: his head was no longer visible. One guard kept watch over several such pits and it was as though he were surrounded by empty space. They kept the accused in this desert with no protection from the Mongolian sun and with no warm clothing against the cold of the night, but no tortures—why waste effort on tortures? The ration they gave was *three and a half ounces of bread* per day and *one glass of water.* Lieutenant Chulpenyev, a giant, a boxer, twenty-one years old, spent *a month* imprisoned this way. Within ten days he was swarming with lice. After fifteen days he was summoned to interrogation for the first time.

18. The accused could be compelled *to stand on his knees*—not in some figurative sense, but literally: on his knees, without sitting back on his heels, and with his back upright. People could be compelled to kneel in the interrogator's office or the corridor for twelve, or even twenty-four or forty-eight hours. (The interrogator himself could go home, sleep, amuse himself in one way or another—this was an organized system; watch was kept over the kneeling prisoner, and the guards worked in shifts.) What kind of prisoner was most vulnerable to such treatment? One already broken, already inclined to surrender. It was also a good method to use with women. Ivanov-Razumnik reports a variation of it: Having set young Lordkipanidze on his knees, the interrogator urinated in his face! And what happened? Unbroken by anything else, Lordkipanidze was broken by this. Which shows that the method also worked well on proud people. . . .

19. Then there is the method of simply compelling a prisoner to *stand there.* This can be arranged so that the accused stands only while being interrogated—because that, too, exhausts and breaks a person down. It can be set up in another way—so that the prisoner sits down during interrogation but is forced to stand up between interrogations. (A watch is set over him, and the guards see to it that he doesn't lean against the wall, and if he goes to sleep and falls over he is given a kick and straightened up.) Sometimes even one day of standing is enough to deprive a person of all his strength and to force him to *testify* to *anything at all.*

20. During all these tortures which involved standing for three, four, and five days, they ordinarily *deprived a person of water.*

The most natural thing of all is to *combine* the psychological and physical methods. It is also natural to combine all the preceding methods with:

21. *Sleeplessness,* which they quite failed to appreciate in medieval times. They did not understand how narrow are the limits within which a human being can preserve his personality intact. Sleeplessness (yes, combined with standing, thirst, bright light, terror, and the unknown —what other tortures are needed!?) befogs the reason, undermines the will, and the human being ceases to be himself, to be his own "I." (As in Chekhov's "I Want to Sleep," but there it was much easier, for there the girl could lie down and slip into lapses of consciousness, which even in just a minute would revive and refresh the brain.) A person deprived

of sleep acts half-unconsciously or altogether unconsciously, so that his testimony cannot be held against him.

They used to say: "You are *not truthful* in your testimony, and *therefore* you will not be allowed to sleep!" Sometimes, as a refinement, instead of making the prisoner stand up, they made him *sit down* on a *soft* sofa, which made him want to sleep all the more. (The jailer on duty sat next to him on the same sofa and kicked him every time his eyes began to shut.) Here is how one victim—who had just sat out days in a box infested with bedbugs—describes his feelings after this torture: "Chill from great loss of blood. Irises of the eyes dried out as if someone were holding a red-hot iron in front of them. Tongue swollen from thirst and prickling as from a hedgehog at the slightest movement. Throat racked by spasms of swallowing."

Sleeplessness was a great form of torture: it left no visible marks and could not provide grounds for complaint even if an inspection— something unheard of anyway—were to strike on the morrow.

"They didn't let you sleep? Well, after all, this is not supposed to be a *vacation resort.* The Security officials were awake too!" (They would catch up on their sleep during the day.) One can say that sleep-lessness became the universal method in the *Organs.* From being one among many tortures, it became *an integral part of the system* of State Security; it was the cheapest possible method and did not require the posting of sentries. In all the interrogation prisons the prisoners were forbidden to sleep even one minute from reveille till taps. (In Su-khanovka and several other prisons used specifically for interrogation, the cot was folded into the wall during the day; in others, the prisoners were simply forbidden to lie down, and even to close their eyes while seated.) Since the major interrogations were all conducted at night, it was automatic: whoever was undergoing interrogation got no sleep for at least five days and nights. (Saturday and Sunday nights, the inter-rogators themselves tried to get some rest.)

22. The above method was further implemented by *an assembly line of interrogators.* Not only were you not allowed to sleep, but for three or four days *shifts of interrogators kept up a continuous interroga-tion.*

23. *The bedbug-infested box* has already been mentioned. In the dark closet made of wooden planks, there were hundreds, maybe even thousands, of bedbugs, which had been allowed to multiply. The guards removed the prisoner's jacket or field shirt, and immediately the hungry bedbugs assaulted him, crawling onto him from the walls or falling off the ceiling. At first he waged war with them strenuously, crushing them

on his body and on the walls, suffocated by their stink. But after several hours he weakened and let them drink his blood without a murmur.

24. *Punishment cells.* No matter how hard it was in the ordinary cell, the punishment cells were always worse. And on return from there the ordinary cell always seemed like paradise. In the punishment cell a human being was systematically worn down by starvation and also, usually, by *cold.* (In Sukhanovka Prison there were also *hot* punishment cells.) For example, the Lefortovo punishment cells were entirely unheated. There were radiators in the corridor only, and in this "heated" corridor the guards on duty *walked* in felt boots and padded jackets. The prisoner was forced to undress down to his underwear, and sometimes to his undershorts, and he was forced to spend from three to five days in the punishment cell without moving (since it was so confining). He received hot gruel on the third day only. For the first few minutes you were convinced you'd not be able to last an hour. But, by some miracle, a human being would indeed sit out his five days, perhaps acquiring in the course of it an illness that would last him the rest of his life.

There were various aspects to punishment cells—as, for instance, dampness and water. In the Chernovtsy Prison after the war, Masha G. was kept barefooted for two hours *and up to her ankles* in icy water —confess! (She was eighteen years old, and how she feared for her feet! She was going to have to live with them a long time.)

25. Should one consider it a variation of the punishment cell when a prisoner was *locked in an alcove?* As long ago as 1933 this was one of the ways they tortured S. A. Chebotaryev in the Khabarovsk GPU. They locked him naked in a concrete alcove in such a way that he could neither bend his knees, nor straighten up and change the position of his arms, nor turn his head. And that was not all! They began to drip cold water onto his scalp—a classic torture—which then ran down his body in rivulets. They did not inform him, of course, that this would go on for only twenty-four hours. It was awful enough at any rate for him to lose consciousness, and he was discovered the next day apparently dead. He came to on a hospital cot. They had brought him out of his faint with spirits of ammonia, caffeine, and body massage. At first he had no recollection of where he had been, or what had happened. For a whole month he was useless even for interrogation.

26. *Starvation* has already been mentioned in combination with other methods. Nor was it an unusual method: to starve the prisoner into confession. Actually, the starvation technique, like interrogation at night, was an integral element in the entire system of coercion. The

miserly prison bread ration, amounting to ten and a half ounces in the peacetime year of 1933, and to one pound in 1945 in the Lubyanka, and permitting or prohibiting food parcels from one's family and access to the commissary, were universally applied to everyone. But there was also the technique of intensified hunger: for example, Chulpenyev was kept for a month on three and a half ounces of bread, after which—when he had just been brought in from the pit—the interrogator Sokol placed in front of him a pot of thick borscht, and half a loaf of white bread sliced diagonally. (What does it matter, one might ask, how it was sliced? But Chulpenyev even today will insist that it was really sliced very attractively.) However, he was not given a thing to eat. How ancient it all is, how medieval, how primitive! The only thing new about it was that it was applied in a socialist society! Others, too, tell about such tricks. They were often tried. But we are going to cite another case involving Chebotaryev because it combined so many methods. They put him in the interrogator's office for seventy-two hours, and the only thing he was allowed was to be taken to the toilet. For the rest, they allowed him neither food nor drink— even though there was water in a carafe right next to him. Nor was he permitted to sleep. Throughout there were three interrogators in the office, working in shifts. One kept writing something—silently, without disturbing the prisoner. The second slept on the sofa, and the third walked around the room, and as soon as Chebotaryev fell asleep, beat him instantly. Then they switched roles. (Maybe they themselves were being punished for failure to deliver.) And then, all of a sudden, they brought Chebotaryev a meal: fat Ukrainian borscht, a chop, fried potatoes, and red wine in a crystal carafe. But because Chebotaryev had had an aversion to alcohol all his life, he refused to drink the wine, and the interrogator couldn't go too far in forcing him to, because that would have spoiled the whole game. After he had eaten, they said to him: "Now here's what you have *testified to in the presence of two witnesses.* Sign here." In other words, he was to sign what had been silently composed by one interrogator in the presence of another, who had been asleep, and a third, who had been actively working. On the very first page Chebotaryev learned he had been on intimate terms with all the leading Japanese generals and that he had received espionage assignments from all of them. He began to cross out whole pages. They beat him up and threw him out. Blaginin, another Chinese Eastern Railroad man, arrested with him, was put through the same thing; but he drank the wine and, in a state of pleasant intoxication, signed the confession—and was shot. (Even one tiny glass can have an enormous effect on a famished man—and that was a whole carafe.)

27. *Beatings*—of a kind that leave no marks. They use rubber truncheons, and they use wooden mallets and small sandbags. It is very, very painful when they hit a bone—for example, an interrogator's jackboot on the shin, where the bone lies just beneath the skin. They beat Brigade Commander Karpunich-Braven for twenty-one days in a row. And today he says: "Even after thirty years all my bones ache—and my head too." In recollecting his own experience and the stories of others, he counts up to fifty-two methods of torture. Here is one: They grip the hand in a special vise so that the prisoner's palm lies flat on the desk—and then they hit the joints with the thin edge of a ruler. And one screams! Should we single out particularly the technique by which teeth are knocked out? They knocked out eight of Karpunich's.

As everyone knows, a blow of the fist in the solar plexus, catching the victim in the middle of a breath, leaves no mark whatever. The Lefortovo Colonel Sidorov, in the postwar period, used to take a "penalty kick" with his overshoes at the dangling genitals of male prisoners. Soccer players who at one time or another have been hit in the groin by a ball know what that kind of blow is like. There is no pain comparable to it, and ordinarily the recipient loses consciousness.

28. In the Novorossisk NKVD they invented a machine for squeezing fingernails. As a result it could be observed later at transit prisons that many of those from Novorossisk had lost their fingernails.

29. And what about the *strait jacket?*

30. And *breaking the prisoner's back?* (As in that same Khabarovsk GPU in 1933.)

31. Or *bridling* (also known as "the swan dive")? This was a Sukhanovka method—also used in Archangel, where the interrogator Ivkov applied it in 1940. A long piece of rough toweling was inserted between the prisoner's jaws like a bridle; the ends were then pulled back over his shoulders and tied to his heels. Just try lying on your stomach like a wheel, with your spine breaking—and without water and food for two days!

Is it necessary to go on with the list? Is there much left to enumerate? What won't idle, well-fed, unfeeling people invent?

Brother mine! Do not condemn those who, finding themselves in such a situation, turned out to be weak and confessed to more than they should have. . . . Do not be the first to cast a stone at them.

■

From childhood on we are educated and trained—for our own profession; for our civil duties; for military service; to take care of our bodily needs; to behave well; even to appreciate beauty (well, this last not really all that much!). But neither our education, nor our upbringing, nor our experience prepares us in the slightest for the greatest trial of our lives: being arrested for nothing and interrogated about nothing. Novels, plays, films (their authors should themselves be forced to drink the cup of Gulag to the bottom!) depict the types one meets in the offices of interrogators as chivalrous guardians of truth and humanitarianism, as our loving fathers. We are exposed to lectures on everything under the sun—and are even herded in to listen to them. But no one is going to lecture to us about the true and extended significance of the Criminal Code; and the codes themselves are not on open shelves in our libraries, nor sold at newsstands; nor do they fall into the hands of the heedless young.

It seems a virtual fairy tale that somewhere, at the ends of the earth, an accused person can avail himself of a lawyer's help. This means having beside you in the most difficult moment of your life a clear-minded ally who knows the law.

The principle of our interrogation consists further in depriving the accused of even a knowledge of the law.

An indictment is presented. And here, incidentally, is how it's presented: "Sign it." "It's not true." "Sign." "But I'm not guilty of anything!" It turns out that you are being indicted under the provisions of Articles 58-10, Part 2, and 58-11 of the Criminal Code of the Russian Republic. "Sign!" "But what do these sections say? Let me read the Code!" "I don't have it." "Well, get it from your department head!" "He doesn't have it either. Sign!" "But I want to see it." "You are not supposed to see it. It isn't written for you but for us. You don't need it. I'll tell you what it says: these sections spell out exactly what you are guilty of. And anyway, at this point your signature doesn't mean that you agree with the indictment but that you've read it, that it's been presented to you."

All of a sudden, a new combination of letters, UPK, flashes by on one of the pieces of paper. Your sense of caution is aroused. What's the difference between the UPK and the UK—the Criminal Code? If you've been lucky enough to catch the interrogator when he is in a good mood, he will explain it to you: the UPK is the Code of Criminal Procedure. What? This means that there are two distinct codes, not just one, of whose contents you are completely ignorant even as you are being trampled under their provisions.

Since that time ten years have passed; then fifteen. The grass has grown thick over the grave of my youth. I served out my term and even "eternal exile" as well. And nowhere—neither in the "cultural education" sections of the camps, nor in district libraries, nor even in medium-sized cities, have I seen with my own eyes, held in my own hands, been able to buy, obtain, or even *ask for* the Code of Soviet law!

And of the hundreds of prisoners I knew who had gone through interrogation and trial, and more than once too, who had served sentences in camp and in exile, none had ever seen the Code or held it in his hand!

It was only when both codes were thirty-five years old and on the point of being replaced by new ones that I saw them, two little paperback brothers, the UK or Criminal Code, and the UPK or Code of Criminal Procedure, on a newsstand in the Moscow subway (because they were outdated, it had been decided to release them for general circulation).

I read them today touched with emotion. For example, the UPK —the Code of Criminal Procedure:

"Article 136: The interrogator does not have the right to extract testimony or a confession from an accused by means of compulsion and threats." (It was as though they had foreseen it!)

"Article 111: The interrogator is obliged to establish clearly all the relevant facts, both those tending toward acquittal and any which might lessen the accused's measure of guilt."

But it was I who helped establish Soviet power in October! It was I who shot Kolchak! I took part in the dispossession of the kulaks! I saved the state ten million rubles in lowered production costs! I was wounded twice in the war! I have three orders and decorations.

"You're not being tried for that!" History . . . the bared teeth of the interrogator: "Whatever good you may have done has nothing to do with the case."

"Article 139: The accused has the right to set forth his testimony in his own hand, and to demand the right to make corrections in the deposition written by the interrogator."

Oh, if we had only known that in time! But what I should say is: If that were only the way it really was! We were always vainly imploring the interrogator not to write "my repulsive, slanderous fabrications" instead of "my mistaken statements," or not to write "our underground weapons arsenal" instead of "my rusty Finnish knife."

If only the defendants had first been taught some prison science! If only interrogation had been run through first in rehearsal, and only afterward for real. . . . They didn't, after all, play that interrogation game with the *second-termers* of 1948: it would have gotten them nowhere. But *newcomers* had no experience, no knowledge! And there was no one from whom to seek advice.

The loneliness of the accused! That was one more factor in the success of unjust interrogation! The entire apparatus threw its full weight on one lonely and inhibited will. From the moment of his arrest and throughout the entire *shock* period of the interrogation the prisoner was, ideally, to be kept entirely alone. In his cell, in the corridor, on the stairs, in the offices, he was not supposed to encounter others like himself, in order to avoid the risk of his gleaning a bit of sympathy, advice, support from someone's smile or glance. The *Organs* did everything to blot out for him his future and distort his present: to lead him to believe that his friends and family had all been arrested and that material proof of his guilt had been found. It was their habit to exaggerate their power to destroy him and those he loved as well as their authority to pardon (which the *Organs* didn't even have). They pretended that there was some connection between the sincerity of a prisoner's "repentance" and a reduction in his sentence or an easing of the camp regimen. (No such connection ever existed.) While the prisoner was still in a state of shock and torment and totally beside himself, they tried to get from him very quickly as many irreparably damaging items of evidence as possible and to implicate with him as many totally innocent persons as possible. Some defendants became so depressed in these circumstances that they even asked not to have the depositions read to them. They could not stand hearing them. They asked merely to be allowed to sign them, just to sign and get it over with. Only after all this was over would the prisoner be released from solitary into a large cell, where, in belated desperation, he would discover and count over his mistakes one by one.

How was it possible not to make mistakes in such a duel? Who could have failed to make a mistake?

We said that "ideally he was to be kept alone." However, in the overcrowded prisons of 1937, and, for that matter, of 1945 as well, this ideal of solitary confinement for a newly arrested defendant could not be attained. Almost from his first hours, the prisoner was in fact in a terribly overcrowded common cell.

But there were virtues to this arrangement, too, which more than made up for its flaws. The overcrowding of the cells not only took the place of the tightly confined solitary "box" but also assumed the character of a first-class *torture* in itself . . . one that was particularly useful because it continued for whole days and weeks—with no effort on the part of the interrogators. The prisoners tortured the prisoners! The jailers pushed so many prisoners into the cell that not every one had even a piece of floor; some were sitting on others' feet, and people walked on people and couldn't even move about at all. Thus, in the Kishinev KPZ's—Cells for Preliminary Detention—in 1945, they pushed *eighteen* prisoners into a cell designed for the solitary confinement of one person; in Lugansk in 1937 it was *fifteen*. And in 1938 Ivanov-Razumnik found *one hundred forty* prisoners in a standard Butyrki cell intended for twenty-five—with toilets so overburdened that prisoners were taken to the toilet only once a day, sometimes at night; and the same thing was true of their outdoor walk as well.

That same year in the Butyrki, those newly arrested, who had already been processed through the bath and the boxes, sat on the stairs for several days at a stretch, waiting for departing prisoner transports to leave and release space in the cells. T——v had been imprisoned in the Butyrki seven years earlier, in 1931, and says that it was overcrowded under the bunks and that prisoners lay on the asphalt floor. I myself was imprisoned seven years later, in 1945, and it was just the same. But recently I received from M. K. B——ch valuable personal testimony about overcrowding in the Butyrki in *1918*. In October of that year—during the second month of the Red Terror —it was so full that they even set up a cell for seventy women in the laundry. When, then, was the Butyrki not crowded?

It was Ivanov-Razumnik who in the Lubyanka reception "kennel" calculated that for weeks at a time there were *three* persons for each square yard of floor space (just as an experiment, try to fit three people into that space!).

But this, too, is no miracle: in the Vladimir *Internal Prison* in 1948, thirty people had to stand in a cell ten feet by ten feet in size! (S. Potapov.)

In this "kennel" there was neither ventilation nor a window, and the prisoners' body heat and breathing raised the temperature to 40 or 45 degrees Centigrade—104 to 113 degrees Fahrenheit—and everyone sat there in undershorts with their winter clothing piled beneath them. Their naked bodies were pressed against one another, and they got eczema from one another's sweat. They sat like that for *weeks at a time,* and were given neither fresh air nor water—except for gruel and tea in the morning.

And if at the same time the latrine bucket replaced all other types of toilet (or if, on the other hand, there was no latrine bucket for use between trips to an outside toilet, as was the case in several Siberian prisons); and if four people ate from one bowl, sitting on each other's knees; and if someone was hauled out for interrogation, and then someone else was pushed in beaten up, sleepless, and broken; and if the appearance of such broken men was more persuasive than any threats on the part of the interrogators; and if, by then, death and any camp whatever seemed easier to a prisoner who had been left unsummoned for months than his tormented current situation—perhaps this really did replace the theoretically ideal isolation in solitary. And you could not always decide in such a porridge of people with whom to be forthright; and you could not always find someone from whom to seek advice. And you would believe in the tortures and beatings not when the interrogator threatened you with them but when you saw their results on other prisoners.

You could learn from those who had suffered that they could give you a salt-water douche in the throat and then leave you in a box for a day tormented by thirst (Karpunich). Or that they might scrape the skin off a man's back with a grater till it bled and then oil it with turpentine. (Brigade Commander Rudolf Pintsov underwent both treatments. In addition, they pushed needles under his nails, and poured water into him to the bursting point—demanding that he confess to having *wanted* to turn his brigade of tanks against the government during the November parade.) And from Aleksandrov, the former head of the Arts Section of the All-Union Society for Cultural Relations with Foreign Countries, who has a broken spinal column which tilts to one side, and who cannot control his tear ducts and thus cannot stop crying, one can learn how *Abakumov* himself could beat —in 1948.

Yes, yes, Minister of State Security Abakumov himself did not by any means spurn such menial labor. He was not averse to taking a rubber truncheon in his hands every once in a while. And his deputy Ryumin was even more willing. He did this at Sukhanovka in the "Generals'" interrogation office. The office had imitation-walnut paneling on the walls, silk portieres at the windows and doors, and a great Persian carpet on the floor. In order not to spoil all this beauty, a dirty runner bespattered with blood was rolled out on top of the carpet when a prisoner was being beaten. When Ryumin was doing the beating, he was assisted not by some ordinary guard but by a colonel. "And so," said Ryumin politely, stroking his rubber truncheon, which was four centimeters—an inch and a half—thick, "you have survived trial by sleeplessness with honor." (Alexander Dolgun had cleverly managed to last a month "without sleep" by sleeping while he was standing up.) "So now we will try the club. Prisoners can't take more than two or three sessions of this. Let down your trousers and lie down on the runner." The colonel sat down on the prisoner's back. Alexander Dolgun was going to *count* the blows. He didn't yet know about a blow from a rubber truncheon on the sciatic nerve when the buttocks have disappeared as a consequence of prolonged starvation. The effect is not felt in the place where the blow is delivered—it explodes inside the head. After the first blow the victim was mad with pain and broke his nails on the carpet. Ryumin beat away, trying to hit accurately. The colonel pressed down on Alexander Dolgun's torso—this was just the right sort of work for three big shoulder-board stars, assisting the all-powerful Ryumin! (After the beating the prisoner could not walk and, of course, was not carried. They just dragged him along the floor. What was left of his buttocks was soon so swollen that he could not button his trousers, and yet there were practically no scars. He was hit by a violent case of diarrhea, and, sitting there on the latrine bucket in solitary, Alexander Dolgun guffawed. He went through a second and a third session, and his skin cracked, and Ryumin went wild, and started to beat him on the stomach, breaking through the intestinal wall and creating an enormous hernia through which Alexander Dolgun's intestines protruded. The prisoner was taken off to the Butyrki hospital with a case of peritonitis, and for the time being their attempts to compel him to commit a foul deed were suspended.)

That is how they can torture you too! After that it could seem a simple fatherly caress when the Kishinev interrogator Danilov beat Father Viktor Shipovalnikov across the back of the head with a poker and pulled him by his long hair. (It is very convenient to drag a priest around in that fashion; ordinary laymen can be dragged by the beard from one corner of the office to the other. And Richard Ohola—a Finnish Red Guard, and a participant in the capture of British agent Sidney Reilly, and commander of a company during the suppression of the Kronstadt revolt—was lifted up with pliers first by one end of his great mustaches and then by the other, and held for ten minutes with his feet off the floor.)

But the most awful thing they can do with you is this: undress you from the waist down, place you on your back on the floor, pull your legs apart, seat assistants on them (from the glorious corps of sergeants!) who also hold down your arms; and then the interrogator (and women interrogators have not shrunk from this) stands between your legs and with the toe of his boot (or of her shoe) gradually, steadily, and with ever greater pressure crushes against the floor those organs which once made you a man. He looks into your eyes and repeats and repeats his questions or the betrayal he is urging on you. If he does not press down too quickly or just a shade too powerfully, you still have fifteen seconds left in which to scream that you will confess to everything, that you are ready to see arrested all twenty of those people he's been demanding of you, or that you will slander in the newspapers everything you hold holy. . . .

And may you be judged by God, but not by people. . . .

"There is no way out! You have to confess to everything!" whisper the stoolies who have been planted in the cell.

"It's a simple question: hang onto your health!" say people with common sense.

"You can't get new teeth," those who have already lost them nod at you.

"They are going to convict you in any case, whether you confess or whether you don't," conclude those who have got to the bottom of things.

"Those who don't sign get shot!" prophesies someone else in the corner. "Out of vengeance! So as not to risk any leaks about how they conduct interrogations."

"And if you die in the interrogator's office, they'll tell your relatives you've been sentenced to camp without the right of correspondence. And then just let them look for you."

If you are an orthodox Communist, then another orthodox Communist will sidle up to you, peering about with hostile suspicion, and he'll begin to whisper in your ear so that the uninitiated cannot overhear:

"It's our duty to support Soviet interrogation. It's a combat situation. We ourselves are to blame. We were too softhearted; and now look at all the rot that has multiplied in the country. There is a vicious secret war going on. Even here we are surrounded by enemies. Just listen to what they are saying! The Party is not obliged to account for what it does to every single one of us—to explain the whys and wherefores. If they ask us to, that means we should sign."

And another orthodox Communist sidles up:

"I signed denunciations against thirty-five people, against all my acquaintances. And I advise you too: Drag along as many names as you can in your wake, as many as you can. That way it will become obvious that the whole thing is an absurdity and they'll let everyone out!"

But that is precisely what the *Organs* need. The conscientiousness of the orthodox Communist and the purpose of the NKVD naturally coincide. Indeed, the NKVD needs just that arched fan of names, that fat multiplication of them. That is the mark of quality of their work, and these are also new patches of woods in which to set out snares. "Your accomplices, accomplices! Others who share your views!" That is what they keep pressing to shake out of everyone. They say that R. Ralov named Cardinal Richelieu as one of his accomplices and that the Cardinal was in fact so listed in his depositions—and no one was astonished by this until Ralov was questioned about it at his rehabilitation proceedings in 1956.

Apropos of the orthodox Communists, Stalin was necessary, for such a *purge* as that, yes, but a Party like that was necessary too: the majority of those in power, up to the very moment of their own arrest, were pitiless in arresting others, obediently destroyed their peers in accordance with those same instructions and handed over to retribution any friend or comrade-in-arms of yesterday. And all the big Bolsheviks, who now wear martyrs' halos, managed to be the executioners of other Bolsheviks (not even taking into account how *all of them* in the first place had been the executioners of non-Communists). Perhaps 1937 was *needed* in order to show how little their whole ideology was worth —that *ideology* of which they boasted so enthusiastically, turning Russia upside down, destroying its foundations, trampling everything it held sacred underfoot, that Russia where *they themselves* had never been threatened by *such* retribution. The victims of the Bolsheviks from 1918 to 1946 never conducted themselves so despicably as the leading Bolsheviks when the lightning struck them. If you study in detail the whole history of the arrests and trials of 1936 to 1938, the principal revulsion you feel is not against Stalin and his accomplices, but against the humiliatingly repulsive defendants—nausea at their spiritual baseness after their former pride and implacability.

So what is the answer? How can you stand your ground when you are weak and sensitive to pain, when people you love are still alive, when you are unprepared?

What do you need to make you stronger than the interrogator and the whole trap?

From the moment you go to prison you must put your cozy past firmly behind you. At the very threshold, you must say to yourself: "My life is over, a little early to be sure, but there's nothing to be done about it. I shall never return to freedom. I am condemned to die—now or a little later. But later on, in truth, it will be even harder, and so the sooner the better. I no longer have any property whatsoever. For me those I love have died, and for them I have died. From today on, my body is useless and alien to me. Only my spirit and my conscience remain precious and important to me."

Confronted by such a prisoner, the interrogator will tremble.

Only the man who has renounced everything can win that victory. But how can one turn one's body to stone?

Well, they managed to turn some individuals from the Berdyayev circle into puppets for a trial, but they didn't succeed with Berdyayev. They wanted to drag him into an open trial; they arrested him twice; and (in 1922) he was subjected to a night interrogation by Dzerzhinsky himself. Kamenev was there too (which means that he, too, was not averse to using the Cheka in an ideological conflict). But Berdyayev did not humiliate himself. He did not beg or plead. He set forth firmly those religious and moral principles which had led him to refuse to accept the political authority established in Russia. And not only did they come to the conclusion that he would be useless for a trial, but they liberated him.

A human being has *a point of view!*

N. Stolyarova recalls an old woman who was her neighbor on the Butyrki bunks in 1937. They kept on interrogating her every night. Two years earlier, a former Metropolitan of the Orthodox Church, who had escaped from exile, had spent a night at her home on his way through Moscow. "But he wasn't the former Metropolitan, he was the Metropolitan! Truly, I was worthy of receiving him." "All right then. To whom did he go when he left Moscow?" "I know, but I won't tell you!" (The Metropolitan had escaped to Finland via an underground railroad of believers.) At first the interrogators took turns, and then they went after her in groups. They shook their fists in the little old woman's face, and she replied: "There is nothing you can do with me even if you cut me into pieces. After all, you are afraid of your bosses, and you are afraid of each other, and you are even afraid of killing me." (They would lose contact with the underground railroad.) "But I am not afraid of anything. I would be glad to be judged by God right this minute."

There were such people in 1937 too, people who did not return to their cell for their bundles of belongings, who chose death, who *signed* nothing denouncing anyone.

One can't say that the history of the Russian revolutionaries has given us any better examples of steadfastness. But there is no comparison anyway, because none of our revolutionaries ever knew what a really *good* interrogation could be, with fifty-two different methods to choose from. Just as oxcart drivers of Gogol's time could not have imagined the speed of a jet plane, those who have never gone through the receiving-line meat grinder of Gulag cannot grasp the true possibilities of interrogation.

We read in *Izvestiya* for May 24, 1959, that Yuliya Rumyantseva was confined in the internal prison of a Nazi camp while they tried to find out from her the whereabouts of her husband, who had escaped from that same camp. She knew, but she refused to tell! For a reader who is not in the know this is a model of heroism. For a reader with a bitter Gulag past it's a model of inefficient interrogation: Yuliya did not die under torture, and she was not driven insane. A month later she was simply released—still very much alive and kicking.

Chapter 4

■

The Bluecaps

Throughout the grinding of our souls in the gears of the great Nighttime Institution, when our souls are pulverized and our flesh hangs down in tatters like a beggar's rags, we suffer too much and are too immersed in our own pain to rivet with penetrating and far-seeing gaze those pale night executioners who torture us. A surfeit of inner grief floods our eyes. Otherwise what historians of our torturers we would be! For it is certain they will never describe themselves as they actually are. But alas! Every former prisoner remembers his own interrogation in detail, how they squeezed him, and what foulness they squeezed out of him—but often he does not even remember their names, let alone think about them as human beings. So it is with me. I can recall much more—and much more that's interesting—about any one of my cellmates than I can about Captain of State Security Yezepov, with whom I spent no little time face to face, the two of us alone in his office.

There is one thing, however, which remains with us all as an accurate, generalized recollection: foul rot—a space totally infected with putrefaction. And even when, decades later, we are long past fits of anger or outrage, in our own quieted hearts we retain this firm impression of low, malicious, impious, and, possibly, muddled people.

There is an interesting story about Alexander II, the Tsar surrounded by revolutionaries, who were to make seven attempts on his life. He once visited the House of Preliminary Detention on Shpalernaya—the uncle of the Big House—where he ordered them to lock him up in solitary-confinement cell No. 227. He stayed in it for more than an hour, attempting thereby to sense the state of mind of those he had imprisoned there.

One cannot but admit that for a monarch this was evidence of moral aspiration, to feel the need and make the effort to take a spiritual view of the matter.

But it is impossible to picture any of our interrogators, right up to Abakumov and Beria, wanting to slip into a prisoner's skin even for one hour, or feeling compelled to sit and meditate in solitary confinement.

Their branch of service does not require them to be educated people of broad culture and broad views—and they are not. Their branch of service does not require them to think logically—and they do not. Their branch of service requires only that they carry out orders exactly and be impervious to suffering—and that is what they do and what they are. We who have passed through their hands feel suffocated when we think of that legion, which is stripped bare of universal human ideals.

Although others might not be aware of it, it was clear to the interrogators at least that the *cases* were fabricated. Except at staff conferences, they could not seriously say to one another or to themselves that they were exposing criminals. Nonetheless they kept right on producing depositions page after page to make sure that we rotted. So the essence of it all turns out to be the credo of the blatnye—the underworld of Russian thieves: "You today; me tomorrow."

They understood that the cases were fabricated, yet they kept on working year after year. How could they? Either they forced themselves *not to think* (and this in itself means the ruin of a human being), and simply accepted that this was the way it had to be and that the person who gave them their orders was always right . . .

But didn't the Nazis, too, it comes to mind, argue that same way?

Or else it was a matter of the Progressive Doctrine, ̣̣ ̣ʈranite ideology. An interrogator in awful Orotukan—sent there ʈo the Kolyma in 1938 as a penalty assignment—was so touched when M. Lurye, former director of the Krivoi Rog Industrial Complex, readily agreed to sign an indictment which meant a second camp term that he used the time they had thus saved to say: "You think we get any satisfaction from using *persuasion?* We have to do what the Party demands of us. You are an old Party member. Tell me what would you do in my place?" Apparently Lurye nearly agreed with him, and it may have been the fact that he had already been thinking in some such terms that led him to sign so readily. It is after all a convincing argument.

But most often it was merely a matter of cynicism. The bluecaps understood the workings of the meat grinder and loved it. In the Dzhida camps in 1944, interrogator Mironenko said to the condemned Babich with pride in his faultless logic: "Interrogation and trial are merely judicial corroboration. They cannot alter your fate, which was *previously* decided. If it is necessary to shoot you, then you will be shot even if you are altogether innocent. If it is necessary to acquit you, then no matter how guilty you are you will be cleared and acquitted."

"Just give us a person—and we'll create the *case!*" That was what many of them said jokingly, and it was their slogan. What we think of as torture they think of as good work. The wife of the interrogator Nikolai Grabishchenko (the Volga Canal Project) said touchingly to her neighbors: "Kolya is a very good worker. One of them didn't confess for a long time—and they gave him to Kolya. Kolya talked with him for one night and he confessed."

What prompted them all to slip into harness and pursue so zealously not truth but *totals* of the processed and condemned? Because it was *most comfortable* for them not to be different from the others. And because these totals meant an easy life, supplementary pay, awards and decorations, promotions in rank, and the expansion and prosperity of the *Organs* themselves. If they ran up high totals, they could loaf when they felt like it, or do poor work or go out and enjoy themselves at night. And that is just what they did. Low totals led to their being kicked out, to the loss of their feedbag. For Stalin could never be convinced that in any district, or city, or military unit, he might suddenly cease to have enemies.

That was why they felt no mercy, but, instead, an explosion of resentment and rage toward those maliciously stubborn prisoners who opposed being fitted into the totals, who would not capitulate to sleeplessness or the punishment cell or hunger. By refusing to confess they menaced the interrogator's personal standing. It was as though they wanted to bring *him* down. In such circumstances all measures were justified! If it's to be war, then war it will be! We'll ram the tube down your throat—swallow that salt water!

Excluded by the nature of their work and by deliberate choice from the *higher* sphere of human existence, the servitors of the Blue Institution lived in their lower sphere with all the greater intensity and avidity. And there they were possessed and directed by the two strongest instincts of the lower sphere, other than hunger and sex: greed for *power* and greed for *gain*. (Particularly for power. In recent decades it has turned out to be more important than money.)

Power is a poison well known for thousands of years. If only no one were ever to acquire material power over others! But to the human being who has faith in some force that holds dominion over all of us, and who is therefore conscious of his own limitations, power is not necessarily fatal. For those, however, who are unaware of any higher sphere, it is a deadly poison. For them there is no antidote.

Here attraction is not the right word—it is *intoxication!* After all, it *is* intoxicating. You are still young—still, shall we say parenthetically, a sniveling youth. Only a little while ago your parents were deeply concerned about you and didn't know where to turn to launch you in life. You were such a fool you didn't even want to study, but you got through three years of *that* school—and then how you took off and flew! How your situation changed! How your gestures changed, your glance, the turn of your head! The learned council of the scientific institute is in session. You enter and everyone notices you and trembles. You don't take the chairman's chair. Those headaches are for the rector to take on. You sit off to one side, but everyone understands that you are head man there. You are the Special Department. And you can sit there for just five minutes and then leave. You have that advantage over the professors. You can be called away by more important business—but later on, when you're considering their decision, you will raise your eyebrows or, better still, purse your lips and say to the rector: "You can't do that. There are *special considerations* involved." That's all! And it won't be done. Or else you are an osobist—a State Security representative in the army—a SMERSH man, and a mere lieutenant; but the portly old colonel, the commander of the unit, stands up when you enter the room and tries to flatter you, to play up to you. He doesn't even have a drink with his chief of staff without inviting you to join them. You have a power over all the people in that military unit, or factory, or district, incomparably greater than that of the military commander, or factory director, or secretary of the district Communist Party. These men control people's military or official duties, wages, reputations, but you control people's freedom. And no one dares speak about you at meetings, and no one will ever dare write about you in the newspaper—not only something bad but anything *good!* They don't dare. Your name, like that of a jealously guarded deity, cannot even be mentioned. You are there; everyone feels your presence; but it's as though you didn't exist. From the moment you don that heavenly blue service cap, you stand higher than the publicly acknowledged power.

No one dares check up on what *you* do. But no one is exempt from your checking up on him. And therefore, in dealing with ordinary so-called citizens, who for you are mere blocks of wood, it is altogether appropriate for you to wear an ambiguous and deeply thoughtful expression. For, of course, you are the one—and no one else—who knows about the *special considerations.* And therefore you are always right.

There is just one thing you must never forget. You, too, would have been just such a poor block of wood if you had not had the luck to become one of the little links in the *Organs*—that flexible, unitary organism inhabiting a nation as a tapeworm inhabits a human body. Everything is yours now! Everything is for you! Just be true to the *Organs!* They will always stand up for you! They will help you swallow up anyone who bothers you! They will help move every obstacle from your path! But—be true to the *Organs!* Do everything they order you to! They will do the thinking for you in respect to your functions too.

The duties of an interrogator require work, of course: you have to come in during the day, at night, sit for hours and hours—but not split your skull over "proof." (Let the prisoner's head ache over that.) And you don't have to worry whether the prisoner is guilty or not but simply do what the *Organs* require. And everything will be all right. It will be up to you to make the interrogation periods pass as pleasurably as possible and not to get overly fatigued. And it would be nice to get some good out of it—at least to amuse yourself. You have been sitting a long time, and all of a sudden a new method of *persuasion* occurs to you! Eureka! So you call up your friends on the phone, and you go around to other offices and tell them about it—what a laugh! Who shall we try it on, boys? It's really pretty monotonous to keep doing the same thing all the time. Those trembling hands, those imploring eyes, that cowardly submissiveness—they are really a bore. If you could just get one of them to resist! "I love strong opponents! *It's such fun to break their backs!*"

And if your opponent is so strong that he refuses to give in, all your methods have failed, and you are in a rage? Then don't control your fury! It's tremendously satisfying, that outburst! Let your anger have its way; don't set any bounds to it! Don't hold yourself back! That's when interrogators spit in the open mouth of the accused! And shove his face into a full cuspidor! That's the state of mind in which they drag priests around by their long hair! Or urinate in a kneeling prisoner's face! After such a storm of fury you feel yourself a real honest-to-God man!

Or else you are interrogating a "foreigner's girl friend." So you curse her out and then you say: "Come on now, does an American have a special kind of ———? Is that it? Weren't there enough Russian ones for you?" And all of a sudden you get an idea: maybe she learned something from those foreigners. Here's a chance not to be missed, like an assignment abroad! And so you begin to interrogate her energetically: *How?* What positions? More! In detail! Every scrap of information! (You can use the information yourself, and you can tell the other boys too!) The girl is blushing all over and in tears. "It doesn't have anything to do with the case," she protests. "Yes, it does, speak up!" That's power for you! She gives you the full details. If you want, she'll draw a picture for you. If you want, she'll demonstrate with her body. She has no way out. In your hands you hold the punishment cell and her *prison term.*

And if you have asked for a stenographer to take down the questions and answers, and they send in a pretty one, you can shove your paw down into her bosom right in front of the boy being interrogated. He's not a human being after all, and there is no reason to feel shy in his presence.

In fact, there's no reason for you to feel shy with anyone. And if you like the broads—and who doesn't?—you'd be a fool not to make use of your position. Some will be drawn to you because of your power, and others will give in out of fear. So you've met a girl somewhere and she's caught your eye? She'll belong to you, never fear; she can't get away! Someone else's wife has caught your eye? She'll be yours too! Because, after all, there's no problem about removing the husband. No, indeed! To know what it meant to be a bluecap one had to experience it! Anything you saw was yours! Any apartment you looked at was yours! Any woman was yours! Any enemy was struck from your path! The earth beneath your feet was yours! The heaven above you was yours —it was, after all, like your cap, sky blue!

The passion for gain was their universal passion. After all, in the absence of any checking up, such power was inevitably used for personal enrichment. One would have had to be *holy* to refrain!

If we were able to discover the hidden motivation behind individual arrests, we would be astounded to find that, granted the rules governing *arrests* in general, 75 percent of the time the particular choice of *whom* to arrest, the personal cast of the die, was determined by human greed and vengefulness; and of that 75 percent, half were the result of material self-interest on the part of the local NKVD (and, of course, the prosecutor too, for on this point I do not distinguish between them).

The motivations and actions of the bluecaps are sometimes so petty that one can only be astounded. Security officer Senchenko took a map case and dispatch case from an officer he'd arrested and started to use them right in his presence, and, by manipulating the documentation, he took a pair of foreign gloves from another prisoner. (When the armies were advancing, the bluecaps were especially irritated because they got only second pick of the booty.) The counterintelligence officer of the Forty-ninth Army who arrested me had a yen for my cigarette case—and it wasn't even a cigarette case but a small German Army box, of a tempting scarlet, however. And because of that piece of shit he carried out a whole maneuver: As his first step, he omitted it from the list of belongings that were confiscated from me. ("You can keep it.") He thereupon ordered me to be searched again, knowing all the time that it was all I had in my pockets. "Aha! what's that? Take it away!" And to prevent my protests: "Put him in the punishment cell!" (What Tsarist gendarme would have dared behave that way toward a defender of the Fatherland?)

Every interrogator was given an allowance of a certain number of cigarettes to encourage those willing to confess and to reward stool pigeons. Some of them kept all the cigarettes for themselves.

Even in accounting for hours spent in interrogating, they used to cheat. They got higher pay for night work. And we used to note the way they wrote down more hours on the night interrogations than they really spent.

Interrogator Fyodorov (Reshety Station, P. O. Box No. 235) stole a wristwatch while searching the apartment of the free person Korzukhin. During the Leningrad blockade Interrogator Nikolai Fyodorovich Kruzhkov told Yelizaveta Viktorovna Strakhovich, wife of the prisoner he was interrogating, K. I. Strakhovich: "I want a quilt. Bring it to me!" When she replied: "All our warm things are in the room they've sealed," he went to her apartment and, without breaking the State Security seal on the lock, unscrewed the entire doorknob. "That's how the MGB works," he explained gaily. And he went in and began to collect the warm things, shoving some crystal in his pocket at the same time. She herself tried to get whatever she could out of the room, but he stopped her. "That's enough for you!"—and he kept on raking in the booty.

There's no end to such cases. One could issue a thousand "White

Papers" (and beginning in 1918 too). One would need only to question systematically former prisoners and their wives. Maybe there are and were bluecaps who never stole anything or appropriated anything for themselves—but I find it impossible to imagine one. I simply do not understand: given the bluecaps' philosophy of life, what was there to restrain them if they liked some particular thing? Way back at the beginning of the thirties, when all of us were marching around in the German uniforms of the Red Youth Front and were building the First Five-Year Plan, they were spending their evenings in salons like the one in the apartment of Konkordiya Iosse, behaving like members of the nobility or Westerners, and their lady friends were showing off their foreign clothes. Where were they getting those clothes?

■

As the folk saying goes: *If you speak for the wolf, speak against him as well.*

Where did this wolf-tribe appear from among our people? Does it really stem from our own roots? Our own blood?

It is our own.

And just so we don't go around flaunting too proudly the white mantle of the just, let everyone ask himself: "If my life had turned out differently, might I myself not have become just such an executioner?"

It is a dreadful question if one really answers it honestly.

I remember my third year at the university, in the fall of 1938. We young men of the Komsomol were summoned before the District Komsomol Committee not once but twice. Scarcely bothering to ask our consent, they shoved an application form at us: You've had enough physics, mathematics, and chemistry; it's more important to your country for you to enter the NKVD school. (That's the way it always is. It isn't just some person who needs you; it is always your Motherland. And it is always some official or other who speaks on behalf of your Motherland and who knows what she needs.)

One year before, the District Committee had conducted a drive among us to recruit candidates for the air force schools. We avoided getting involved that time too, because we didn't want to leave the university—but we didn't sidestep recruitment then as stubbornly as we did this time.

Twenty-five years later we could think: Well, yes, we understood the sort of arrests that were being made at the time, and the fact that they were torturing people in prisons, and the slime they were trying to drag us into. But it isn't true! After all, the Black Marias were going through the streets at night, and we were the same young people who were parading with banners during the day. How could we know anything about those arrests and why should we think about them? All the provincial leaders had been removed, but as far as we were concerned it didn't matter. Two or three professors had been arrested, but after all they hadn't been our dancing partners, and it might even be easier to pass our exams as a result. Twenty-year-olds, we marched in the ranks of those born the year the Revolution took place, and because we were the same age as the Revolution, the brightest of futures lay ahead.

It would be hard to identify the exact source of that inner intuition, not founded on rational argument, which prompted our refusal to enter the NKVD schools. It certainly didn't derive from the lectures on historical materialism we listened to: it was clear from them that the struggle against the internal enemy was a crucial battlefront, and to share in it was an honorable task. Our decision even ran counter to our material interests: at that time the provincial university we attended could not promise us anything more than the chance to teach in a rural school in a remote area for miserly wages. The NKVD school dangled before us special rations and double or triple pay. Our feelings could not be put into words—and even if we had found the words, fear would have prevented our speaking them aloud to one another. It was not our minds that resisted but something inside our breasts. People can shout at you from all sides: "You must!" And your own head can be saying also: "You must!" But inside your breast there is a sense of revulsion, repudiation. I don't want to. *It makes me feel sick.* Do what you want without me; I want no part of it.

Still, some of us were recruited at that time, and I think that if they had really put the pressure on, they could have broken everybody's resistance. So I would like to imagine: if, by the time war broke out, I had already been wearing an NKVD officer's insignia on my blue tabs, what would I have become? What do shoulder boards do to a human being? And where have all the exhortations of grandmother, standing before an ikon, gone? And where the young Pioneer's daydreams of future sacred Equality?

And at the moment when my life was turned upside down and the SMERSH officers at the brigade command point tore off those cursed shoulder boards, and took my belt away and shoved me along to their automobile, I was pierced to the quick by worrying how, in my stripped and sorry state, I was going to make my way through the telephone operator's room. The rank and file must not see me in that condition!

So let the reader who expects this book to be a political exposé slam its covers shut right now.

If only it were all so simple! If only there were evil people somewhere insidiously committing evil deeds, and it were necessary only to separate them from the rest of us and destroy them. But the line dividing good and evil cuts through the heart of every human being. And who is willing to destroy a piece of his own heart?

During the life of any heart this line keeps changing place; sometimes it is squeezed one way by exuberant evil and sometimes it shifts to allow enough space for good to flourish. One and the same human being is, at various ages, under various circumstances, a totally different human being. At times he is close to being a devil, at times to sainthood. But his name doesn't change, and to that name we ascribe the whole lot, good and evil.

Socrates taught us: *Know thyself!*

Confronted by the pit into which we are about to toss those who have done us harm, we halt, stricken dumb: it is after all only because of the way things worked out that they were the executioners and we weren't.

From good to evil is one quaver, says the proverb.

And correspondingly, from evil to good.

Whoever got in by mistake either adjusted to the milieu or else was thrown out, or eased out, or even fell across the rails himself. Still . . . were there no good people left there?

In Kishinev, a young lieutenant gaybist went to Father Viktor Shipovalnikov a full month before he was arrested: "Get away from here, go away, they plan to arrest you!" (Did he do this on his own, or did his mother send him to warn the priest?) After the arrest, this young man was assigned to Father Viktor as an escort guard. And he grieved for him: "Why didn't you go away?"

When the interrogator Goldman gave Vera Korneyeva the "206" form on nondisclosure to sign, she began to catch on to her rights, and then she began to go into the *case* in detail, involving as it did all seventeen members of their "religious group." Goldman raged, but he had to let her study the file. In order not to be bored waiting for her, he led her to a large office, where half a dozen employees were sitting, and left her there. At first she read quietly, but then a conversation began—perhaps because the others were bored—and Vera launched aloud into a real religious sermon. (One would have had to know her to appreciate this to the full. She was a luminous person, with a lively mind and a gift of eloquence, even though in freedom she had been no more than a lathe operator, a stable girl, and a housewife.) They listened to her impressively, now and then asking questions in order to clarify something or other. It was catching them from an unexpected side of things. People came in from other offices, and the room filled up. Even though they were only typists, stenographers, file clerks, and not interrogators, in 1946 this was still their milieu, the *Organs.* It is impossible to reconstruct her monologue. She managed to work in all sorts of things, including the question of "traitors of the Motherland." Why were there no traitors in the 1812 War of the Fatherland, when there was still serfdom? It would have been natural to have traitors then! But mostly she spoke about religious faith and religious believers. *Formerly,* she declared, unbridled passions were the basis for everything—"Steal the stolen goods"—and, in that state of affairs, religious believers were naturally a hindrance to you. But now, when you want to *build* and prosper in this world, why do you persecute your best citizens? They represent your most precious material: after all, believers don't need to be watched, they do not steal, and they do not shirk. Do you think you can build a just society on a foundation of self-serving and envious people? Everything in the country is falling apart. Why do you spit in the hearts of your best people? Separate church and state properly and do not touch the church; you will not lose a thing thereby. Are you materialists? In that case, put your faith in education—in the possibility that it will, as they say, disperse religious faith. But why arrest people? At this point Goldman came in and started to interrupt rudely. But everyone shouted at him: "Oh, shut up! Keep quiet! Go ahead, woman, talk." (And how should they have addressed her? Citizeness? Comrade? Those forms of address were forbidden, and these people were bound by the conventions of Soviet life. But "woman"—that was how Christ had spoken, and you couldn't go wrong there.) And Vera continued in the presence of her interrogator.

So there in the MGB office those people listened to Korneyeva—and why did the words of an insignificant prisoner touch them so near the quick?

And why is it that for nearly two hundred years the Security forces have hung onto the color of the heavens? That was what they wore in Lermontov's lifetime—"and you, blue uniforms!" Then came blue service caps, blue shoulder boards, blue tabs, and then they were ordered to make themselves less conspicuous, and the blue brims were hidden from the gratitude of the people and everything blue on heads and shoulders was made narrower—until what was left was piping, narrow rims . . . but still blue.

Is this only a masquerade?

Or is it that even blackness must, every so often, however rarely, partake of the heavens?

It would be beautiful to think so. But when one learns, for example, the nature of Yagoda's striving toward the sacred . . . An eyewitness from the group around Gorky, who was close to Yagoda at the time, reports that in the vestibule of the bathhouse on Yagoda's estate near Moscow, ikons were placed so that Yagoda and his comrades, after undressing, could use them as targets for revolver practice before going in to take their baths.

Just how are we to understand that? As the act of an *evildoer?* What sort of behavior is it? Do such people really exist?

We would prefer to say that such people cannot exist, that there aren't any. It is permissible to portray evildoers in a story for children, so as to keep the picture simple. But when the great world literature of the past—Shakespeare, Schiller, Dickens—inflates and inflates images of evildoers of the blackest shades, it seems somewhat farcical and clumsy to our contemporary perception. The trouble lies in the way these classic evildoers are pictured. They recognize themselves as evildoers, and they know their souls are black. And they reason: "I cannot live unless I do evil. So I'll set my father against my brother! I'll drink the victim's sufferings until I'm drunk with them!" Iago very precisely identifies his purposes and his motives as being black and born of hate.

But no; that's not the way it is! To do evil a human being must first of all believe that what he's doing is good, or else that it's a well-considered act in conformity with natural law. Fortunately, it is in the nature of the human being to seek a *justification* for his actions.

Macbeth's self-justifications were feeble—and his conscience devoured him. Yes, even Iago was a little lamb too. The imagination and the spiritual strength of Shakespeare's evildoers stopped short at a dozen corpses. Because they had no *ideology.*

Ideology—that is what gives evildoing its long-sought justification and gives the evildoer the necessary steadfastness and determination. That is the social theory which helps to make his acts seem good instead of bad in his own and others' eyes, so that he won't hear reproaches and curses but will receive praise and honors. That was how the agents of the Inquisition fortified their wills: by invoking Christianity; the conquerors of foreign lands, by extolling the grandeur of their Motherland; the colonizers, by civilization; the Nazis, by race; and the Jacobins (early and late), by equality, brotherhood, and the happiness of future generations.

Thanks to *ideology,* the twentieth century was fated to experience evildoing on a scale calculated in the millions. This cannot be denied, nor passed over, nor suppressed. How, then, do we dare insist that evildoers do not exist? And who was it that destroyed these millions? Without evildoers there would have been no Archipelago.

There was a rumor going the rounds between 1918 and 1920 that the Petrograd Cheka, headed by Uritsky, and the Odessa Cheka, headed by Deich, did not shoot all those condemned to death but fed some of them alive to the animals in the city zoos. I do not know whether this is truth or calumny, or, if there were any such cases, how many there were. But I wouldn't set out to look for proof, either. Following the practice of the bluecaps, I would propose that they prove to us that this was impossible. How else could they get food for the zoos in those famine years? Take it away from the working class? Those enemies were going to die anyway, so why couldn't their deaths support the zoo economy of the Republic and thereby assist our march into the future? Wasn't it *expedient?*

That is the precise line the Shakespearean evildoer could not cross. But the evildoer with ideology does cross it, and his eyes remain dry and clear.

Physics is aware of phenomena which occur only at *threshold* magnitudes, which do not exist at all until a certain *threshold* encoded by and known to nature has been crossed. No matter how intense a yellow light you shine on a lithium sample, it will not emit electrons. But as soon as a weak bluish light begins to glow, it does emit them. (The threshold of the photoelectric effect has been crossed.) You can cool oxygen to 100 degrees below zero Centigrade and exert as much pressure as you want; it does not yield, but remains a gas. But as soon as minus 183 degrees is reached, it liquefies and begins to flow.

Evidently evildoing also has a threshold magnitude. Yes, a human being hesitates and bobs back and forth between good and evil all his life. He slips, falls back, clambers up, repents, things begin to darken again. But just so long as the threshold of evildoing is not crossed, the possibility of returning remains, and he himself is still within reach of our hope. But when, through the density of evil actions, the result either of their own extreme degree or of the absoluteness of his power, he suddenly crosses that threshold, he has left humanity behind, and without, perhaps, the possibility of return.

■

From the most ancient times justice has been a two-part concept: virtue triumphs, and vice is punished.

We have been fortunate enough to live to a time when virtue, though it does not triumph, is nonetheless not always tormented by attack dogs. Beaten down, sickly, virtue has now been allowed to enter in all its tatters and sit in the corner, as long as it doesn't raise its voice.

However, no one dares say a word about vice. Yes, they did mock virtue, but there was no vice in that. Yes, so-and-so many millions did get mowed down—but no one was to blame for it. And if someone pipes up: "What about *those who* . . ." the answer comes from all sides, reproachfully and amicably at first: "What are you talking about, comrade! Why *open* old *wounds?*" Then they go after you with an oaken club: "Shut up! Haven't you had enough yet? You think you've been rehabilitated!"

In that same period, by 1966, *eighty-six thousand* Nazi criminals had been convicted in West Germany. And still we choke with anger here. We do not hesitate to devote to the subject page after newspaper page and hour after hour of radio time. We even stay after work to attend protest meetings and vote: *"Too few!* Eighty-six thousand are too few. And twenty years is too little! It must go on and on."

And during the same period, in our own country (according to the reports of the Military Collegium of the Supreme Court) about *ten men* have been convicted.

What takes place beyond the Oder and the Rhine gets us all worked up. What goes on in the environs of Moscow and behind the green fences near Sochi, or the fact that the murderers of our husbands and fathers ride through our streets and we make way for them as they pass, doesn't get us worked up at all, doesn't touch us. That would be "digging up the past."

Meanwhile, if we translate 86,000 West Germans into our own terms, on the basis of comparative population figures, it would become *one-quarter of a million.*

But in a quarter-century we have not tracked down anyone. We have not brought anyone to trial. It is their wounds we are afraid to reopen. And as a symbol of them all, the smug and stupid Molotov lives on at Granovsky No. 3, a man who has learned nothing at all, even now, though he is saturated with our blood and nobly crosses the sidewalk to seat himself in his long, wide automobile.

Here is a riddle not for us contemporaries to figure out: *Why* is

Germany allowed to punish its evildoers and Russia is not? What kind of disastrous path lies ahead of us if we do not have the chance to purge ourselves of that putrefaction rotting inside our body? What, then, can Russia teach the world?

In the German trials an astonishing phenomenon takes place from time to time. The defendant clasps his head in his hands, refuses to make any defense, and from then on asks no concessions from the court. He says that the presentation of his crimes, revived and once again confronting him, has filled him with revulsion and he no longer wants to live.

That is the ultimate height a trial can attain: when evil is so utterly condemned that even the criminal is revolted by it.

A country which has condemned evil 86,000 times from the rostrum of a court and irrevocably condemned it in literature and among its young people, year by year, step by step, is purged of it.

What are we to do? Someday our descendants will describe our several generations as generations of driveling do-nothings. First we submissively allowed them to massacre us by the millions, and then with devoted concern we tended the murderers in their prosperous old age.

What are we to do if the great Russian tradition of penitence is incomprehensible and absurd to them? What are we to do if the animal terror of hearing even one-hundredth part of all they subjected others to outweighs in their hearts any inclination to justice? If they cling greedily to the harvest of benefits they have watered with the blood of those who perished?

It is clear enough that those men who turned the handle of the meat grinder even as late as 1937 are no longer young. They are fifty to eighty years old. They have lived the best years of their lives prosperously, well nourished and comfortable, so that it is too late for any kind of *equal* retribution as far as they are concerned.

But let us be generous. We will not shoot them. We will not pour salt water into them, nor bury them in bedbugs, nor bridle them into a "swan dive," nor keep them on sleepless "stand-up" for a week, nor kick them with jackboots, nor beat them with rubber truncheons, nor squeeze their skulls in iron rings, nor push them into a cell so that they lie atop one another like pieces of baggage—we will not do any of the things they did! But for the sake of our country and our children we have the duty to *seek them all out and bring them all to trial!* Not to put them on trial so much as their crimes. And to compel each one of them to announce loudly:

"Yes, I was an executioner and a murderer."

And if these words were spoken in our country *only* one-quarter of a million times (a just proportion, if we are not to fall behind West Germany), would it, perhaps, be enough?

It is unthinkable in the twentieth century to fail to distinguish between what constitutes an abominable atrocity that must be prosecuted and what constitutes that "past" which "ought not to be stirred up."

We have to condemn publicly the very *idea* that some people have the right to repress others. In keeping silent about evil, in burying it so deep within us that no sign of it appears on the surface, we are *implanting* it, and it will rise up a thousandfold in the future. When we neither punish nor reproach evildoers, we are not simply protecting their trivial old age, we are thereby ripping the foundations of justice from beneath new generations. It is for this reason, and not because of the "weakness of indoctrinational work," that they are growing up "indifferent." Young people are acquiring the conviction that foul deeds are never punished on earth, that they always bring prosperity.

It is going to be uncomfortable, horrible, to live in such a country!

Chapter 5

■

First Cell, First Love

How is one to take the title of this chapter? A cell and love in the same breath? Ah, well, probably it has to do with Leningrad during the blockade—and you were imprisoned in the Big House. In that case it would be very understandable. That's why you are still alive—because they shoved you in there. It was the best place in Leningrad—not only for the interrogators, who even lived there and had offices in the cellars in case of shelling. Joking aside, in Leningrad in those days no one washed and everyone's face was covered with a black crust, but in the Big House prisoners were given a hot shower every tenth day. Well, it's true that only the corridors were heated—for the jailers. The cells were left unheated, but after all, there were water pipes in the cells that worked and a toilet, and where else in Leningrad could you find that? And the bread ration was just like the ration outside—barely four and a half ounces. In addition, there was broth made from slaughtered horses once a day! And thin gruel once a day as well!

It was a case of the cat's being envious of the dog's life!

You sit down and half-close your eyes and try to remember them all. How many different cells you were imprisoned in during your term! It is difficult even to count them. And in each one there were people, people. There might be two people in one, 150 in another. You were imprisoned for five minutes in one and all summer long in another.

But in every case, out of all the cells you've been in, your first cell is a very special one, the place where you first encountered others like yourself, doomed to the same fate. All your life you will remember it with an emotion that you otherwise experience only in remembering your first love. And those people, who shared with you the floor and air of that stone cubicle during those days when you rethought your entire life, will from time to time be recollected by you as members of your own family.

Yes, in those days they were your only family.

What you experience in your first interrogation cell parallels nothing in your entire *previous* life or your whole *subsequent* life. No doubt prisons have stood for thousands of years before you came along, and may continue to stand after you too—longer than one would like to think—but that first interrogation cell is unique and inimitable.

Maybe it was a terrible place for a human being. A lice-laden, bedbug-infested lock-up, without windows, without ventilation, without bunks, and with a dirty floor. . . . Or maybe it was "solitary" in the Archangel prison, where the glass had been smeared over with red lead so that the only rays of God's maimed light which crept in to you were crimson, and where a 15-watt bulb burned constantly in the ceiling, day and night. Or "solitary" in the city of Choibalsan, where, for six months at a time, fourteen of you were crowded onto seven square yards of floor space in such a way that you could only shift your bent legs in unison. Or it was one of the Lefortovo "psychological" cells, like No. 111, which was painted black and also had a day-and-night 25-watt bulb, but was in all other respects like every other Lefortovo cell: asphalt floor; the heating valve out in the corridor where only the guards had access to it; and, above all, that interminable irritating roar from the wind tunnel of the neighboring Central Aero- and Hydro-dynamics Institute—a roar one could not believe was unintentional, a roar which would make a bowl or cup vibrate so violently that it would slip off the edge of the table, a roar which made it useless to converse and during which one could sing at the top of one's lungs and the jailer wouldn't even hear. And then when the roar stopped, there would ensue a sense of relief and felicity superior to freedom itself.

But it was not the dirty floor, nor the murky walls, nor the odor of the latrine bucket that you loved—but those fellow prisoners with whom you about-faced at command, and that something which beat between your heart and theirs, and their sometimes astonishing words, and then, too, the birth within you, on that very spot, of free-floating thoughts you had so recently been unable to leap up or rise to.

And how much it had cost you to last out until that first cell! You had been kept in a pit, or in a box, or in a cellar. No one had addressed a human word to you. No one had looked at you with a human gaze. All they did was to peck at your brain and heart with iron beaks, and when you cried out or groaned, they laughed.

For a week or a month you had been an abandoned waif, alone among enemies, and you had already said good-bye to reason and to life; and you had already tried to kill yourself by "falling" from the radiator in such a way as to smash your brains against the iron cone of the valve. Then all of a sudden you were alive again, and were brought in to your friends. And reason returned to you.

That's what your first cell is!

You waited for that cell. You dreamed of it almost as eagerly as of freedom. Meanwhile, they kept shoving you around between cracks in the wall and holes in the ground, from Lefortovo into some legendary, diabolical Sukhanovka.

Sukhanovka was the most terrible prison the MGB had. Its very name was used to intimidate prisoners; interrogators would hiss it threateningly. And you'd not be able to question those who had been there: either they were insane and talking only disconnected nonsense, or they were dead.

Sukhanovka was a former monastery, dating back to Catherine the Great. It consisted of two buildings—one in which prisoners served out their terms, and the other a structure that contained sixty-eight monks' cells and was used for interrogations. The journey there in a Black Maria took two hours, and only a handful of people knew that the prison was really just a few miles from Lenin's Gorki estate and near the former estate of Zinaida Volkonskaya. The countryside surrounding it was beautiful.

There they stunned the newly arrived prisoner with a stand-up punishment cell again so narrow that when he was no longer able to stand he had to sag, supported by his bent knees propped against the wall. There was no alternative. They kept prisoners thus for more than a day to break their resistance. But they ate tender, tasty food at Sukhanovka, which was like nothing else in the MGB—because it was brought in from the Architects' Rest Home. They didn't maintain a separate kitchen to prepare hogwash. However, the amount one architect would eat—including fried potatoes and meatballs—was divided among twelve prisoners. As a result the prisoners were not only always hungry but also exceedingly irritable.

The cells were all built for two, but prisoners under interrogation were usually kept in them singly. The dimensions were five by six and a half feet.

To be absolutely precise, they were 156 centimeters by 209 centimeters. How do we know? Through a triumph of engineering calculation and a strong heart that even

Sukhanovka could not break. The measurements were the work of Alexander Dolgun, who would not allow them to drive him to madness or despair. He resisted by striving to use his mind to calculate distances. In Lefortovo he counted steps, converted them into kilometers, remembered from a map how many kilometers it was from Moscow to the border, and then how many across all Europe, and how many across the Atlantic Ocean. He was sustained in this by the hope of returning to America. And in one year in Lefortovo solitary he got, so to speak, halfway across the Atlantic. Thereupon they took him to Sukhanovka. Here, realizing how few would survive to tell of it—and all our information about it comes from him—he invented a method of measuring the cell. The numbers 10/22 were stamped on the bottom of his prison bowl, and he guessed that "10" was the diameter of the bottom and "22" the diameter of the outside edge. Then he pulled a thread from a towel, made himself a tape measure, and measured everything with it. Then he began to invent a way of sleeping *standing up,* propping his knees against the small chair, and of deceiving the guard into thinking his eyes were open. He succeeded in this deception, and that was how he managed not to go insane when Ryumin kept him sleepless for a month.

Two little round stools were welded to the stone floor, like stumps, and at night, if the guard unlocked a cylinder lock, a shelf dropped from the wall onto each stump and remained there for seven hours (in other words, during the hours of interrogation, since there was no daytime interrogation at Sukhanovka at all), and a little straw mattress large enough for a child also dropped down. During the day, the stool was exposed and free, but one was forbidden to sit on it. In addition, a table lay, like an ironing board, on four upright pipes. The "fortochka" in the window—the small hinged pane for ventilation—was always closed except for ten minutes in the morning when the guard cranked it open. The glass in the little window was reinforced. There were never any exercise periods out of doors. Prisoners were taken to the toilet at 6 A.M. only—i.e., when no one's stomach needed it. There was no toilet period in the evening. There were two guards for each block of seven cells, so that was why the prisoners could be under almost constant inspection through the peephole, the only interruption being the time it took the guard to step past two doors to a third. And that was the purpose of silent Sukhanovka: to leave the prisoner not a single moment for sleep, not a single stolen moment for privacy. You were always being watched and always in their power.

But if you endured the whole duel with insanity and all the trials of loneliness, and had stood firm, you deserved your first cell! And now when you got into it, your soul would heal.

If you had surrendered, if you had given in and betrayed everyone, you were also ready for your first cell. But it would have been better for you not to have lived until that happy moment and to have died a victor in the cellar, without having signed a single sheet of paper.

Now for the first time you were about to see people who were not your enemies. Now for the first time you were about to see others who were alive, who were traveling your road, and whom you could join to yourself with the joyous word "we."

Yes, that word which you may have despised out in freedom, when they used it as a substitute for your own individuality ("All of us, like one man!" Or: "We are deeply angered!" Or: "We demand!" Or: "We swear!"), is now revealed to you as something sweet: you are not alone in the world! Wise, spiritual beings—*human beings*—still exist.

■

I had been dueling for four days with the interrogator, when the jailer, having waited until I lay down to sleep in my blindingly lit box, began to unlock my door. I heard him all right, but before he could say: "Get up! Interrogation!" I wanted to lie for another three-hundredths of a second with my head on the pillow and pretend I was sleeping. But, instead of the familiar command, the guard ordered: "Get up! Pick up your bedding!"

Uncomprehending, and unhappy because this was my most precious time, I wound on my footcloths, put on my boots, my overcoat, my winter cap, and clasped the government-issue mattress in my arms. The guard was walking on tiptoe and kept signaling me not to make any noise as he led me down a corridor silent as the grave, through the fourth floor of the Lubyanka, past the desk of the section supervisor, past the shiny numbers on the cells and the olive-colored covers of the peepholes, and unlocked Cell 67. I entered and he locked it behind me immediately.

Even though only a quarter of an hour or so had passed since the signal to go to sleep had been given, the period allotted the prisoners for sleeping was so fragile, and undependable, and brief that, by the time I arrived, the inhabitants of Cell 67 were already asleep on their metal cots with their hands on top of the blankets.

At the sound of the door opening, all three started and raised their heads for an instant. They, too, were waiting to learn which of them might be taken to interrogation.

And those three lifted heads, those three unshaven, crumpled pale faces, seemed to me so human, so dear, that I stood there, hugging my mattress, and smiled with happiness. And they smiled. And what a forgotten look that was—after only one week!

"Are you from freedom?" they asked me. (That was the question customarily put to a newcomer.)

"Nooo," I replied. And that was a newcomer's usual first reply. They had in mind that I had probably been arrested recently, which meant that I came *from freedom.* And I, after ninety-six hours of interrogation, hardly considered that I was from "freedom." Was I not already a veteran prisoner? Nonetheless I was *from freedom.* The beardless old man with the black and very lively eyebrows was already asking me for military and political news. Astonishing! Even though it was late February, they knew nothing about the Yalta Conference, nor the encirclement of East Prussia, nor anything at all about our own attack below Warsaw in mid-January, nor even about the woeful December retreat of the Allies. According to regulations, those under interrogation were not supposed to know anything about the outside world. And here indeed they didn't!

I was prepared to spend half the night telling them all about it—with pride, as though all the victories and advances were the work of my own hands. But at this point the duty jailer brought in my cot, and I had to set it up without making any noise. I was helped by a young fellow my own age, also a military man. His tunic and aviator's cap hung on his cot. He had asked me, even before the old man spoke, not for news of the war but for tobacco. But although I felt openhearted toward my new friends, and although not many words had been exchanged in the few minutes since I joined them, I sensed something alien in this front-line soldier who was my contemporary, and, as far as he was concerned, I clammed up immediately and forever.

(I had not yet even heard the word "nasedka"—"stool pigeon"—nor learned that there had to be one such "stool pigeon" in each cell. And I had not yet had time to think things over and conclude that I did not like this fellow, Georgi Kramarenko. But a spiritual relay, a sensor relay, had clicked inside me, and it had closed him off from me for good and all. I would not bother to recall this event if it had been the only one of its kind. But soon, with astonishment, and alarm, I became aware of the work of this internal sensor relay as a constant, inborn trait. The years passed and I lay on the same bunks, marched in the same formations, and worked in the same work brigades with hundreds of others. And always that secret sensor relay, for whose creation I deserved not the least bit of credit, worked even before I remembered it was there, worked at the first sight of a human face and eyes, at the first sound of a voice—so that I opened my heart to that person either fully or just the width of a crack, or else shut myself off from him completely. This was so consistently unfailing that all the efforts of the State Security officers to employ stool pigeons began to seem to me as insignificant as being pestered by gnats: after all, a person who has undertaken to be a traitor always betrays the fact in his face and in his voice, and even though some were more skilled in pretense, there was always something fishy about them. On the other hand, the sensor relay helped me distinguish those to whom I could from the very beginning of our acquaintance completely disclose my most precious depths and secrets—secrets for which heads roll. Thus it was that I got through eight years of imprisonment, three years of exile, and another six years of underground authorship, which were in no wise less dangerous. During all those seventeen years I recklessly revealed myself to dozens of people—and didn't make a misstep even once. (I have never read about this trait anywhere, and I mention it here for those interested in psychology. It seems to me that such spiritual sensors exist in many of us, but because we live in too technological and rational an age, we neglect this miracle and don't allow it to develop.)

We set up the cot, and I was then ready to talk—in a whisper, of course, and lying down, so as not to be sent from this cozy nest into a punishment cell. But our third cellmate, a middle-aged man whose cropped head already showed the white bristles of imminent grayness, peered at me discontentedly and said with characteristic northern severity: "Tomorrow! Night is for sleeping."

That was the most intelligent thing to do. At any minute, one of us could have been pulled out for interrogation and held until 6 A.M., when the interrogator would go home to sleep but we were forbidden to.

One night of undisturbed sleep was more important than all the fates on earth!

One more thing held me back, which I didn't quite catch right away but had felt nonetheless from the first words of my story, although I could not at this early date find a name for it: As each of us had been arrested, everything in our world had switched places, a 180-degree shift in all our concepts had occurred, and the good news I had begun to recount with such enthusiasm might not be good news for *us* at all.

My cellmates turned on their sides, covered their eyes with their handkerchiefs to keep out the light from the 200-watt bulb, wound towels around their upper arms, which were chilled from lying on top of the blankets, hid their lower arms furtively beneath them, and went to sleep.

And I lay there, filled to the brim with the joy of being among them. One hour ago I could not have counted on being with anyone. I could have come to my end with a bullet in the back of my head—which was what the interrogator kept promising me—without having seen anyone at all. Interrogation still hung over me, but how far it had retreated! Tomorrow I would be telling them my story (though not talking about my *case,* of course) and they would be telling me their stories too. How interesting tomorrow would be, one of the best days of my life! (Thus, very early and very clearly, I had this consciousness that prison was not an abyss for me, but the most important turning point in my life.)

■

And there was no reason to be bored with my companions in my new cell. They were people to listen to and people with whom to compare notes.

The old fellow with the lively eyebrows—and at sixty-three he in no way bore himself like an old man—was Anatoly Ilyich Fastenko. He was a big asset to our Lubyanka cell—both as a keeper of the old Russian prison traditions and as a living history of Russian revolutions. Thanks to all that he remembered, he somehow managed to put in perspective everything that had taken place in the past and everything that was taking place in the present. Such people are valuable not only in a cell. We badly need them in our society as a whole.

Right there in our cell we read Fastenko's name in a book about the 1905 Revolution. He had been a Social Democrat for such a long, long time that in the end, it seemed, he had ceased to be one.

He had been sentenced to his first prison term in 1904 while still a young man, but he had been freed outright under the "manifesto" proclaimed on October 17, 1905. . . .

There was much about Fastenko I could not yet understand. In my eyes, perhaps the main thing about him, and the most surprising, was that he had known Lenin personally. Yet he was quite cool in recalling this. (Such was my attitude at the time that when someone in the cell called Fastenko by his patronymic alone, without using his given name—in other words simply "Ilyich," asking: "Ilyich, is it your turn to take out the latrine bucket?"—I was utterly outraged and offended because it seemed sacrilege to me not only to use Lenin's patronymic in the same sentence as "latrine bucket," but even to call anyone on earth "Ilyich" except that one man, Lenin.) For this reason, no doubt, there was much that Fastenko would have liked to explain to me that he still could not bring himself to.

Nonetheless, he did say to me, in the clearest Russian: "Thou shalt not make unto thee any graven image!" But I failed to understand him!

Observing my enthusiasm, he more than once said to me insistently: "You're a mathematician; it's a mistake for you to forget that maxim of Descartes: 'Question everything!' Question *everything!*" What did this mean—"everything"? Certainly not *everything!* It seemed to me that I had questioned enough things as it was, and that was enough of that!

Or he said: "Hardly any of the old hard-labor political prisoners of Tsarist times are left. I am one of the last. All the hard-labor politicals have been destroyed, and they even dissolved our society in the thirties." "Why?" I asked. "So we would not get together and discuss things." And although these simple words, spoken in a calm tone, should have been shouted to the heavens, should have shattered windowpanes, I understood them only as indicating one more of Stalin's evil deeds. It was a troublesome fact, but without roots.

One thing is absolutely definite: not everything that enters our ears penetrates our consciousness. Anything too far out of tune with our attitude is lost, either in the ears themselves or somewhere beyond, but it is lost. And even though I clearly remember Fastenko's many stories, I recall his opinions but vaguely. He gave me the names of various books which he strongly advised me to read whenever I got back to freedom. In view of his age and his health, he evidently did not count on getting out of prison alive, and he got some satisfaction from hoping that I would someday understand his ideas. . . .

Fastenko was the most cheerful person in the cell, even though, in view of his age, he was the only one who could not count on surviving and returning to freedom. Flinging an arm around my shoulders, he would say:

> To *stand up* for the truth is nothing!
> For truth you have to *sit* in jail!

Or else he taught me to sing this song from Tsarist hard-labor days:

> And if we have to perish
> In mines and prisons wet,
> Our cause will ever find renown
> In future generations yet.

And I believe this! May these pages help his faith come true!

■

Spring promises everyone happiness—and tenfold to the prisoner. Oh, April sky! It didn't matter that I was in prison. Evidently, they were not going to shoot me. And in the end I would become wiser here. I would come to understand many things here, Heaven! I would correct my mistakes yet, O Heaven, not for *them* but for you, Heaven! I had come to understand those mistakes here, and I would correct them!

The walk in the fresh air lasted only twenty minutes, but how much there was about it to concern oneself with; how much one had to accomplish while it lasted.

During that outdoor walk you concentrated on breathing as much fresh air as possible.

There, too, alone beneath that bright heaven, you had to imagine your bright future life, sinless and without error.

There, too, was the best place of all to talk about the most dangerous subjects. It didn't matter that conversation during the walk was forbidden. One simply had to know how to manage it. The compensation was that in all likelihood you could not be overheard either by a stoolie or by a microphone.

During these walks I tried to get into a pair with Susi, once a leading lawyer in Estonia. We talked together in the cell, but we liked to try talking about the main things here. We hadn't come together quickly. It took some time. But he had already managed to tell me a great deal. I acquired a new capability from him: to accept patiently and purposefully things that had never had any place in my own plans and had, it seemed, no connection at all with the clearly outlined direction of my life. From childhood on, I had somehow known that my objective was the history of the Russian Revolution and that nothing else concerned me. To understand the Revolution I had long since required nothing beyond Marxism. I cut myself off from everything else that came up and turned my back on it. And now fate brought me together with Susi. He breathed a completely different sort of air. And he would tell me passionately about his own interests, and these were Estonia and democracy. And although I had never expected to become interested in Estonia, much less bourgeois democracy, I nevertheless kept listening and listening to his loving stories of twenty free years in that modest, work-loving, small nation of big men whose ways were slow and set. I listened to the principles of the Estonian constitution, which had been borrowed from the best of

European experience, and to how their hundred-member, one-house parliament had worked. And, though the *why* of it wasn't clear, I began to like it all and store it all away in my experience. I listened willingly to their fatal history: the tiny Estonian anvil had, from way, way back, been caught between two hammers, the Teutons and the Slavs. Blows showered on it from East and West in turn; there was no end to it, and there still isn't. And there was the well-known (totally unknown) story of how we Russians wanted to take them over in one fell swoop in 1918, but they refused to yield. And how, later on, Yudenich spoke contemptuously of their Finnish heritage, and we ourselves christened them "White Guard Bandits." Then the Estonian gymnasium students enrolled as volunteers. We struck at Estonia again in 1940, and again in 1941, and again in 1944. Some of their sons were conscripted by the Russian Army, and others by the German Army, and still others ran off into the woods. The elderly Tallinn intellectuals discussed how they might break out of that iron ring, break away somehow, and live for themselves and by themselves. Their Premier might, possibly, have been Tief, and their Minister of Education, say, Susi. But neither Churchill nor Roosevelt cared about them in the least; but "Uncle Joe" did. And during the very first nights after the Soviet armies entered Tallinn, all these dreamers were seized in their Tallinn apartments. Fifteen of them were imprisoned in various cells of the Moscow Lubyanka, one in each, and were charged under Article 58-2 with the criminal desire for national self-determination.

Each time we returned to the cell from our walk was like being arrested again. Even in our very special cell the air seemed stifling after the outdoors. And it would have been good to have a snack afterward too. But it was best not to think about it—not at all.

■

I often argued with Yuri Yevtukhovich. Yuri spoke German fluently; in 1941 they dressed him as a German POW officer, provided him with the necessary documents, and sent him on a reconnaissance mission. He fulfilled his mission and on his way back changed into a Soviet uniform, which he took off a dead officer. Then he was taken prisoner by the Germans. They sent him to a concentration camp near Vilnius.

In every life there is one particular event that is decisive for the entire person—for his fate, his convictions, his passions. Two years in that camp shook Yuri up once and for all. It is impossible to catch with words or to circumvent with syllogisms what that camp was. That was a camp to die in—and whoever did not die was compelled to reach certain conclusions.

The slops for which the POW officers stood in line with their mess tins from 6 A.M. on, while the Ordners beat them with sticks and the cooks with ladles, were not enough to sustain life. At evening, Yuri could see from the windows of their room the one and only picture for which his artistic talent had been given him: the evening mist hovering above a swampy meadow encircled by barbed wire; a multitude of bonfires; and, around the bonfires, beings who had once been Russian officers but had now become beastlike creatures who gnawed the bones of dead horses, who baked patties from potato rinds, who smoked manure and were all swarming with lice. Not all those two-legged creatures had died as yet. Not all of them had yet lost the capacity for intelligible speech, and one could see in the crimson reflections of the bonfires how a belated understanding was dawning on those faces which were descending to the Neanderthal.

Wormwood on the tongue! That life which Yuri had preserved was no longer precious to him for its own sake. He was not one of those who easily agree to forget. No, if he was going to survive, he was obliged to draw certain conclusions.

It was already clear to them that the Germans were not the heart of the matter, or at least not the Germans alone; that among the POW's of many nationalities only the Soviets lived like this and died like this. None were worse off than the Soviets. Even the Poles, even the Yugoslavs, existed in far more tolerable conditions; and as for the English and the Norwegians, they were inundated by the International Red Cross with parcels from home. They didn't even bother to line up for the German rations. Wherever there were Allied POW camps next door, their prisoners, out of kindness, threw our men handouts over the fence, and our prisoners jumped on these gifts like a pack of dogs on a bone.

The Russians were carrying the whole war on their shoulders—and this was the Russian lot. Why?

What is the right course of action if our mother has sold us to the gypsies? No, even worse, thrown us to the dogs? Does she really remain our mother? If a wife has become a whore, are we really still bound to her in fidelity? A Motherland that betrays its soldiers—is that really a Motherland?

When, in the spring of 1943, recruiters from the first Byelorussian "legions" put in an appearance, some POW's signed up with them to escape starvation. Yuri went with them out of conviction, with a clear mind. Just where could one draw the line? Which step was the fatal one? Yuri became a lieutenant in the German Army.

In all, Yuri spent three weeks in our cell. I argued with him during all those weeks. I said that our Revolution was magnificent and just; that only its 1929 distortion was terrible. He looked at me regretfully, compressing his nervous lips.

■

Just before May 1 they took down the blackout shade on the window. The war was perceptibly coming to an end.

That evening it was quieter than ever before in the Lubyanka. It was, I remember, almost like the second day of Easter, since May Day and Easter came one after the other that year. All the interrogators were out in Moscow celebrating. No one was taken to interrogation. In the silence we could hear someone across the corridor protesting. They took him from the cell and into a box. By listening, we could detect the location of all the doors. They left the door of the box open, and they kept beating him a long time. In the suspended silence every blow on his soft and choking mouth could be heard clearly.

On May 2 a thirty-gun salute roared out. That meant a European capital. Only two had not yet been captured—Prague and Berlin. We tried to guess which it was.

On the ninth of May they brought us our dinner at the same time as our lunch—which was done at the Lubyanka only on May 1 and November 7.

And that is how we guessed that the war had ended.

That evening they shot off another thirty-gun salute. We then knew that there were no more capitals to be captured. And later that same evening one more salute roared out—forty guns, I seem to remember. And that was the end of all the ends.

Above the muzzle of our window, and from all the other cells of the Lubyanka, and from all the windows of all the Moscow prisons, we, too, former prisoners of war and former front-line soldiers, watched the Moscow heavens, patterned with fireworks and crisscrossed by the beams of searchlights.

Boris Gammerov, a young antitank man, already demobilized because of wounds, with an incurable wound in his lung, having been arrested with a group of students, was in prison that evening in an overcrowded Butyrki cell, where half the inmates were former POW's and front-line soldiers. He described this last salute of the war in a terse eight-stanza poem, in the most ordinary language: how they were already lying down on their board bunks, covered with their overcoats; how they were awakened by the noise; how they raised their heads; squinted up at the muzzle—"Oh, it's just a salute"—and then lay down again:

And once again covered themselves with their coats.

With those same overcoats which had been in the clay of the trenches, and the ashes of bonfires, and been torn to tatters by German shell fragments.

That victory was not for us. And that spring was not for us either.

Chapter 6

■

That Spring

Through the windows of the Butyrki Prison every morning and evening in June, 1945, we could hear the brassy notes of bands not far away—coming from either Lesnaya Street or Novoslobodskaya. They kept playing marches over and over.

Behind the murky green "muzzles" of reinforced glass, we stood at the wide-open but impenetrable prison windows and listened. Were they military units that were marching? Or were they workers cheerfully devoting their free time to marching practice? We didn't know, but the rumor had already gotten through to us that preparations were under way for a big Victory Parade on Red Square on June 22—the fourth anniversary of the beginning of the war.

The foundation stones of a great building are destined to groan and be pressed upon; it is not for them to crown the edifice. But even the honor of being part of the foundation was denied those whose doomed heads and ribs had borne the first blows of this war and thwarted the foreigners' victory, and who were now abandoned for no good reason.

"Joyful sounds mean nought to the traitor."

That spring of 1945 was, in our prisons, predominantly the spring of the Russian *prisoners of war.* They passed through the prisons of the Soviet Union in vast dense gray schools like ocean herring. The first trace of those schools I glimpsed was Yuri Yevtukhovich. But I was soon entirely surrounded by their purposeful motion, which seemed to know its own fated design.

Not only war prisoners passed through those cells. A wave of those who had spent any time in Europe was rolling too: émigrés from the Civil War; the "ostovtsy"—workers recruited as laborers by the Germans during World War II; Red Army officers who had been too astute and farsighted in their conclusions, so that Stalin feared they might bring European freedom back from their European crusade, like the Decembrists 120 years before. And yet it was the war prisoners who constituted the bulk of the wave. And among the war prisoners of various ages, most were of my own age—not precisely my age, but *the twins of October,* those born along with the Revolution, who in 1937 had poured forth undismayed to celebrate the twentieth anniversary of the Revolution, and whose age group, at the beginning of the war, made up the standing army—which had been scattered in a matter of weeks.

That tedious prison spring had, to the tune of the victory marches, become the spring of reckoning for my whole generation.

Over our cradles the rallying cry had resounded: "All power to the Soviets!" It was we who had reached out our suntanned childish hands to clutch the Pioneers' bugle, and who in response to the Pioneer challenge, "Be prepared," had saluted and answered: "We are always prepared!" It was we who had smuggled weapons into Buchenwald and joined the Communist Party there. And it was we who were now in disgrace, only because we had survived.

Back when the Red Army had cut through East Prussia, I had seen downcast columns of returning war prisoners—the only people around who were grieving instead of celebrating. Even then their gloom had shocked me, though I didn't yet grasp the reason for it. I jumped down and went over to those voluntarily formed-up columns. (Why were they marching in columns? Why had they lined themselves up in ranks? After all, no one had compelled them to, and the war prisoners of all other nations went home as scattered individuals. But ours wanted to return as submissively as possible.) I was wearing a captain's shoulder boards, and they, plus the fact that I was moving forward, helped prevent my finding out why our POW's were so sad. But then fate turned me around and sent me in the wake of those prisoners along the same path they had taken. I had already marched with them from army counterintelligence headquarters to the headquarters at the front, and when we got there I had heard their first stories, which I didn't yet understand; and then Yuri Yetukhovich told me the whole thing. And here beneath the domes of the brick-red Butyrki castle, I felt that the story of these several million Russian prisoners had got me in its grip once and for all, like a pin through a specimen beetle. My own story of landing in prison seemed insignificant. I stopped regretting my torn-off shoulder boards. It was mere chance that had kept me from ending up exactly where these contemporaries of mine had ended. I came to understand that it was my duty to take upon my shoulders a share of their common burden—and to bear it to the last man, until it crushed us.

Sometimes we try to lie but our tongue will not allow us to. These people were labeled traitors, but a remarkable slip of the tongue occurred—on the part of the judges, prosecutors, and interrogators. And the convicted prisoners, the entire nation, and the newspapers repeated and reinforced this mistake, involuntarily letting the truth out of the bag. They intended to declare them "traitors *to* the Motherland." But they were universally referred to, in speech and in writing, even in the court documents, as "traitors *of* the Motherland."

You said it! They were not traitors *to her.* They were *her* traitors. It was not they, the unfortunates, who had betrayed the Motherland, but their calculating Motherland who had betrayed them, and not just once but *thrice.*

The first time she betrayed them was on the battlefield, through ineptitude—when the government, so beloved by the Motherland, did everything it could to lose the war: destroyed the lines of fortifications; set up the whole air force for annihilation; dismantled the tanks and artillery; removed the effective generals; and forbade the armies to resist. And the war prisoners were the men whose bodies took the blow and stopped the Wehrmacht.

The second time they were heartlessly betrayed by the Motherland was when she abandoned them to die in captivity.

And the third time they were unscrupulously betrayed was when, with motherly love, she coaxed them to return home, with such phrases as "The Motherland has forgiven you! The Motherland calls you!" and snared them the moment they reached the frontiers.

It would appear that during the one thousand one hundred years of Russia's existence as a state there have been, ah, how many foul and terrible deeds! But among them was there ever so multimillioned foul a deed as this: to betray one's own soldiers and proclaim them traitors?

How many wars Russia has been involved in! (It would have been better if there had been fewer.) And were there many traitors in all those wars? Had anyone observed that treason had become deeply rooted in the hearts of Russian soldiers? Then, under the most just social system in the world, came the most just war of all—and out of nowhere millions of traitors appeared, from among the simplest, lowliest elements of the population. How is this to be understood and explained?

Capitalist England fought at our side against Hitler; Marx had eloquently described the poverty and suffering of the working class in that same England. Why was it that in this war only one traitor could be found among *them,* the businessman "Lord Haw Haw"—but in our country millions?

It is frightening to open one's trap about this, but might the heart of the matter not be in the political system?

All the Western peoples behaved the same in our war: parcels, letters, all kinds of assistance flowed freely through the neutral countries. The Western POW's did not have to lower themselves to accept ladlefuls from German soup kettles. They talked back to the German guards. Western governments gave their captured soldiers their seniority rights, their regular promotions, even their pay.

The only soldier in the world *who cannot surrender* is the soldier of the world's one and only Red Army. That's what it says in our military statutes. (The Germans would shout at us from their trenches: "Ivan plen nicht!"—"Ivan no prisoner!") Who can picture all that means? There is war; there is death—but there is no surrender! What a discovery! What it means is: Go and die; we will go on living. And if you lose your legs, yet manage to return from captivity on crutches, we will convict you.

Our soldiers alone, renounced by their Motherland and degraded to nothing in the eyes of enemies and allies, had to push their way to the swine swill being doled out in the backyards of the Third Reich. Our soldiers alone had the doors shut tight to keep them from returning to their homes, although their young souls tried hard not to believe this. There was something called Article 58-1b—and, in wartime, it provided only for execution by shooting! For not wanting to die from a German bullet, the prisoner had to die from a Soviet bullet for having been a prisoner of war! Some get theirs from the enemy; we get it from our own!

Very few of the war prisoners returned across the Soviet border as free men, and if one happened to get through by accident because of the prevailing chaos, he was seized later on, even as late as 1946 or 1947. Some were arrested at assembly points in Germany. Others weren't arrested openly right away but were transported from the border in freight cars, under convoy, to one of the numerous Identification and Screening Camps (PFL's) scattered throughout the country. These camps differed in no way from the common run of Corrective Labor Camps (ITL's) except that their prisoners had not yet been sentenced but would be sentenced there. All these PFL's were also attached to some kind of factory, or mine, or construction project, and the former

POW's, looking out on the Motherland newly restored to them through the same barbed wire through which they had seen Germany, could begin work from their first day on a ten-hour work day. Those under suspicion were questioned during their rest periods, in the evenings, and at night, and there were large numbers of Security officers and interrogators in the PFL's for this purpose. As always, the interrogation began with the hypothesis that you were obviously guilty. And you, without going outside the barbed wire, had to prove that you were *not* guilty. Your only available means to this end was to rely on witnesses who were exactly the same kind of POW's as you. Obviously they might not have turned up in your own PFL; they might, in fact, be at the other end of the country; in that case, the Security officers of, say, Kemerovo would send off inquiries to the Security officers of Solikamsk, who would question the witnesses and send back their answers along with new inquiries, and you yourself would be questioned as a witness in some other case. True, it might take a year or two before your fate was resolved, but after all, the Motherland was losing nothing in the process. You were out mining coal every day. And if one of your witnesses gave the wrong sort of testimony about you, or if none of your witnesses was alive, you had only yourself to blame, and you were sure to be entered in the documents as a traitor *of* the Motherland. And the visiting military court would rubber-stamp your *tenner*. And if, despite all their twisting things about, it appeared that you really hadn't worked for the Germans, and if—and this was the main point—you had not had the chance to see the Americans and English with your own eyes (to have been liberated from captivity by *them* instead of by us was a gravely aggravating circumstance), then the Security officers would decide the degree of isolation in which you were to be held.

"Oh, if I had only known!" That was the refrain in the prison cells that spring. If I had only known that this was how I would be greeted! That they would deceive me so! That this would be my fate! Would I have really returned to my Motherland? Not for anything!

The only ones who did not sigh: "Oh, if I had only known"— because they knew very well what they were doing—and the only ones who did not expect any mercy and did not expect an amnesty—were the Vlasov men.

■

I had known about them and been perplexed about them long before our unexpected meeting on the board bunks of prison.

First there had been the leaflets, reporting the creation of the

ROA, the "Russian Liberation Army." Not only were they written in bad Russian, but they were imbued with an alien spirit that was clearly German and, moreover, seemed little concerned with their presumed subject; besides, and on the other hand, they contained crude boasting about the plentiful chow available and the cheery mood of the soldiers. Somehow one couldn't believe in that army, and, if it really did exist, what kind of cheery mood could it be in? Only a German could lie like that.

Actually, no Russian Liberation Army ever existed until almost the very end of the war. During all those years several *hundred thousand* voluntary helpers—the *Hilfswillige*—were scattered throughout all sorts of German units as enlisted men or in even more subordinate positions. In addition, there were a few volunteer anti-Soviet units, made up of former Soviet citizens but under the command of German officers. The Lithuanians were the first to start supporting the Germans (understandably so: we had really hurt them beyond endurance in just one year!). Then the Ukrainians formed a voluntary SS division, and the Estonians joined a few SS units. In Byelorussia there was a people's militia fighting against the partisans: 100,000 men! There was a Turkestan battalion, and in Crimea a Tartar one. (All this was the harvest of what the Soviets had sowed, like the senseless persecution of Islam in Crimea, whose farsighted conqueror, Catherine II, had assigned state funds to build new mosques and enlarge others. Hitler's military units occupying the area had also had enough common sense to protect the mosques.) When the Germans conquered our southern regions, the number of volunteer battalions increased: there was a Georgian one, an Armenian one, a battalion of the Northern Caucasus peoples, and sixteen Kalmyk battalions. (And there were almost no Soviet partisans in the South.) During the German retreat from the Don region, about fifteen thousand Cossacks followed the German army; half of them were able to fight. In the Briansk region, near Lokot, in 1941, before the arrival of the Germans, the local population dissolved the kolkhozes and readied itself to fight Soviet partisans; the autonomous region that was then created remained in existence until 1943, headed by an engineer, Voskoboynikov. It had twenty thousand armed men, whose flag bore the image of St. George. They called themselves "The Russian National Liberation Army."

In the fall of 1942 Vlasov allowed the use of his name in order to unite all the anti-Bolshevik units, and during that same fall Hitler's headquarters turned down a proposal from middle rank army officers that Germany should renounce all plans for eastern colonization and substitute for them the creation of Russian national military units. Vlasov had only just made his fatal choice and taken the first step on this road when he became entirely useless except for propaganda purposes. The situation never changed until the very end.

Wearing a homemade brown uniform which did not belong to any army, with the red lapels of a general's coat but without any insignia of rank, Vlasov made his first trip in March, 1943 (Smolensk-Mogilev-Bobruysk), and a second one in April (Riga-Pechory-Pskov-Gdov-Luga). These trips caused much enthusiasm among the Russian population; they seemed to prove that a Russian national movement was being born and that an independent Russia could be resurrected. Vlasov made public appearances in the theaters of Smolensk and Pskov, both filled to capacity; he spoke about the goals of the liberation movement and then proceeded to declare openly that national socialism was unacceptable for Russia but that, on the other hand, it was impossible to overthrow the Bolsheviks without the Germans. Just as openly, people asked him whether it was true that the Germans intended to turn Russia into a colony and the Russian people into beasts of burden. They asked: Why has nobody so far stated clearly what will be Russia's future after the war? Why don't the Germans allow Russian self-government in the occupied regions? Why must the anti-Stalin volunteers fight under the command of German officers? Vlasov answered these questions with some embarrassment, displaying more optimism than he could truly have felt at that time. The German General Headquarters reacted with an order issued by Marshall Keitel: "In view of the incompetent and shameless declarations made by the prisoner of war, Russian General Vlasov, during his trip to the Northern Army Group, which was undertaken without the Führer's and my knowledge, he is to be immediately transferred to a POW camp." The general's name could be used only for propaganda purposes, and if he were ever to make a public statement again, he would be turned over to the Gestapo and rendered harmless.

Those were the last months during which millions of Soviet people were still out of Stalin's reach and could fight against Bolshevik slavery and organize their own independent existence. But the German leadership had no hesitations: on June 8, 1943, on the eve of the Kursk-Orlov battle, Hitler confirmed that a Russian independent army would never be created and that Germany needed Russians only as manpower. Hitler was unable to understand the historical fact that the opportunity to overthrow a Communist regime can come only from a popular movement, from an uprising of the long-suffering population. But Hitler was more afraid of *such* a Russia and *such* a victory than of a defeat. Even after Stalingrad, even after he had lost the Caucasus, Hitler did not notice anything new. While Stalin was assuming the role of the supreme defender of the nation, reintroducing the old Russian epaulettes, restoring the Orthodox Church, and dissolving the Comintern, Hitler was helping him as much as he could by ordering, in September, 1943, that all volunteer units be disarmed and assigned to guarding coal mines. Later, he changed his mind and transferred them to the Western front, to fight against the Allies.

Such was, fundamentally, the end of the entire project of an independent Russian army. So what did Vlasov do? He did not quite know how bad things really were (he did not know that after his March and April journeys he was again considered a prisoner of war and was in danger); he adopted the irreparable and fatal course of hoping and to a certain extent attempting to reach an agreement with the Beast, whereas in dealing with apocalyptic beasts there is only one way to safety: unswerving firmness from the first minute to the last. But here one must ask whether such an avenue to safety ever existed for the liberation movement of the citizens of Russia. It was doomed right from the start to be one more victim on the 1917 sacrificial altar, which had not yet cooled off completely. The first war winter of 1941–1942, which destroyed several million Soviet prisoners of war, extended the long chain made out of victims' bones—the chain begun in the summer, when unarmed people's militia units had been sent to save Bolshevism.

Thus, the already fading significance of this bitter volunteer struggle was altogether lost. These people were sent as cannon fodder in the fight against the Allies and against the French Resistance, that is, against the only forces with whom the Russians in Germany could have had a genuine feeling of solidarity, having experienced both the cruelty and the self-satisfaction of the Germans. This was the end of the secret hope cherished by those around Vlasov: If the British and Americans support Communists against Hitler, they simply must help a democratic non-Communist Russia in the same struggle. . . . At the downfall of the Third Reich, when it will become quite clear that the Soviet Union is increasing the pressure to extend its regime to all Europe and to the whole world, how could the West continue to support the Bolshevik dictatorship?

But there was a gap between the Russian and the Western conscience which exists to this day. The West was fighting *only* against Hitler, and for this purpose *all* means and *all* allies were good, the Soviets above all. Not only could the West not concede that the Soviet people might have their own purposes which did not coincide with the goals of the Communist government; it did not want to admit any such thought, because it would have been embarrassing and difficult to live with. It is a tragicomic fact that on the leaflets which the Western allies were distributing among the anti-Bolshevik volunteer battalions on the Western front, they wrote: "We *promise* all defectors that they will be immediately sent back to the Soviet Union (to prison. . . .)."

We soon discovered that there really were Russians fighting against us and that they fought harder than any SS men. In July, 1943, for example, near Orel, a platoon of Russians in German uniform defended Sobakinskiye Vyselki. They fought with the desperation that might have been expected if they had built the place themselves. One of them was driven into a root cellar. They threw hand grenades in after him and he fell silent. But they had no more than stuck their heads in than he let them have another volley from his automatic pistol. Only when they lobbed in an antitank grenade did they find out that, within the root cellar, he had another foxhole in which he had taken shelter from the infantry grenades. Just try to imagine the degree of shock, deafness, and hopelessness in which he had kept on fighting.

In East Prussia, a trio of captured Vlasov men was being marched along the roadside a few steps away from me. At that moment a T-34 tank thundered down the highway. Suddenly one of the captives twisted around and dived underneath the tank. The tank veered, but the edge of its track crushed him nevertheless. The broken man lay writhing, bloody foam coming from his mouth. And one could certainly understand him! He preferred a soldier's death to being hanged in a dungeon.

They had no choice. There was no other way for them to fight. They had no chance to find a way out, to safeguard their lives, by some more cautious mode of fighting. If "pure" surrender was considered unforgivable treason to the Motherland, then what about those who had taken up enemy arms? Our propaganda, in all its crudity, explained their conduct as: (1) treason (was it biologically based? carried in the bloodstream?); or (2) cowardice—which it certainly was not! A coward tries to find a spot where things are easy, soft, safe. And men could be induced to enter the Wehrmacht's Vlasov detachments only in the last extremity, only at the limit of desperation, only out of inexhaustible hatred of the Soviet regime, only with total contempt for their own safety. For they knew they would never have the faintest glimpse of mercy! When we captured them, we shot them as soon as the first intelligible Russian word came from their mouths. In Russian captivity, as in German captivity, the worst lot of all was reserved for the Russians.

In general, this war revealed to us that the worst thing in the world was to be a Russian.

The Vlasov men had a presentiment of all this; they knew it ahead of time; nevertheless, on the left sleeve of their German uniforms they sewed the shield with the white-blue-red edging, the field of St. Andrew, and the letters "ROA."

However, by February, 1945, the First Division of the Russian Liberation Army had been formed, and the formation of the Second Division had begun. It was too late even to hypothesize that these divisions would ever fight together with the Germans. The old secret hope of the Vlasov leadership that a conflict would arise between the Soviets and the Allies was now gaining strength. This hope was reflected in a report by the German Ministry of Propaganda (in February, 1945): "The Vlasov movement does not consider itself bound for life and for death to Germany; there are within it strong pro-English feelings, and they are thinking about a change of course. It is not a national socialist movement, and they simply do not recognize the existence of the Jewish problem."

The breakdown of Germany, which by that time was total, made it possible for the commander of the division, Buniachenko, to take it out of the front line by his own decision; despite the opposition of the German generals, the division started fighting its way into Czechoslovakia. (On the way it freed Soviet prisoners of war, who joined it "so that Russians may all be together.") The men reached the outskirts of Prague at the beginning of May. The Czechs had started an uprising in the capital on May 5 and asked them for help. On May 6 Buniachenko's division entered Prague and, in a violent battle on May 7, saved both the uprising and the city. As if ironically, as if to confirm the farsightedness of the most shortsighted Germans, the first and last independent action of the First Vlasov Division was a blow dealt the Germans; it must have been a relief for all those Russian hearts which during the past three senseless and cruel years had accumulated so much bitterness and anger against them. (In those days the Czechs welcomed the Russians with flowers; they understood. But who knows whether all of them remembered later *which* Russians had saved their city? The official Soviet version is that Prague was liberated by the Soviet army. It is true that, in accordance with Stalin's wishes, Churchill was in no hurry in those days to arm the inhabitants of Prague; and as to the U.S. army, it slowed down its advance in order to allow the Soviets to enter the city first. Joseph Smrkovsky, who at that time was a leading Czech Communist, did not foresee the distant future and insulted the "traitors" of the Vlasov units. The only freedom which he wanted had to come from the Soviets.)

Vlasov consistently refused to escape alone (a plane was ready to take him to Spain); in what must have been a paralysis of will, he gave up and accepted the end. During these last weeks his activity was limited to dispatching secret delegations in an effort to establish a contact with the British and Americans.

The only sense the Vlasovites could see in all these events, so as to justify somehow their long dangling in the German noose, was in getting a chance to be useful to the Allies now that everything was finished. The hope kept glimmering, or rather burning high, that at this time, after the end of the war, the powerful English and American Allies would ask Stalin to change his domestic policy. The armies coming from the West and from the East were getting closer and closer; they might well clash over Hitler's crushed remains, and that would be the time when the West could gain by saving and using the anti-Bolshevik Russians. The West simply had to understand that Bolshevism is an enemy for all mankind.

But the West did not understand at all. The democratic West simply could not understand: What do you mean when you call yourselves a political opposition? An opposition exists inside your country? Why has it never publicly declared its existence? If you are dissatisfied with Stalin, go back home and, in the first subsequent election, do not re-elect him. That would be the honest course. But why did you have to take up arms, and, what is worse, German arms? No, we have to extradite you; it would be terribly bad form to act otherwise, and we might spoil our relations with a gallant ally.

In World War II the West kept defending its *own* freedom and defended it for *itself.* As for us and as for Eastern Europe, it buried us in an even more absolute and hopeless slavery.

Vlasov's last effort was his statement that the leadership of the ROA was ready to appear before an international court and that turning the army over to the Soviets for extermination contravened international law, since it would involve extraditing an opposition movement. But nobody heard that squeaking. Most of the American military commanders were amazed to learn about the existence of Russians who were not Soviets; they thought it quite natural to hand them all over to the Soviet state.

The ROA not only surrendered to the Americans; it *implored* them to accept its capitulation and begged for one thing only: the promise that the Americans would not extradite them to the Soviets. Midlevel American officers who did not know anything about big politics sometimes naïvely gave such promises. (But all of them were broken; the ROA soldiers were deceived.) The First Division (on May 11 near Pilsen) found itself facing an armed wall of American military men; it was almost the same with the Second Division. The Americans *refused* to consider them prisoners of war and refused to let them into their zone. In Yalta Churchill and Roosevelt had signed the agreement to repatriate all Soviet citizens, and especially the military, without specifying whether the repatriation was to be voluntary or enforced: How could any people on earth not be willing to return to their homes? The nearsightedness of the West was condensed in what was written at Yalta.

At the same time, in May, 1945, Great Britain also acted as a loyal ally of the Soviets; the usual modesty of the Soviet leadership prevented this action from being publicized. The English turned over to the Soviet army command a Cossack corps of forty to forty-five thousand men which had fought its way to Austria from Yugoslavia. The extradition was carried out with a perfidy which is characteristic of British diplomatic tradition. The gist of the matter is that the Cossacks meant to fight to the death or emigrate overseas, maybe to Paraguay, maybe to Indochina, anywhere—as long as they would not have to surrender to the Soviets alive. The British provided the Cossacks with military food rations of extra quality, dressed them in fine British uniforms, promised them that they could serve in the British army, and even held military reviews. Therefore, the Cossacks did not grow suspicious when they were asked to turn in their weapons, on the grounds that this was necessary in order to standardize their equipment. On May 28 all officers, from squadron commanders upward, were summoned separately from their soldiers to the town of Judenburg, on the pretext that they would confer with Field Marshal Alexander about the future fate of the army. En route the officers were surreptitiously placed under a strong escort (the British beat them until they bled), and the whole motorcade was gradually surrounded by Soviet tanks. When they arrived in Judenburg, police vans were waiting, as were armed guards holding lists of names. They could not even shoot or stab themselves to death, since all their weapons had been taken away. Some jumped off the high viaduct into the river or onto the stones. Among the generals thus turned over to the Soviets, the majority were émigrés who had fought as allies of the British during World War I. During the Civil

War the British had not had enough time to show their gratitude; now they were paying their debt. In the following days the British extradited the enlisted men as treacherously, in trains which were covered with barbed wire.

In the meantime, a Cossack transport had arrived from Italy, carrying 35,000 people. They stopped in the Drava Valley near Lienz. There were Cossack soldiers among them, but also many old people, children, and women; none of them wanted to go back to their beloved Cossack rivers. The hearts of the British were not troubled, nor were their democratic minds. The British commanding officer, Major Davies, whose name will certainly survive from now on in Russian history at least, could be exuberantly friendly or merciless, as needed. After the surreptitious extradition of the officers, he openly announced on June 1 that there would be a compulsory extradition. Thousands of voices yelled: "We will not go!" Black flags appeared over the refugees' camp, where church services were being celebrated non-stop: people arranging their own funeral services while they were still alive! . . . British tanks and soldiers arrived. The order was given through loudspeakers for everybody to get into the trucks. The crowd was singing hymns from the requiem service; the priests lifted their crosses high above their heads; the young people formed a chain around the elderly, the women, and the children. Then British soldiers started beating them with rifle butts and clubs, grabbing them and throwing them onto the trucks, including the wounded, as if they were packages. As the crowd retreated, first the platform on which the priests were standing broke down under their weight; then the camp fence collapsed. The crowd rushed to the bridge over the Drava; British tanks rolled on to stop them, büt entire families sought death by throwing themselves into the river. Meanwhile, the British units in the neighborhood pursued and shot at the fugitives. (The cemetery where the people who were shot or trampled to death were buried still exists in Lienz.)

In those same days, just as treacherously and mercilessly, the British extradited to the Yugoslav Communists thousands of their regime's enemies who had been Great Britain's allies in 1941! They, too, were to be shot and exterminated without trial.

But even that was only the beginning. During all of 1946 and 1947 the Western allies, faithful to Stalin, continued to turn over to him Soviet citizens, former soldiers as well as civilians. It did not really matter who they were as long as the West could get rid of this human confusion as quickly as possible. People were extradited from Austria,

Germany, Italy, France, Denmark, Norway, and Sweden, from the American occupation zones, and from the territory of the United States as well.

I myself fell under Vlasov fire a few days before my arrest. There were Russians in the East Prussian "sack" which we had surrounded, and one night at the end of January their unit tried to break through our position to the west, without artillery preparation, in silence. There was no firmly delineated front in any case, and they penetrated us in depth, catching my sound-locator battery, which was out in front, in a pincers. I just barely managed to pull it back by the last remaining road. But then I went back for a piece of damaged equipment, and, before dawn, I watched as they suddenly rose from the snow where they'd dug in, wearing their winter camouflage cloaks, hurled themselves with a cheer on the battery of a 152-millimeter gun battalion at Adlig Schwenkitten, and knocked out twelve heavy cannon with hand grenades before they could fire a shot. Pursued by their tracer bullets, our last little group ran almost two miles in fresh snow to the bridge across the Passarge River. And there they were stopped.

Soon after that I was arrested. And now, on the eve of the Victory Parade, here we all were sitting together on the board bunks of the Butyrki. I took puffs from their cigarettes and they took puffs from mine. And paired with one or another of them, I used to carry out the six-bucket tin latrine barrel.

Now, a quarter of a century later, when most of the Vlasov men have perished in camps and those who have survived are living out their lives in the Far North, I would like to issue a reminder, through these pages, that this was a phenomenon totally unheard of in all world history: that several hundred thousand young men, aged twenty to thirty, took up arms against their Fatherland as allies of its most evil enemy. Perhaps there is something to ponder here: Who was more to blame, those youths or the gray Fatherland? One cannot explain this treason biologically. It has to have had a social cause.

Because, as the old proverb says: *Well-fed horses don't rampage.*

Then picture to yourself a field in which starved, neglected, crazed horses are rampaging back and forth.

■

That same spring many Russian émigrés were also in those cells.

It was very like a dream: the resurrection of buried history. The weighty tomes on the Civil War had long since been completed and their covers shut tight. The causes for which people fought in it had been decided. The chronology of its events had been set down in textbooks. The leaders of the White movement were, it appeared, no longer our contemporaries on earth but mere ghosts of a past that had melted away. The Russian émigrés had been more cruelly dispersed than the tribes of Israel. And, in our Soviet imagination, if they were still dragging out their lives somewhere, it was as pianists in stinking little restaurants, as lackeys, laundresses, beggars, morphine and cocaine addicts, and virtual corpses. Right up to 1941, when the war came, it would have been impossible to find out from any hints in our newspapers, our lofty literature, our criticism of the arts (nor did our own well-fed masters of art and literature help us find out) that Russia Abroad was a great spiritual world, that in it Russian philosophy was living and developing; that out there were philosophers like Bulgakov, Berdyayev, and Lossky; that Russian art had enchanted the world; that Rachmaninoff, Chaliapin, Benois, Diaghilev, Pavlova, and the Don Cossack Chorus of Jaroff were out there; that profound studies of Dostoyevsky were being undertaken (at a time when he was anathema in the Soviet Union); that the incredible writer Nabokov-Sirin also existed out there; that Bunin himself was still alive and had been writing for all these twenty years; that journals of the arts were being published; that theatrical works were being produced; that Russians from the same areas of Russia came together in groups where their mother tongue could be heard; and that émigré men had not given up marrying émigré women, who in turn presented them with children, which meant young people our own age.

The picture of emigration presented in our country was so falsified that if one had conducted a mass survey to ask which side the Russian émigrés were on in the Spanish Civil War, or else, perhaps, what side they were on in the Second World War, with one voice everyone would have replied: For Franco! For Hitler! Even now people in our country do not know that many White émigrés fought on the Republican side in Spain. The émigrés did not support Hitler. They ostracized Merezhkovsky and Gippius, who took Hitler's part, leaving them to alienated loneliness. There was a joke—except it wasn't a joke—to the effect that Denikin wanted to fight for the Soviet Union against Hitler, and that at one time Stalin planned to arrange his return to the Motherland, not for military reasons, obviously, but as a symbol of national unity. During the German occupation of France, a horde of Russian émigrés, young and old, joined the Resistance. And after the liberation of Paris they swarmed to the Soviet Embassy to apply for permission to return to the Motherland. No matter what kind of Russia it was—it was still Russia! That was their slogan, and that is how they proved they had not been lying previously about their love for her. (Imprisoned in 1945 and 1946, they were almost happy that these prison bars and these jailers were their own, Russian. And they observed with surprise the Soviet boys scratching their heads and saying: "Why the hell did we come back? Wasn't there room enough for us in Europe?")

But, given that Stalinist logic which said that every Soviet person who had lived abroad had to be imprisoned in camp, how could the émigrés possibly escape the same lot? In the Balkans, Central Europe, Harbin, they were arrested as soon as the Soviet armies arrived. They were arrested in their apartments and on the street, just like Soviet citizens. For a while State Security arrested only men, and not all of them, only those who had in one or another way revealed a political bias. Later on, their families were transported to exile in Russia, but some were left where they were in Bulgaria and Czechoslovakia. In France they were welcomed into Soviet citizenship with honors and flowers and sent back to the Motherland in comfort; and only when they got to the U.S.S.R. were they raked in. Things dragged out longer for the Shanghai émigrés. In 1945 Russian hands didn't reach that far. But a plenipotentiary from the Soviet government went to Shanghai and announced a decree of the Presidium of the Supreme Soviet extending forgiveness to all émigrés. Well, now, how could one refuse to believe that? The government certainly couldn't lie! Whether or not there actually was such a decree, it did not, in any case, tie the hands of the *Organs.* The Shanghai Russians expressed their delight. They were told they could take with them as many possessions as they wanted and whatever they wanted. They went home with automobiles —the country could put them to good use. They were told they could settle wherever they wanted to in the Soviet Union and, of course, work at any profession or trade. They were transported from Shanghai in steamships. The fate of the passengers varied. On some of the ships, for some reason, they were given no food at all. They also suffered various fates after reaching the port of Nakhodka (which was, incidentally, one of the main transit centers of Gulag). Almost all of them were loaded into freight cars, like prisoners, except that they had, as yet, no strict convoy, and there were no police dogs. Some of them were actually delivered to inhabited places, to cities, and allowed to live there for two or three years. Others were delivered in trainloads straight to their camps and were dumped out somewhere off a high embankment into the forest beyond the Volga, together with their white pianos and their jardinieres. In 1948–1949, the former Far Eastern émigrés who had until then managed to stay out of camps were scraped up to the last man.

Chapter 7

■

In the Engine Room

There was a box at the so-called Butyrki "station": the famous *frisking* box, where new arrivals were searched. It had space enough for five or six jailers to process up to twenty zeks in one batch. Now, however, it was empty and the rough-hewn search tables had nothing on them. Over at one side of the room, seated behind a small nondescript table beneath a small lamp, was a neat, black-haired NKVD major. Patient boredom was what his face chiefly revealed. The intervals during which the zeks were brought in and led out one by one were a waste of his time. Their signatures could have been collected much, much faster.

He indicated that I was to sit down on the stool opposite him, on the other side of his table. He asked my name. To the right and left of the inkwell lay two piles of white papers the size of a half-sheet of typewriter paper, all looking much the same. In format they were just like the fuel requisitions handed out in apartment-house management offices, or warrants in official institutions for purchase of office supplies. Leafing through the pile on the right, the major found the paper which referred to me. He pulled it out and read it aloud to me in a bored patter. (I understood I had been sentenced to eight years.) Immediately, he began to write a statement on the back of it, with a fountain pen, to the effect that the text had been read to me on the particular date.

My heart didn't give an extra half-beat—it was all so everyday and routine. Could this really be my sentence—the turning point in my life? I would have liked to feel nervous, to experience this moment to the full, but I just couldn't. And the major had already pushed the sheet over to me, the blank side facing up. And a schoolchild's seven-kopeck pen, with a bad point that had lint on it from the inkwell, lay there in front of me.

"No, I have to read it myself."

"Do you really think I would deceive you?" the major objected lazily. "Well, go ahead, read it."

Unwillingly, he let the paper out of his hand. I turned it over and began to look through it with deliberate slowness, not just word by word but letter by letter. It had been typed, but what I had in front of me was not the original but a carbon:

EXTRACT
from a decree of the OSO of the NKVD of the U.S.S.R.
of July 7, 1945, No. ——.

All of this was underscored with a dotted line and the sheet was vertically divided with a dotted line:

Case heard:	Decreed:
Accusation of so-and-so (name, year of birth, place of birth)	To designate for so-and-so (name) for anti-Soviet propaganda, and for an attempt to create an anti-Soviet organization, 8 (eight) years in corrective labor camps.
Copy verified. Secretary _____	

Was I really just supposed to sign and leave in silence? I looked at the major—to see whether he intended to say something to me, whether he might not provide some clarification. No, he had no such intention. He had already nodded to the jailer at the door to get the next prisoner ready.

To give the moment at least a little importance, I asked him, with a tragic expression: "But, really, this is terrible! Eight years! What for?"

And I could hear how false my own words sounded. Neither he nor I detected anything terrible.

"Right there." The major showed me once again where to sign.

I signed. I could simply not think of anything else to do.

"In that case, allow me to write an appeal right here. After all, the sentence is unjust."

"As provided by regulations," the major assented with a nod, placing my sheet of paper on the left-hand pile.

"Let's move along," commanded the jailer.

And I *moved along.*

(I had not really shown much initiative. Georgi Tenno, who, to be sure, had been handed a paper worth twenty-five years, answered: "After all, this is a life sentence. In olden times they used to beat the drums and assemble a crowd when a person was given a life sentence. And here it's like being on a list for a soap ration—twenty-five years and run along!"

Arnold Rappoport took the pen and wrote on the back of the verdict: "I protest categorically this terroristic, illegal sentence and demand immediate release." The officer who had handed it to him had at first waited patiently, but when he read what Rappoport had written, he was enraged and tore up the paper with the note on it. So what! The term remained in force anyway. This was just a copy.

Vera Korneyeva was expecting *fifteen* years and she saw with delight that there was a typo on the official sheet—it read only *five*. She laughed her luminous laugh and hurried to sign before they took it back. The officer looked at her dubiously: "Do you really understand what I read to you?" "Yes, yes, thank you very much. Five years in corrective-labor camps."

The ten-year sentence of Janos Rozsas, a Hungarian, was read to him in the corridor in Russian, without any translation. He signed it, not knowing it was his sentence, and he waited a long time afterward for his trial. Still later, when he was in camp, he recalled the incident very vaguely and realized what had happened.)

■

The OSO was nowhere mentioned in either the Constitution or the Code. However, it turned out to be the most convenient kind of hamburger machine—easy to operate, undemanding, and requiring no legal lubrication. The Code existed on its own, and the OSO existed on its own, and it kept on deftly grinding without all the Code's 205 articles, neither invoking them nor even mentioning them.

As they used to joke in camp: "There is no court for nothing—for that there is an OSO."

Of course, the OSO itself also needed for convenience some kind of operational shorthand, but for that purpose it worked out on its own a dozen "letter" articles which made operations very much simpler. It wasn't necessary, when they were used, to cudgel your brains trying to make things fit the formulations of the Code. And they were few enough to be easily remembered by a child. Some of them we have already described:

ASA —Anti-Soviet Agitation
KRD —Counter-Revolutionary Activity
KRTD—Counter-Revolutionary Trotskyite Activity (And that "T" made the life of a zek in camp much harder.)
PSh —Suspicion of Espionage (Espionage that went beyond the bounds of suspicion was handed over to a tribunal.)
SVPSh—Contacts Leading (!) to Suspicion of Espionage
KRM —Counter-Revolutionary Thought
VAS —Dissemination of Anti-Soviet Sentiments
SOE —Socially Dangerous Element
SVE —Socially Harmful Element
PD —Criminal Activity (a favorite accusation against former camp inmates if there was nothing else to be used against them)

And then, finally, there was the very expansive category:

ChS —Member of a Family (of a person convicted under one of the foregoing "letter" categories)

It has to be remembered that these categories were not applied uniformly and equally among different groups and in different years. But, as with the articles of the Code and the sections in special decrees, they broke out in sudden epidemics.

There is one more qualification. The OSO did not claim to be handing down a *sentence.* It did not sentence a person but, instead, *imposed an administrative penalty.* And that was the whole thing in a nutshell. Therefore it was, of course, natural for it to have juridical independence!

But even though they did not claim that the administrative penalty was a court sentence, it could be up to twenty-five years and include:

• Deprivation of titles, ranks, and decorations
• Confiscation of all property
• Imprisonment
• Deprivation of the right to correspond

Thus a person could disappear from the face of the earth with the help of the OSO even more reliably than under the terms of some primitive court sentence.

The OSO enjoyed another important advantage in that its penalty could not be appealed. There was nowhere to appeal to. There was no appeals jurisdiction above it, and no jurisdiction beneath it. It was subordinate only to the Minister of Internal Affairs, to Stalin, and to Satan.

Another big advantage the OSO had was speed. This speed was limited only by the technology of typewriting.

And, last but not least, not only did the OSO not have to confront the accused face to face, which lessened the burden on interprison transport: it didn't even have to have his photograph. At a time when the prisons were badly overcrowded, this was a great additional advantage because the prisoner did not have to take up space on the prison floor, or eat free bread once his interrogation had been completed. He could be sent off to camp immediately and put to honest work. The copy of the sentence could be read to him much later.

■

All the articles of the Code had become encrusted with interpretations, directions, instructions. And if the actions of the accused are not covered by the Code, he can still be convicted:

- By analogy (What opportunities!)
- Simply because of *origins* (7-35: belonging to a socially dangerous milieu)
- For *contacts with dangerous persons* (Here's scope for you! Who is "dangerous" and what "contacts" consist of only the judge can say.)

But one should not complain about the precise wording of our published laws either. On January 13, 1950, a decree was issued reestablishing capital punishment. (One is bound, of course, to consider that capital punishment never did depart from Beria's cellars.) And the decree stated that the death sentence could be imposed on *subversives* —diversionists. What did that mean? It didn't say. Iosif Vissarionovich loved it that way: not to say all of it, just to hint. Did it refer only to someone who blew up rails with TNT? It didn't say. We had long since come to know what a "diversionist" was: someone who produced goods of poor quality was a diversionist. But what was a *subversive?* Was someone *subverting* the authority of the government, for example, in a conversation on a streetcar? Or if a girl married a foreigner—wasn't she *subverting* the majesty of our Motherland?

But it is not the judge who judges. The judge only takes his pay. The directives did the judging. The directive of 1937: ten years; twenty years; execution by shooting. The directive of 1943: twenty years at hard labor; hanging. The directive of 1945: ten years for everyone, plus five of disenfranchisement (manpower for three Five-Year Plans). The directive of 1949: everyone gets twenty-five.

The machine stamped out the sentences. The prisoner had already been deprived of all rights when they cut off his buttons on the threshold of State Security, and he couldn't avoid a stretch. The members of the legal profession were so used to this that they fell on their faces in 1958 and caused a big scandal. The text of the projected new "Fundamental Principles of Criminal Prosecution of the U.S.S.R." was published in the newspapers, and they'd *forgotten* to include any reference to *possible* grounds for acquittal. The government newspaper issued a mild rebuke: *"The impression might be created that our courts only bring in convictions."*

But just take the jurists' side for a moment: why, in fact, should a trial be supposed to have *two* possible outcomes when our general *elections* are conducted on the basis of *one* candidate? An acquittal is, in fact, unthinkable from the economic point of view! It would mean that the informers, the Security officers, the interrogators, the prosecutor's staff, the internal guard in the prison, and the convoy had all worked to no purpose.

Here is one straightforward and typical case that was brought before a military tribunal. In 1941, the Security operations branch of our inactive army stationed in Mongolia was called on to show its activity and vigilance. The military medical assistant Lozovsky, who was jealous of Lieutenant Pavel Chulpenyev because of some woman, realized this. He addressed three questions to Chulpenyev when they were alone: 1. "Why, in your opinion, are we retreating from the Germans?" (Chulpenyev's reply: "They have more equipment and they were mobilized earlier." Lozovsky's counter: "No, it's a *maneuver*. We're *decoying* them.") 2. "Do you believe the Allies will help?" (Chulpenyev: "I believe they'll help, but not from unselfish motives." Lozovsky's counter: "They are deceiving us. They won't help us at all.") 3. "Why was Voroshilov sent to command the Northwest Front?"

Chulpenyev answered and forgot about them. And Lozovsky wrote a denunciation. Chulpenyev was summoned before the Political Branch of the division and expelled from the Komsomol: for a defeatist attitude, for praising German equipment, for belittling the strategy of our High Command. The loudest voice raised against him belonged to the Komsomol organizer Kalyagin, who had behaved like a coward at the battle of Khalkhin-Gol, in Chulpenyev's presence, and therefore found it convenient to get rid of the witness once and for all.

Chulpenyev's arrest followed. He had one confrontation with Lozovsky. Their previous conversation *was not even brought up* by the interrogator. One question was asked: "Do you know this man?" "Yes." "Witness, you may leave." (The interrogator was afraid the charge might fall through.)

Depressed by his month's incarceration in the sort of hole in the ground we have already described, Chulpenyev appeared before a military tribunal of the 36th Motorized Division. Present were Lebedev, the Divisional Political Commissar, and Slesarev, the Chief of the Political Branch. The witness Lozovsky was not even summoned to testify. However, after the trial, to document the false testimony, they got Lozovsky's signature and that of Political Commissar Seryegin. The questions the tribunal asked were: Did you have a conversation with Lozovsky? What did he ask you about? What were your answers? Naïvely, Chulpenyev told them. He still couldn't understand what he was guilty of. "After all, many people talk like that!" he innocently exclaimed. The tribunal was interested: "Who? Give us their names." But Chulpenyev was not of their breed! He had the last word. "I beg the court to give me an assignment that will mean my death so as to assure itself once more of my patriotism"—and, like a simplehearted warrior of old—"Me and the person who slandered me—both of us together."

Oh, no! Our job is to kill off all those chivalrous sentiments in the people. Lozovsky's duty was to hand out pills and Seryegin's duty was to indoctrinate the soldiers. Whether or not you died wasn't important. What was important was that *we* were on guard.

Chapter 8

∎

The Law as a Child

We forget everything. What we remember is not what actually happened, not history, but merely that hackneyed dotted line they have chosen to drive into our memories by incessant hammering.

I do not know whether this is a trait common to all mankind, but it is certainly a trait of our people. And it is a vexing one. It may have its source in goodness, but it is vexing nonetheless. It makes us an easy prey for liars.

Therefore, if they demand that we forget even the public trials, we forget them. The proceedings were open and were reported in our newspapers, but they didn't drill a hole in our brains to make us remember—and so we've forgotten them. Only things repeated on the radio day after day drill holes in the brain. I am not even talking about young people, since they, of course, know nothing of all this, but about people who were alive at the time of those trials. Ask any middle-aged person to enumerate the highly publicized open trials. He will remember those of Bukharin and Zinoviev. And, knitting his brow, that of the Promparty too. And that's all. There were no other public trials.

Yet in actual fact they began right after the October Revolution. In 1918, quantities of them were taking place, in many different tribunals. They were taking place before there were either laws or codes, when the judges had to be guided solely by the requirements of the revolutionary workers' and peasants' power. At the same time, they were regarded as blazing their own trail of bold legality. Their detailed history will someday be written by someone, and it's not for us even to attempt to include it in our present investigation.

This chapter is concerned with the public trials conducted in the first few years following the success of the Bolshevik Revolution. It reviews five specific trials from 1918 to 1920. One can already observe the indiscriminate character of the accusations and the collaboration of prosecution and defense attorneys against the accused.

Chapter 9

■

The Law Becomes a Man

This chapter describes the law "while it is still in its Boy Scout stage," in the early 1920s. Five trials are traced in detail, among them the 1922 Moscow and Leningrad trials against prominent Church leaders, leading to the execution of the defendants. The objective of these trials is to speed up the repression and the looting of the Church. Lenin draws up the political section of the Criminal Code. Its scope is unlimited.

Chapter 10

■

The Law Matures

This chapter continues the theme of the two preceding chapters, with special attention to three trials of the late 1920s and early 1930s. At this stage, loyal engineers and even fellow Communists come under attack.

■

And may my compassionate reader now have mercy on me! Until now my pen sped on untrembling, my heart didn't skip a beat, and we slipped along unconcerned, because for these fifteen years we have been firmly protected either by legal revolutionality or else by revolutionary legality. But from now on things will be painful: as the reader will recollect, as we have had explained to us dozens of times, beginning with Khrushchev, "from approximately 1934, violations of Leninist norms of legality began." And how are we to enter this abyss of illegality now? How are we to drag our way along yet another bitter stretch of the road?

However, *these* trials which follow were, because of the fame of the defendants, a cynosure for the whole world. They did not escape the attention of the public. They were written about. They were interpreted and they will be interpreted again and again. It is for us merely to touch lightly on their *riddle.*

Let us make one qualification, though not a big one; the published stenographic records did not coincide completely with what was said at the trials. One writer who received an entrance pass—they were given out only to selected individuals—took running notes and subsequently discovered these differences. All the correspondents also noted the snag with Krestinsky, which made a recess necessary in order to get him back on the track of his assigned testimony. (Here is how I picture it. Before the trial a chart was set up for emergencies: in the first column was the name of the defendant; in the second, the method to be used during the recess if he should depart from his text during the open trial; in the third column, the name of the Chekist responsible for applying the indicated method. So if Krestinsky departed from his text, then who would come on the run and what that person would do had already been arranged.)

But the inaccuracies of the stenographic record do not change or lighten the picture. Dumfounded, the world watched three plays in a row, three wide-ranging and expensive dramatic productions in which the powerful leaders of the fearless Communist Party, who had turned the entire world upside down and terrified it, now marched forth like doleful, obedient goats and bleated out everything they had been ordered to, vomited all over themselves, cringingly abased themselves and their convictions, and confessed to crimes they could not in any wise have committed.

This was unprecedented in remembered history. It was particularly astonishing in contrast with the recent Leipzig trial of Dimitrov. Dimitrov had answered the Nazi judges like a roaring lion, and, immediately afterward, his comrades in Moscow, members of that same unyielding cohort which had made the whole world tremble—and the greatest of them at that, those who had been called the "Leninist guard" —came before the judges drenched in their own urine.

And even though much appears to have been clarified since then —with particular success by Arthur Koestler—the *riddle* continues to circulate as durably as ever.

People have speculated about a Tibetan potion that deprives a man of his will, and about the use of hypnosis. Such explanations must by no means be rejected: if the NKVD possessed such methods, clearly *there were no moral rules* to prevent resorting to them. Why not weaken or muddle the will? And it is a known fact that in the twenties some leading hypnotists gave up their careers and entered the service of the GPU. It is also reliably known that in the thirties a school for hypnotists existed in the NKVD. Kamenev's wife was allowed to visit her husband before his trial and found him not himself, his reactions retarded. (And she managed to communicate this to others before she herself was arrested.)

But why was neither Palchinsky nor Khrennikov broken by the Tibetan potion or hypnosis?

The fact is that an explanation on a higher, psychological plane is called for.

One misunderstanding in particular results from the image of these men as old revolutionaries who had not trembled in Tsarist dungeons —seasoned, tried and true, hardened, etc., fighters. But there is a plain and simple mistake here. These defendants were *not those* old revolutionaries. They had acquired that glory by inheritance from and association with the Narodniks, the SR's, and the Anarchists. They were the ones, the bomb throwers and the conspirators, who had known hard-labor imprisonment and real prison *terms*—but even *they* had never in their lives experienced a *genuinely merciless interrogation* (because such a thing did not exist at all in Tsarist Russia). And *these others,* the Bolshevik defendants at the treason trials, had never known either interrogation or real prison terms. The Bolsheviks had never been sentenced to special "dungeons," any Sakhalin, any special hard labor in Yakutsk. It is well known that Dzerzhinsky had the hardest time of them all, that he had spent all his life in prisons. But, according to our yardstick, he had served *just a normal "tenner,"* just a simple *"ten-ruble bill,"* like any ordinary collective farmer in our time. True, included in that tenner were three years in the hard-labor central prison, but that is nothing special either.

The Party leaders who were the defendants in the trials of 1936 to 1938 had, in their revolutionary pasts, known short, easy imprisonment, short periods in exile, and had never even had a whiff of hard labor. Bukharin had many petty arrests on his record, but they amounted to nothing. Apparently, he was never imprisoned anywhere for a whole year at a time, and he had just a wee bit of exile on Onega. Kamenev, despite long years of propaganda work and travel to all the cities of Russia, spent only two years in prison and one and a half years in exile. In our time, even sixteen-year-old kids got *five* right off. Zinoviev, believe it or not, *never spent as much as three months in prison.* He never received *even one sentence!* In comparison with the ordinary natives of our Archipelago they were all *callow youths;* they didn't know what prison was like. Rykov and I. N. Smirnov had been arrested several times and had been imprisoned for five years, but somehow they went through prison very easily, and they either escaped from exile without any trouble at all or were released because of an amnesty. Until they were arrested and imprisoned in the Lubyanka, they hadn't the slightest idea what a real prison was nor what the jaws of unjust interrogation were like. (There is no basis for assuming that if Trotsky had fallen into those jaws, he would have conducted himself with any less self-abasement, or that his resistance would have proved stronger than theirs. He had had no occasion to prove it. He, too, had known only easy imprisonment, no serious interrogations, and a mere two years of exile in Ust-Kut. The terror Trotsky inspired as Chairman of the Revolutionary Military Council was something he acquired very cheaply, and does not at all demonstrate any true strength of character or courage. Those who have condemned many others to be shot often wilt at the prospect of their own death. The two kinds of toughness are not connected.)

And, after all, our entire failure to understand derives from our belief in the unusual nature of these people. We do not, after all, where ordinary confessions signed by ordinary citizens are concerned, find their reasons for denouncing themselves and others so fulsomely baffling. We accept it as something we understand: a human being is weak; a human being gives in. But we consider Bukharin, Zinoviev, Kamenev, Pyatakov, I. N. Smirnov to be supermen to begin with—and, in essence, our failure to understand is due to that fact alone.

True, the directors of this dramatic production seem to have had a harder task in selecting the performers than they'd had in the earlier trials of the engineers: in those trials they had forty barrels to pick from, so to speak, whereas here the available troupe was small. Everyone knew who the chief performers were, and the audience wanted to see them in the roles and them only.

Yet there was a choice! The most farsighted and determined of those who were doomed did not allow themselves to be arrested. They committed suicide first (Skrypnik, Tomsky, Gamarnik). It was the ones who *wanted to live* who allowed themselves to be arrested. And one could certainly braid a rope from the ones who wanted to live! But even among them some behaved differently during the interrogations, realized what was happening, turned stubborn, and died silently but at least not shamefully. For some reason, they did not, after all, put on public trial Rudzutak, Postyshev, Yenukidze, Chubar, Kosior, and, for that matter, Krylenko himself, even though their names would have embellished the trials.

They put on trial the most compliant. A selection was made after all.

The men selected were drawn from a lower order, but, on the other hand, the mustached Producer knew each of them very well. He also knew that on the whole they were *weaklings,* and he knew, one by one, the particular weaknesses of each. Therein lay his dark and special talent, his main psychological bent and his life's achievement: to see people's weaknesses on the lowest plane of being.

And the man who seems, in the perspective of time, to have embodied the highest and brightest intelligence of all the disgraced and executed leaders (and to whom Arthur Koestler apparently dedicated his talented inquiry) was N. I. Bukharin. Stalin saw through him, too, at that lowest stratum at which the human being unites with the earth; and Stalin held him in a long death grip, playing with him as a cat plays with a mouse, letting him go just a little, and then catching him again. Bukharin wrote every last word of our entire existing—in other words, nonexistent—Constitution, which is so beautiful to listen to. And he flew about up there, just below the clouds, and thought that he had outplayed Koba: that he had thrust a constitution on him that would compel him to relax the dictatorship. And at that very moment, he himself had already been caught in those jaws. . . .

Bukharin in his last days began to compose his "Letter to the Future Central Committee." Committed to memory and thereby preserved, it recently became known to the whole world. However, it did not shake the world to its foundations. For what were the last words this brilliant theoretician decided to hand down to future generations? Just one more cry of anguish and a plea to be restored to the Party. (He paid dearly in shame for that devotion!) And one more affirmation that he "fully approved" everything that had happened up to and including 1937. And that included not only all the previous jeeringly mocking trials, but also all the foul-smelling waves of our great prison sewage disposal system.

There remained an easy dialogue with Vyshinsky along set lines: "Is it true that every opposition to the Party is a struggle against the Party?" "In general it is, factually it is." "But a struggle against the Party cannot help but grow into a war against the Party." "According to the logic of things—yes, it must." "And that means that in the end, given the existence of oppositionist beliefs, any foul deeds whatever might be perpetrated against the Party [espionage, murder, sellout of the Motherland]?" "But wait a minute, none were actually committed." "But they *could have been?*" "Well, theoretically speaking." (Those are your theoreticians for you!) "But for us the highest of all interests are those of the Party?" "Yes, of course, of course!" "So you see, only a very fine distinction separates us. We are required to concretize the eventuality: in the interest of discrediting for the future any idea of opposition, we are required to accept as *having taken place* what *could* only theoretically have taken place. After all, it *could* have, couldn't it?" "It could have." "And so it is necessary to recognize as actual what was possible; that's all. It's a small philosophical transition. Are we in agreement? ... Yes, and one thing more, and it's not for me to explain to you, but if you retreat and say something different during the trial, you understand that it will only play into the hands of the world bourgeoisie and will only do the Party harm. Well, and it's clear that in that case you yourself will not die an easy death. But if everything goes off all right, we will, of course, allow you to go on living. We'll send you in secret to the island of Monte Cristo, and you can work on the economics of socialism there." "But in previous trials, as I understand it, you did shoot them all?" "But what comparison is there between *you* and *them!* And then, we also left many of them alive too. They were shot only in the newspapers."

And so perhaps there isn't any insoluble riddle in those trials?

Chapter 11

■

The Supreme Measure

Capital punishment has had an up-and-down history in Russia. In the Code of the Tsar Aleksei Mikhailovich Romanov there were fifty crimes for which capital punishment could be imposed. By the time of the Military Statutes of Peter the Great there were two hundred. Yet the Empress Elizabeth, while she did not repeal those laws authorizing capital punishment, never once resorted to it. They say that when she ascended the throne she swore an oath never to execute anyone—and for all twenty years of her reign she kept that oath. She fought the Seven Years' War! Yet she still got along without capital punishment. It was an astounding record in the mid-eighteenth century—fifty years before the guillotine of the Jacobins. True, we have taught ourselves to ridicule all our past; we never acknowledge a good deed or a good intention in our history. And one can very easily blacken Elizabeth's reputation too; she replaced capital punishment with flogging with the knout; tearing out nostrils; branding with the word "thief"; and eternal exile in Siberia. But let us also say something on behalf of the Empress: how could she have changed things more radically than she did in contravention of the social concepts of her time? And perhaps the prisoner condemned to death today would voluntarily consent to that whole complex of punishments if only the sun would continue to shine on him; but we, in our humanitarianism, don't offer him that chance. And perhaps the reader will come to feel in the course of this book that twenty or even ten years in our camps are harder to bear than were the punishments of Elizabeth?

In today's terms, Elizabeth had a universally human point of view on all this, while the Empress Catherine the Great had, on the contrary, a class point of view (which was consequently more correct). Not to execute anyone at all seemed to her appalling and indefensible. She found capital punishment entirely appropriate to defending herself, her throne, and her system—in other words, in political cases, such as those of Mirovich, the Moscow plague mutiny, and Pugachev. But for *habitual criminals,* for *nonpolitical offenders,* why not consider capital punishment abolished?

Under Paul, the abolition of capital punishment was confirmed. (Despite his many wars, there were no military tribunals attached to military units.) And during the whole long reign of Alexander I, capital punishment was introduced only for war crimes that took place during a campaign (1812). (Right at this point, some people will say to us: What about deaths from running the gantlet? Yes, indeed, there were, of course, hidden executions—for that matter, one can literally drive a person to death with a trade-union meeting!) But the yielding up of one's God-given life because others, sitting in judgment, have so voted simply did not take place in our country even for *crimes* of state for an entire half-century—from Pugachev to the Decembrists.

The blood of the five Decembrists whetted the appetite of our state. From then on, execution for crimes of state was no longer prohibited nor was it forgotten, right up to the February Revolution in 1917. It was confirmed by the Statutes of 1845 and 1904, and further reinforced by the criminal statutes of the army and navy.

And how many people were executed in Russia during that period? We have already, in Chapter 8 above, cited the figures given by liberal leaders of 1905–1907. Let us add to them the verified figures of N. S. Tagantsev, the expert on Russian criminal law. Up until 1905, the death penalty was an exceptional measure in Russia. For a period of thirty years—from 1876 to 1904 (the period of the Narodnaya Volya revolutionaries and the use of terrorism—a terrorism which did not consist merely of *intentions* murmured in the kitchen of a communal apartment—a period of mass strikes and peasant revolts; the period when the parties of the future revolution were created and grew in strength)— 486 people were executed; in other words, about seventeen people per year for the whole country. (This figure includes executions of ordinary, nonpolitical criminals!) During the years of the first revolution (1905) and its suppression, the number of executions rocketed upward, astounding Russian imaginations, calling forth tears from Tolstoi and indignation from Korolenko and many, many others: from 1905 through 1908 about 2,200 persons were executed—forty-five a month. This, as Tagantsev said, was an *epidemic of executions.* It came to an abrupt end.

When the Provisional Government came to power, it abolished capital punishment entirely. In July, 1917, however, it was reinstated in the active army and front-line areas for military crimes, murder, rape, assault, and pillage (very widespread in those areas at that time). This was one of the most unpopular of the measures which destroyed the Provisional Government. The Bolsheviks' slogan before the Bolshevik coup d'état was: "Down with capital punishment, reinstated by Kerensky!"

If we are to judge by official documents, capital punishment was restored in all its force in June, 1918. No, it was not "restored"; instead, a *new* era of executions was inaugurated. If one takes the view that Latsis is not deliberately understating the real figures but simply lacks complete information, and that the Revtribunals carried on approximately the same amount of judicial work as the Cheka performed in an extrajudicial way, one concludes that in the twenty central provinces of Russia in a period of sixteen months (June, 1918, to October, 1919) more than sixteen thousand persons were shot, which is to say *more than one thousand a month.*

However, it may not even have been these individual executions, with or without formally pronounced death sentences, which added up to thousands and inaugurated the new era of executions in 1918 that stunned and froze Russia. Still more terrible to us was the practice—initially followed by both warring sides and, later, by the victors only—of *sinking barges* loaded with uncounted, unregistered hundreds, unidentified even by a roll call. (Naval officers in the Gulf of Finland, in the White, Caspian, and Black seas, and, as late as 1920, hostages in Lake Baikal.) This is outside the scope of our narrow history of courts and trials, but it belongs to the history of *morals,* which is where everything else originates as well. In all our centuries, from the first Ryurik on, had there ever been a period of such cruelties and so much killing as during the post-October Civil War?

At one time 265 condemned prisoners were *awaiting* execution in Leningrad's Kresty Prison alone. And during the whole year, it would certainly seem that more than a thousand were shot in Kresty alone.

And what kind of evildoers were these condemned men? Where did so many plotters and troublemakers come from? Among them, for example, were six collective farmers from nearby Tsarskoye Selo who were guilty of the following crime: After they had finished mowing the collective farm with their own hands, they had gone back and mowed a second time along the hummocks to get a little hay for their own cows. The All-Russian Central Executive Committee *refused to pardon all six of these peasants, and the sentence of execution was carried out.*

What cruel and evil Saltychikha, what utterly repulsive and infamous serf-owner would have *killed* six peasants for their miserable little clippings of hay? If one had dared to beat them with birch switches even once, we would know about it and read about it in school and curse that name. But now, heave the corpses into the water, and pretty soon the surface is all smooth again and no one's the wiser. And one must cherish the hope that someday documents will confirm the report of my witness, who is still alive. Even if Stalin had killed no others, I believe he deserved to be drawn and quartered just for the lives of those six Tsarskoye Selo peasants! And yet they still dare shriek at us (from Peking, from Tirana, from Tbilisi, yes, and plenty of big-bellies in the Moscow suburbs are doing it too): "How could you dare expose him?" "How could you dare disturb his great shade?" "Stalin belongs to the world Communist movement!" But in my opinion all he belongs to is the Criminal Code. "The peoples of all the world remember him as a friend." But not those on whose backs he rode, whom he slashed with his knout.

However, let us return to being dispassionate and impartial once more. Of course, the All-Russian Central Executive Committee would certainly have "completely abolished" the supreme measure, as promised, but unfortunately what happened was that in 1936 the Father and Teacher "completely abolished" the All-Russian Central Executive Committee itself. And the *Supreme Soviet* that succeeded it had an eighteenth-century ring. "The supreme measure" became a *punishment* once again, and ceased to be some kind of incomprehensible "social defense."

As for the executions of 1937–1938, what legal expert, what criminal historian, will provide us with verified statistics? Where is that *Special Archive* we might be able to penetrate in order to read the figures? There is none. There is none and there never will be any. Therefore we dare report only those figures mentioned in rumors that were quite fresh in 1939–1940, when they were drifting around under the Butyrki arches, having emanated from the high- and middle-ranking Yezhov men of the NKVD who had been arrested and had passed through those cells not long before. (And they really knew!) The Yezhov men said that during those two years of 1937 and 1938 a *half-*

million "political prisoners" had been shot throughout the Soviet Union, and 480,000 *blatnye*—habitual thieves—in addition. According to the testimony from Krasnodar, in 1937–1938 in the main building of the GPU on Proletarskaya Street they shot more than two hundred people every night.

In May, 1947, Iosif Vissarionovich inspected his new starched dickey in his mirror, liked it, and dictated to the Presidium of the Supreme Soviet the Decree on the Abolition of Capital Punishment in peacetime (replacing it with a new maximum term of twenty-five years —it was a good pretext for introducing the so-called *quarter*).

But our people are ungrateful, criminal, and incapable of appreciating generosity. Therefore, after the rulers had creaked along and eked out two and a half years without the death penalty, on January 12, 1950, a new decree was published that constituted an about-face: "In view of petitions pouring in from the national republics [the Ukraine?], from the trade unions [oh, those lovely trade unions; they always know what's needed], from peasant organizations [this was dictated by a sleepwalker: the Gracious Sovereign had stomped to death all peasant organizations way back in the Year of the Great Turning Point], and also from cultural leaders [now, *that* is quite likely]," capital punishment was restored for a conglomeration of "traitors of the Motherland, spies, and subversives-diversionists." (And, of course, they forgot to repeal the *quarter,* the twenty-five-year sentence, which remained in force.)

And once this return to our familiar friend, to our beheading blade, had begun, things went further with no effort at all: in 1954, for premeditated murder; in May, 1961, for theft of state property, and counterfeiting, and terrorism in places of imprisonment (this was directed especially at prisoners who killed informers and terrorized the camp administration); in July, 1961, for violating the rules governing foreign currency transactions; in February, 1962, for threatening the lives of (shaking a fist at) policemen or Communist vigilantes; then for rape; and immediately thereafter for bribery.

But all of this is simply temporary—until complete abolition. And that's how it's described today too.

And so it turns out that Russia managed longest of all without capital punishment in the reign of the Empress Elizabeth Petrovna.

■

Thus many were shot—thousands at first, then hundreds of thousands. We divide, we multiply, we sigh, we curse. But still and all, these are

Viktor Petrovich Pokrovsky

Aleksandr Shtrobinder

Vasily Ivanovich Anichkov

Aleksandr Andreyevich Svechin

Mikhail Aleksandrovich Reformatsky

Yelizaveta Yevgenyevna Anichkova

just numbers. They overwhelm the mind and then are easily forgotten. And if someday the relatives of those who had been shot were to send one publisher photographs of their executed kin, and an album of those photographs were to be published in several volumes, then just by leafing through them and looking into the extinguished eyes we would learn much that would be valuable for the rest of our lives. Such reading, almost without words, would leave a deep mark on our hearts for all eternity.

In one household I am familiar with, where some former zeks live, the following ceremony takes place: On March 5, the day of the death of the Head Murderer, they spread out on the table all the photographs of those who were shot and those who died in camps that they have been able to collect—several dozen of them. And throughout the day solemnity reigns in the apartment—somewhat like that of a church, somewhat like that of a museum. There is funeral music. Friends come to visit, to look at the photographs, to keep silent, to listen, to talk softly together. And then they leave without saying good-bye.

And that is how it ought to be everywhere. At least these deaths would have left a small scar on our hearts.

So that they should not have died *in vain!*

And I, too, have a few such chance photographs. Look at these at least:

Viktor Petrovich Pokrovsky—shot in Moscow in 1918.

Aleksandr Shtrobinder, a student—shot in Petrograd in 1918.

Vasily Ivanovich Anichkov—shot in the Lubyanka in 1927.

Aleksandr Andreyevich Svechin, a professor of the General Staff—shot in 1935.

Mikhail Aleksandrovich Reformatsky, an agronomist—shot in Orel in 1938.

Yelizaveta Yevgenyevna Anichkova—shot in a camp on the Yenisei in 1942.

They say that Konstantin Rokossovsky, the future marshal, was twice taken into the forest at night for a supposed execution. The firing squad leveled its rifles at him, and then they dropped them, and he was taken back to prison. And this was also making use of "the supreme measure" as an interrogator's trick. But it was all right; nothing happened; and he is alive and healthy and doesn't even cherish a grudge about it.

And almost always a person obediently allows himself to be killed. Why is it that the death penalty has such a hypnotic effect? Those pardoned recall hardly anyone in their cell who offered any resistance.

Chapter 12

■

Tyurzak

Tyurzak = TYURemnoye ZAKlyucheniye, "prison confinement." Tyurzak is an official term.

In December, 1917, it had already become clear that it was altogether impossible to do without prisons, that some people simply couldn't be left anywhere except behind bars (see Chapter 2, above), because—well, simply because there was no place for them in the new society. The attention of the new prison authorities was directed toward the combat readiness of the prison guards outside the walls and the takeover of the stock of prisons inherited from the Tsar. Fortunately, it turned out that the Civil War had not resulted in the destruction of all the principal *central prisons* and jails.

So we recall the Solovetsky Islands (nicknamed Solovki): it was such a good place, cut off from communication with the outside world for half a year at a time. You couldn't be heard from there no matter how loud you shouted, and you could even burn yourself up for all anyone would know.

■

From our experience of the past and our literature of the past we have derived a naïve faith in the power of a hunger strike. But the hunger strike is a purely moral weapon. It presupposes that the jailer has not entirely lost his conscience. Or that the jailer is afraid of public opinion. Only in such circumstances can it be effective.

The Tsarist jailers were still inexperienced. They got nervous if one of their prisoners went on a hunger strike; they exclaimed over it; they looked after him; they put him in the hospital. There are many examples, but this work is not about them. It is even humorous to note that it was enough for Valentinov to go on a hunger strike for twelve days: as a result, he not only achieved some relaxation in the regimen but was *totally released* from interrogation—whereupon he went to Lenin in Switzerland. Even in the Orel central hard-labor prison the strikers always won. They got the regimen relaxed in 1912 and further relaxed in 1913, to the point of general access to outdoor walks for all political hard-labor prisoners—who were obviously so unrestricted by their supervisors that they managed to compose and send out to freedom their appeal "to the Russian people." (And this from the hard-labor prisoners of a central prison!) Furthermore, it was *published*. (It's enough to make one's eyes pop out of one's head! Someone has to have been crazy!)

In the Revolution of 1905 and the years following it, the prisoners felt themselves to be masters of the prison to such an extent that they did not even go to the trouble of declaring a hunger strike; they simply destroyed prison property (so-called "obstructions"), or went so far as to declare a *strike,* although it might seem that for prisoners this would have hardly any meaning.

In the twenties, the lively picture of hunger strikes grows clouded (though that depends, of course, on the point of view . . .). Still, it was possible in those years to achieve at least one's personal demands by this means.

From the thirties on, state thinking about hunger strikes took a new turn. What did the state want with even such watered-down, isolated, half-suppressed hunger strikes? Wasn't the ideal picture one of prisoners who had no will of their own, nor the capacity to make their own decisions—and of a prison administration that did their thinking and their deciding for them? These are, if you will, the only prisoners who can exist in the new society. And so from the beginning of the thirties, they stopped accepting declarations of hunger strikes as legal. "The hunger strike as a method of resistance *no longer exists,*" they proclaimed to Yekaterina Olitskaya in 1932, and they said the same thing to many others. The government has abolished your hunger strikes—and that's that. But Olitskaya refused to obey and began to fast. They let her go on fasting in solitary for *fifteen* days. Then they took her to the hospital and put milk and dried crusts in front of her to tempt her. But she stood firm, and on the *nineteenth* day she won her victory: she got an extended outdoor period and newspapers and parcels from the Political Red Cross. (That's how one had to moan and groan in order to receive those legitimate relief parcels!) Overall, however, it was an insignificant victory and paid for too dearly. Olitskaya recalls such foolish hunger strikes on the part of others too: people starved up to twenty days in order to get delivery of a parcel or a change of companions for their outdoor walk. Was it worth it? After all, in the *New Type Prison* one's strength, once lost, could not be restored. The religious-sect member Koloskov fasted until he died on the twenty-fifth day. Could one in general permit oneself to fast in the New Type Prison? After all, the new prison heads, operating in secrecy and silence, had acquired several powerful methods of combating hunger strikes:

1. Patience on the part of the administration. (We have seen enough of what this meant from preceding examples.)

2. Deception. This, too, can be practiced thanks to total secrecy. When every step is reported by the newspapers, you aren't going to do much deceiving. But in our country, why not? In 1933, in the Khabarovsk Prison, S. A. Chebotaryev, demanding that his family be informed of his whereabouts, fasted for seventeen days. (He had come from the Chinese Eastern Railroad in Manchuria and then suddenly disappeared, and he was worried about what his wife might be thinking.) On the seventeenth day, Zapadny, the Deputy Chief of the Provincial GPU, and the Khabarovsk Province prosecutor (their ranks indicate that lengthy hunger strikes were really not so frequent) came to see him and showed him a telegraph receipt (There, they said, they had informed his wife!), and thus persuaded him to take some broth. And the receipt was a fake!

3. Forced artificial feeding. This method was adapted, without any question, from experience with wild animals in captivity. And it could be employed only in total secrecy. By 1937 artificial feeding was, evidently, already in wide use.

4. A new view of the hunger strike: that hunger strikes are a continuation of counterrevolutionary activity in prison, and must be punished with a new *prison term.*

Approximately in the middle of 1937, a new directive came: From now on the prison administration *will not in any respect be responsible for those dying on hunger strikes!* The last vestige of personal responsibility on the part of the jailers had disappeared! (In these circumstances, the prosecutor of the province would not have come to visit Chebotaryev!) Furthermore, so that the interrogator shouldn't get disturbed, it was also announced that days spent on hunger strike by a prisoner under interrogation should be crossed off the official interrogation period. In other words, it should not only be considered that the *hunger strike had not taken place,* but the prisoner should be regarded as not having been in prison at all during the period of the strike. Thus the interrogator would not be to blame for being behind schedule. Let the only perceptible result of the hunger strike be the prisoner's exhaustion!

Decades passed and time produced its own results. The hunger strike—the first and most natural weapon of the prisoner—in the end became alien and incomprehensible to the prisoners themselves. Fewer and fewer desired to undertake them. And to prison administrations the whole thing began to seem either plain stupidity or else a malicious violation.

■

Even though the enormous Archipelago was already spreading across the land, the prisons for long-termers didn't fall into decay. The old jail tradition was being zealously carried on. Everything new and invaluable which the Archipelago had contributed to the indoctrination of the masses was still not enough in itself. The deficiency was provided for by the complementary existence of the TON's—the Special Purpose Prisons—and prisons for long-termers in general.

Not everyone swallowed up by the Great Machine was allowed to mingle with the natives of the Archipelago. Well-known foreigners, individuals who were too famous or who were being held secretly, purged gaybisty, could not by any means be seen openly in camps; their hauling a barrow did not compensate for the disclosure and the consequent *moral-political* damage. In the same way, the socialists, who were engaged in a continuous struggle for their prison rights, could not conceivably be permitted to mingle with the masses but had to be kept separately and, in fact; suffocated separately—in view of their special privileges and rights. Much later on, in the fifties, as we shall learn later in this work, the Special Purpose Prisons were also needed to isolate camp rebels. And in the last years of his life, disappointed in the possibilities of "reforming" thieves, Stalin gave orders that various *ringleaders of the thieves* should also get *tyurzak* rather than camp. And then, to be sure, it was necessary for the state to support free of charge in prison those prisoners who because of their feebleness would have immediately died off in camp and would thus have shirked their duty to serve out their terms.

The inventory of old jails, inherited from the Romanov dynasty, was, of necessity, looked after, remodeled, strengthened, and perfected. Certain central prisons, like the one in Yaroslavl, were so well and suitably appointed (doors plated with iron; table, stool, and cot permanently anchored in each cell) that the only thing required to bring them up to date was the installation of "muzzles" on the windows and the fencing in of the courtyards where the prisoners walked in order to reduce them to the size of a cell (by 1937 all the trees on prison grounds had been cut down, all vegetable gardens plowed under, and all grassy areas paved with asphalt). Others, like the one in Suzdal, required new equipment, and the monastery arrangement had to be remodeled, but, after all, self-incarceration of a body in a monastery and its incarceration in a prison by the state serve physically similar purposes, and therefore the buildings were always easy to adapt. One of the buildings of the Sukhanovka Monastery was adapted for use as a prison for long-termers. . . .

During the twenties the prisoner's *food* was very decent in the isolators for politicals: the lunches always included some meat; fresh vegetables were served; milk could be bought in the commissary. In 1931–1933 the food deteriorated sharply, but things were no better out in freedom at that time. Both scurvy and dizziness from lack of food were no rarity in the prisons for politicals in those years. Later on the food improved, but it was never the same as before. The *light* in cells was always "rationed," so to speak, in both the thirties and the forties: the "muzzles" on the windows and the frosted reinforced glass created a permanent twilight in the cells (darkness is an important factor in causing depression). They often stretched netting above the window "muzzle," and in the winter it was covered with snow, which cut off this last access to the light. Reading became no more than a way of ruining one's eyes. In the Vladimir TON, they made up for this lack of light at night: bright electric lights burned all night long, preventing sleep. And in the Dmitrovsk Prison in 1938 (N. A. Kozyrev), there was light in the evenings and at night—a kerosene lamp on a little shelf way up near the ceiling, that burned away and smoked up the last air; in 1939 there were electric lights that glowed red at half-voltage. *Air* was "rationed" too. The hinged panes for ventilation were kept locked, and opened only during the interval of the prisoners' trip to the toilet. In Vladimir in 1948 there was no lack of air, because the transom was open permanently. *Walks outdoors* ranged from fifteen to forty-five minutes at various hours in various prisons. There was no such thing as the communication with the soil that had existed in Schlüsselburg or Solovki; everything that grew had been torn up by the roots, trampled, covered with concrete and asphalt. They even forbade lifting up one's head to the heavens during the walks: "Look at your feet!" This was the command both Kozyrev and Adamova remember from the Kazan

Prison. *Visits* from relatives were forbidden in 1937 and never renewed. *Letters* could be sent to close relatives twice a month and could be received from them in most years. (But in Kazan they had to be returned to the administration the day after they had been read.) Access to the *commissary* to make purchases with the money sent in specifically limited amounts was usually permitted. *Furniture* was no unimportant part of the prison regimen. Adamova wrote eloquently of her happiness at finding a simple wooden cot with a straw mattress and a simple wooden table in her cell in Suzdal, after having had only cots that folded into the wall and chairs anchored to the floor. In the Vladimir TON, I. Korneyev experienced two different prison *regimens:* Under one, in 1947–1948, personal articles were not removed from the cell; one could lie down during the day; and the turnkey very seldom looked through the peephole. But under the other, in 1949–1953, the cell was locked with two locks (the responsibility of the turnkey and duty officer respectively); one was forbidden to lie down, forbidden to talk in a normal voice (in Kazan, only in a whisper); personal articles were all taken away; a uniform of striped mattress ticking was issued; correspondence was permitted only twice a year and only on those days announced without warning by the chief of the prison (anyone who missed that day couldn't write), and only a sheet of paper half the size of a postal sheet could be used; violent *searches* and unscheduled visits were frequent, requiring the complete turning out of one's belongings and undressing down to one's skin. Communication between cells was prohibited to such an extent that the jailers went through the toilets with a portable lantern after each toilet visit and searched in each hole. The entire cell would get *punishment cells* for graffiti in the toilets. The punishment cells were a scourge in the Special Purpose Prisons. One could get into a punishment cell for coughing. ("Cover your head with your blanket. Then you can cough!") Or for *walking around the cell* (Kozyrev: "It was considered to be rebellious"); for the noise made by one's shoes. (In the Kazan Prison women had been issued men's shoes that were much too large for women's feet—size 10½.) Incidentally, Ginzburg was correct in concluding that periods in a punishment cell were meted out not for any particular misdemeanor but *according to a schedule:* every prisoner was required to spend some time there in order to learn what it was like. And the rules included another generally applicable point: "In the event of any display of unruliness in a punishment cell [?], the chief of the prison has the right to extend the term of incarceration there *to twenty days.*" Just what was meant by unruliness? Here's what happened to Kozyrev. (The descriptions of the punishment cell and much else in the prison regimen tally to such an extent among all sources that the stamp of a single system of administrative rules can be detected.) He was given another five days in the punishment cell for pacing back and forth. In the autumn, the building containing the punishment cells was unheated, and it was very cold. They forced prisoners to undress down to their underwear and to take off their shoes. The floor was bare earth and dust (it might be wet dirt; and in the Kazan Prison it might even be covered with water). Kozyrev had a stool in his. (Ginzburg had none in hers.) He immediately concluded that he would perish, that he would freeze to death. But some kind of mysterious inner warmth gradually made itself felt, and it was his salvation. He learned to sleep sitting on his stool. They gave him a mug of hot water three times a day; it made him drunk. One of the duty officers, in violation of the rules, pressed a piece of sugar into his ten-and-a-half-ounce bread ration. On the basis of the rations issued him, and by observing the light from some faraway, tiny, labyrinthine window, Kozyrev kept count of the days. His five days had come to an end, but he had not been released. His sense of hearing had become extremely acute and he heard whispers in the corridor—having to do with either "the sixth" or "six days." This was a provocation: they were waiting for him to say that his five days were over and that it was time to let him out. That would have constituted unruliness, for which his stay in the punishment cell would have been prolonged. But he sat silent and obedient for another day, and then they let him out, just as if everything had been the way it was supposed to be. (Perhaps the chief of the prison used this method for testing all the prisoners in turn for submissiveness? And then he could sentence all those who weren't yet submissive enough to further terms in the punishment cell.) After the punishment cell the ordinary cell seemed like a palace. Kozyrev became deaf for half a year, and he began to get abscesses in his throat. His cellmate went insane from frequent imprisonment in the punishment cell, and Kozyrev was kept locked up with an insane man for more than a year, with just the two of them there.

■

And only here, right here, is where our chapter ought to have begun. It ought to have examined that glimmering light which, in time, the soul of the lonely prisoner begins to emit, like the halo of a saint. Torn from the hustle-bustle of everyday life in so absolute a degree that even counting the passing minutes puts him intimately in touch with the Universe, the lonely prisoner has to have been purged of every imperfection, of everything that has stirred and troubled him in his former life, that has prevented his muddied waters from settling into transparency. How gratefully his fingers reach out to feel and crumble the lumps of earth in the vegetable garden (but, alas, it is all asphalt). How his head rises of itself toward the Eternal Heavens (but, alas, this is forbidden). And how much touching attention the little bird on the window sill arouses in him (but, alas, there is that "muzzle" there, and the netting as well, and the hinged ventilation pane is locked). And what clear thoughts, what sometimes surprising conclusions, he writes down on the paper issued him (but, alas, only if you buy it in the commissary, and only if you turn it in to the prison office when you have used it up —for eternal safekeeping . . .).

But our peevish qualifications somehow interrupt our line of thought. The plan of our chapter creaks and cracks, and we no longer know the answer to the question: Is the soul of a person in the New Type Prison, in the Special Purpose Prison (the TON), purified or does it perish once and for all?

If the first thing you see each and every morning is the eyes of your cellmate who has gone insane, how then shall you save yourself during the coming day? Nikolai Aleksandrovich Kozyrev, whose brilliant career in astronomy was interrupted by his arrest, saved himself only by thinking of the eternal and infinite: of the order of the Universe—and of its Supreme Spirit; of the stars; of their internal state; and what Time and the passing of Time really are.

And in this way he began to discover a new field in physics. And only in this way did he succeed in surviving in the Dmitrovsk Prison. But his line of mental exploration was blocked by forgotten figures. He could not build any further—he had to have a lot of figures. Now just where could he get them in his solitary-confinement cell with its overnight kerosene lamp, a cell into which not even a little bird could enter? And the scientist prayed: "Please, God! I have done everything I could. Please help me! Please help me continue!"

At this time he was entitled to receive one book every ten days (by then he was alone in the cell). In the meager prison library were several different editions of Demyan Bedny's *Red Concert*, which kept coming around to each cell again and again. Half an hour passed after his prayer; they came to exchange his book; and as usual, without asking anything at all, they pushed a book at him. It was entitled *A Course in Astrophysics!* Where had it come from? He simply could not imagine such a book in the prison library. Aware of the brief duration of this coincidence, Kozyrev threw himself on it and began to memorize everything he needed immediately, and everything he might need later on. In all, just two days had passed, and he had eight days left in which to keep his book, when there was an unscheduled inspection by the chief of the prison. His eagle eye noticed immediately. "But you are an astronomer?" "Yes." "Take this book away from him!" But its mystical arrival had opened the way for his further work, which he then continued in the camp in Norilsk.

And so now we should begin the chapter on the conflict between the soul and the bars.

But what is this? The jailer's key is rattling brazenly in the lock. The gloomy block superintendent is there with a long list. "Last name, first name, patronymic? Date of birth? Article of the Code? Term? End of term? Get your *things* together. Be quick about it!"

Well, brothers, a prisoner transport! A prisoner transport! We're off to somewhere! Good Lord, bless us! Shall we gather up our bones?

Well, here's what: If we are still alive, then we'll finish this story another time. In Part IV. If we are still alive . . .

PART II

Perpetual Motion

■

Chapter 1

■

The Ships of the Archipelago

Scattered from the Bering Strait almost to the Bosporus are thousands of islands of the spellbound Archipelago. They are invisible, but they exist. And the invisible slaves of the Archipelago, who have substance, weight, and volume, have to be transported from island to island just as invisibly and uninterruptedly.

And by what means are they to be transported? On what?

Great ports exist for this purpose—transit prisons; and smaller ports—camp transit points. Sealed steel ships also exist: railroad cars especially christened *zak cars* ("prisoner cars"). And out at the anchorages, they are met by similarly sealed, versatile *Black Marias* rather than by sloops and cutters. The *zak cars* move along on regular schedules. And, whenever necessary, whole caravans—trains of red cattle cars—are sent from port to port along the routes of the Archipelago.

All this is a thoroughly developed system! It was created over dozens of years—not hastily. Well-fed, uniformed, unhurried people created it. The Kineshma convoy waits at the Moscow Northern Station at 1700 hours on odd-numbered days to accept Black Marias from the Butyrki, Krasnaya Presnya, and Taganka prisons. The Ivanovo convoy has to arrive at the station at 0600 hours on even-numbered days to receive and hold in custody transit prisoners for Nerekhta, Bezhetsk, and Bologoye.

All this is happening right next to you, you can almost touch it, but it's invisible (and you can shut your eyes to it too). At the big stations the loading and unloading of the dirty faces takes place far, far from the passenger platform and is seen only by switchmen and roadbed inspectors. At smaller stations a blind alleyway between two warehouses is preferred, into which the Black Marias can back so that their steps are flush with the steps of the zak car. The convict doesn't have time to look at the station, to see you, or to look up and down the train. He gets to look only at the steps. (And sometimes the lower step is waist-high, and he hasn't the strength to climb up on it.) And the convoy guards, who have blocked off the narrow crossing from the Black Maria to the zak car, growl and snarl: "Quick, quick! Come on, come on!" And maybe even brandish their bayonets.

And you, hurrying along the platform with your children, your suitcases, and your string bags, are too busy to look closely: Why is that second baggage car hitched onto the train? There is no identification on it, and it is very much like a baggage car—and the gratings have diagonal bars, and there is darkness behind them. But then why are soldiers, defenders of the Fatherland, riding in it, and why, when the train stops, do two of them march whistling along on either side and peer down under the car?

The train starts—and a hundred crowded prisoner destinies, tormented hearts, are borne along the same snaky rails, behind the same smoke, past the same fields, posts, and haystacks as you, and even a few seconds sooner than you. But outside your window even less trace of the grief which has flashed past is left in the air than fingers leave in water. And in the familiar life of the train, which is always exactly the same—with its slit-openable package of bed linen, and tea served in glasses with metal holders—could you possibly grasp what a dark and suppressed horror has been borne through the same sector of Euclidean space just three seconds ahead of you? You are dissatisfied because there are four of you in your compartment and it is crowded. And could you possibly believe—and will you possibly believe when reading these lines—that in the same size compartment as yours, but up ahead in that zak car, there are fourteen people? And if there are twenty-five? And if there are thirty?

Probably this type of railroad car really was first used under Stolypin, in other words before 1911. And in the general Cadet revolutionary embitterment, they christened it with his name. However, it really became the favorite means of prisoner transport only in the twenties; and it became the universal and exclusive means only from 1930 on, when everything in our life became uniform. Therefore it would be more correct to call it a *Stalin* car rather than a *Stolypin* car. But we aren't going to argue with the Russian language here.

The Stolypin car is an ordinary passenger car divided into compartments, except that five of the nine compartments are allotted to the prisoners (here, as everywhere in the Archipelago, half of everything goes to the auxiliary personnel, the guards), and compartments are separated from the corridor not by a solid barrier but by a grating which leaves them open for inspection. This grating consists of intersecting diagonal bars, like the kind one sees in station parks. It rises the full height of the car, and because of it there are not the usual baggage racks projecting from the compartments over the corridor. The windows on the corridor sides are ordinary windows, but they have the same diagonal gratings on the outside. There are no windows in the prisoners' compartments—only tiny, barred blinds on the level of the second sleeping shelves. That's why the car has no exterior windows and looks like a baggage car. The door into each compartment is a sliding door: an iron frame with bars.

From the corridor side all this is very reminiscent of a menagerie: pitiful creatures resembling human beings are huddled there in cages, the floors and bunks surrounded on all sides by metal grilles, looking out at you pitifully, begging for something to eat and drink. Except that in menageries they never crowd the wild animals in so tightly.

According to the calculations of nonprisoner engineers, six people can sit on the bottom bunks of a Stolypin compartment, and another three can lie on the middle ones (which are joined in one continuous bunk, except for the space cut out beside the door for climbing up and getting down), and two more can lie on the baggage shelves above. Now if, in addition to these eleven, eleven more are pushed into the compartment (the last of whom are shoved out of the way of the door by the jailers' boots as they shut it), then this will constitute a normal complement for a Stolypin prisoners' compartment. Two huddle, half-sitting, on each of the upper baggage shelves; another five lie on the joined middle level (and they are the lucky ones—these places are won in battle, and if there are any prisoners present from the underworld companionship of thieves—the blatnye—then it is they who are lying there); and this leaves thirteen down below: five sit on each of the bunks and three are in the aisle between their legs. Somewhere, mixed up with the people, on the people and under the people, are their belongings. And that is how they sit, their crossed legs wedged beneath them, day after day.

No, it isn't done especially to torture people. A sentenced prisoner is a laboring soldier of socialism, so why should he be tortured? They need him for construction work. But, after all, you will agree he is not off on a jaunt to visit his mother-in-law, and there is no reason in the world to treat him so well that people out in *freedom* would envy him. We have problems with our transportation: he'll get there all right, and he won't die on the way either.

Since the fifties, when railroad timetables were actually straightened out, the prisoners haven't had to travel in this fashion for very long at a time—say, a day and a half or two days. During and after the war, things were worse. From Petropavlovsk (in Kazakhstan) to Karaganda, a Stolypin car might be *seven days* en route (with twenty-five people in a compartment). From Karaganda to Sverdlovsk it could be *eight days* (with up to twenty-six in a compartment). Even just going from Kuibyshev to Chelyabinsk in August, 1945, Susi traveled in a Stolypin car for several days, and their compartment held *thirty-five* people lying on top of one another, floundering, fighting. And in the autumn of 1946 N. V. Timofeyev-Ressovsky traveled from Petropavlovsk to Moscow in a compartment that had *thirty-six* people in it! For several days he *hung* suspended between other human beings and his legs did not touch the floor. Then they started to die off—and the guards hauled the corpses out from under their feet. (Not right away, true; only on the second day.) That way things became less crowded. The whole trip to Moscow continued in this fashion for *three weeks.*

Was thirty-six the upper limit for a Stolypin compartment? I have no evidence available on thirty-seven or higher, and yet, adhering to our one-and-only scientific method, and remembering the necessity to struggle against "the limiters," we are compelled to reply: No, no, no! It is not a limit! Perhaps in some other country it would be an upper limit, but not here! As long as there are any cubic centimeters of unbreathed air left in the compartment, even if it be beneath the upper shelves, even if between shoulders, legs, and heads, the compartment is ready to take additional prisoners. One might, however, conditionally accept as the upper limit the number of unremoved corpses which can be contained in the total volume of the compartment, given the possibility of packing them in at leisure.

V. A. Korneyeva traveled from Moscow in a compartment that held *thirty women*—most of them withered old women, exiled for their religious beliefs (on arrival *all* these women, except two, were immediately put in the hospital). Nobody died in the compartment because several of the prisoners were young, well-developed, good-looking girls, arrested "for going out with foreigners." These girls took it upon themselves to shame the convoy: "You ought to be ashamed to transport them this way! These are your own mothers!" It probably wasn't so much their moral argument as their attractive appearance which produced a reaction in the convoy guards, and they did move several of the old women out—*to the punishment cell.* But the punishment cell in a Stolypin car is no punishment; it is a blessing. Of five prisoner compartments, four are used as general cells, and the fifth is set aside and divided in two halves—two narrow half-compartments with one lower and one upper berth, like those the conductors have. These punishment cells serve to isolate prisoners; three or four travel in them at a time, and this gives both comfort and space.

No, it is not intentionally to torture them with thirst that the exhausted and overcrowded prisoners are fed not soup but salt herring or dry smoked Caspian carp for the whole of their trip in the Stolypin car. (This was exactly how it was in *all* the years, the thirties and the fifties, winter and summer, in Siberia and the Ukraine, and it isn't even necessary to cite examples.) It was not to torture them with thirst—but just you tell me what these ragamuffins were to be fed anyway while being moved around. They were not supposed to get hot meals in prisoner-transport railroad cars. (True, there was a kitchen in one of the Stolypin car compartments, but that was only for the convoy.) You couldn't just give the prisoners raw grits, and you couldn't give them raw codfish either, nor could you give them canned meat because they might stuff themselves. Herring was just the thing, with a piece of bread —and what else did they need?

Go ahead, take your half a herring while they are handing it out, and be glad you got it. If you're smart, you aren't going to eat that herring; just be patient, wait, hide it in your pocket, and you can eat it at the next transit point where there is water to be had. It's worse when they issue you wet Sea of Azov anchovies, covered with coarse salt. You can't keep them in your pocket; so scoop them up in the flaps of your pea jacket, or in your handkerchief, in the palm of your hand —and eat them. They divide up these Azov anchovies on somebody's pea jacket, whereas the convoy guards dump the dried carp right on the floor of the compartment, and it is divided up on the benches, on the prisoners' knees.

But once they've given you a fish, they aren't going to hold back on the bread, and maybe they'll even throw in a bit of sugar. Things are much worse when the convoy comes over and announces: "We aren't going to be feeding you today; *nothing was issued* for you." And it could very well be that nothing was actually issued: someone in one or another prison accounting office made a mistake in the figures. And it could also be that it was issued but that the convoy was short on rations—after all, they aren't exactly overfed either—and so they decided *to snag* a bit of your bread for themselves; and in that case to hand over half a herring by itself would seem suspicious.

And, of course, it is not for the purpose of intentionally torturing the prisoner that after his herring he is given neither hot water (and he never gets that here in any case) nor even plain, unboiled water. One has to understand the situation: The convoy staff is limited; some of them have to be on watch in the corridor; some are on duty on the platform; at the stations they clamber all over the car, under it, on top of it, to make sure that there aren't any holes in it. Others are kept busy cleaning guns, and then, of course, there has to be time for political indoctrination and their catechism on the articles of war. And the third shift is sleeping. They insist on their full eight hours—for, after all, the war is over. And then, to go carry water in pails—it has to be hauled a long way, too, and it's insulting: why should a Soviet soldier have to carry water like a donkey for enemies of the people? And there are also times when they spend half a day hauling the Stolypin cars way out from the station in order to reshuffle or recouple the cars (it will be farther away from prying eyes), and the result is that you can't get water even for your own Red Army mess. True, there is one way out. You can go dip up some water from the locomotive tender. It's yellow and murky, with some lubricating grease mixed in with it. But the zeks will drink it willingly. It doesn't really matter that much anyway, since it isn't as if they could see what they are drinking in the semidarkness of their compartment. They don't have their own window, and there isn't any light bulb there either, and what light they get comes from the corridor. And there's another thing too: it takes a long time to dole out that water. The zeks don't have their own mugs. Whoever did have one has had it taken away from him—so what it adds up to is that they have to be given the two government issue mugs to drink out of, and while they are drinking up you have to keep standing there and standing, and dipping it out and dipping it out some more and handing it to them. (Yes, and then, too, the prisoners argue about who's to drink first; they want the healthy prisoners to drink first, and only then those with tuberculosis, and last of all those with syphilis! Just as if it wasn't going to begin all over again in the next cell: first the healthy ones . . .)

But the convoy could have borne with all that, hauled the water, and doled it out, if only those pigs, after slurping up the water, didn't ask to go to the toilet. So here's the way it all works out: if you don't give them water for a day, then they don't ask to go to the toilet. Give them water once, and they go to the toilet once; take pity on them and give them water twice—and they go to the toilet twice. So it's pure and simple common sense: just don't give them anything to drink.

And it isn't that one is stingy about taking them to the toilet because one wants to be stingy about the use of the toilet itself, but because taking prisoners to the toilet is a responsible—even, one might say, a combat—operation: it takes a long, long time for one private first class and two privates. Two guards have to be stationed, one next to the toilet door, the other in the corridor on the opposite side (so that no one tries to escape in that direction), while the private first class has to push open and then shut the door to the compartment, first to admit the returning prisoner, and then to allow the next one out. The statutes permit letting out only one at a time, so that they don't try to escape and so that they can't start a rebellion. Therefore, the way it works out is that the one prisoner who has been let out to go to the toilet is holding up 30 others in his own compartment and 120 in the whole car, not to mention the convoy detail! And so the command resounds: "Come on there, come on! Get a move on, get a move on!" The private first class and the soldiers keep hurrying him all the way there and back and he hurries so fast that he stumbles, and it's as though they think he is going to steal that shithole from the state. (In 1949, traveling in a Stolypin car between Moscow and Kuibyshev, the one-legged German Schultz, having understood the Russian hurry-up by this time, jumped to the toilet and back on his one leg while the convoy kept laughing and ordering him to go faster. During one such trip, one of the convoy guards pushed him when he reached the platform at the end of the corridor, and Schultz fell down on the floor in front of the toilet. The convoy guard went into a rage and began to beat him, while Schultz, who couldn't get up because of the blows raining down on him, crawled and crept into the dirty toilet. The rest of the convoy roared with laughter.)

So that the prisoner shouldn't attempt to escape during the moment he was in the toilet, and also for a faster turnaround, the door to the toilet was not closed, and the convoy guard, watching the process from the platform of the car, could encourage it: "Come on, come on now! That's plenty, that's enough for you!" Sometimes the orders came before you even started: "All right, number one only!" And that meant that from the platform they'd prevent your doing anything else. And then, of course, you couldn't wash your hands. There was never enough water in the tank there, and there wasn't enough time either. The toilet was filthy. Quicker, quicker! And tracking back the liquid mess on his shoes, the prisoner would be shoved back into the compartment, where he would climb up over somebody's arms and shoulders, and then, from the top row, his dirty shoes would dangle to the middle row and drip.

When women were taken to the toilet, the statutes of the convoy service, and common sense as well, required that the toilet door be kept open, but not every convoy insisted on this and some allowed the door to be shut: Oh, all right, go ahead and shut it. (Later on one of the women was sent in to wash out the toilet, and the guard again had to stand right there beside her so that she didn't try to escape.)

And even at this fast tempo, visits to the toilet for 120 people would take more than two hours—more than a quarter of the entire shift for three convoy guards! And in spite of that, you still couldn't make them happy. In spite of that, some old sandpiper or other would begin to cry half an hour later and ask to go to the toilet, and, of course, he wouldn't be allowed to go, and then he would soil himself right there in the compartment, and once again that meant trouble for the private first class: the prisoner had to be forced to pick it up in his hands and carry it away.

So that was all there was to it: fewer trips to the toilet! And that meant less water, and less food too—because then they wouldn't complain of loose bowels and stink up the air; after all, how bad could it be? A man couldn't even breathe.

Less water! But they had to hand out the herring anyway, just as the regulations required! No water—that was a reasonable measure. No herring—that was a service crime.

No one, no one at all, ever set out to torture us on purpose! The convoy's actions were quite reasonable! But, like the ancient Christians, we sat there in the cage while they poured salt on our raw and bleeding tongues.

Also the prisoner-transport convoys did not often deliberately (though sometimes they did) mix the thieves—blatari—and nonpolitical offenders in with Article 58 politicals in the same compartment. But a particular situation existed: There were a great many prisoners and very few railroad cars and compartments, and time was always short, and so when was there time enough to sort them out? One of the four compartments was kept for women, and if the prisoners in the other three were to be sorted out on one basis or another, the most logical basis would be by destination so that it would be easier to unload them.

After all, was it because Pontius Pilate wanted to humiliate him that Christ was crucified between two thieves? It just happened to be crucifixion day that day—and there was only one Golgotha, and time was short. And so *he was numbered with the transgressors.*

■

But it is better still to stop as soon as possible being a *sucker*—that ridiculous greenhorn, that prey, that victim. *You will never return* to your former world. And the sooner you get used to being without your near and dear ones, and the sooner they get used to being without you, the better it will be. And the easier!

And keep as few things as possible, so that you don't have to fear for them. Don't take a suitcase for the convoy guard to crush at the door of the car (when there are twenty-five people in a compartment, what else could he figure out to do with it?). And don't wear new boots, and don't wear fashionable oxfords, and don't wear a woolen suit: these things are going to be stolen, taken away, swept aside, or switched, either in the Stolypin car, or in the Black Maria, or in the transit prison. Give them up without a struggle—because otherwise the humiliation will poison your heart. They will take them away from you in a fight, and trying to hold onto your property will only leave you with a bloodied mouth. All those brazen snouts, those jeering manners, those two-legged dregs, are repulsive to you. But by owning things and trembling about their fate aren't you forfeiting the rare opportunity of observing and understanding? And do you think that the freebooters, the pirates, the great privateers, painted in such lively colors by Kipling and Gumilyev, were not simply these same blatnye, these same thieves? That's just what they were. Fascinating in romantic literary portraits, why are they so repulsive to you here?

Understand them too! To them prison is *their native home.* No matter how fondly the government treats them, no matter how it softens their punishments, no matter how often it amnesties them, their inner destiny brings them back again and again. Was not the first word in the legislation of the Archipelago for them? In our country, the right to own private property was at one time just as effectively banished out *in freedom* too. (And then those who had banished it began to enjoy *possessing* things.) So why should it be tolerated in prison? You were too slow about it; you didn't eat up your fat bacon; you didn't share your sugar and tobacco with your friends. And so now the thieves empty your *bindle* in order to correct your moral error. Having given you their pitiful worn-out boots *in exchange* for your fashionable ones, their soiled coveralls in return for your sweater, they won't keep these things for long: your boots were merely something to lose and win back five times at cards, and they'll *hawk* your sweater the very next day for a liter of vodka and a round of salami. They, too, will have nothing left of them in one day's time—just like you. This is the principle of the second law of thermodynamics: all differences tend to level out, to disappear. . . .

Own nothing! Possess nothing! Buddha and Christ taught us this, and the Stoics and the Cynics. Greedy though we are, why can't we seem to grasp that simple teaching? Can't we understand that with property we destroy our soul?

Own only what you can always carry with you: know languages, know countries, know people. Let your memory be your travel bag. Use your memory! Use your memory! It is those bitter seeds alone which might sprout and grow someday.

Look around you—there are people around you. Maybe you will remember one of them all your life and later eat your heart out because you didn't make use of the opportunity to ask him questions. And the less you talk, the more you'll hear. Thin strands of human lives stretch from island to island of the Archipelago. They intertwine, touch one another for one night only in just such a clickety-clacking half-dark car as this and then separate once and for all. Put your ear to their quiet humming and the steady clickety-clack beneath the car. After all, it is the spinning wheel of life that is clicking and clacking away there.

Chapter 2

■

The Ports of the Archipelago

Spread out on a large table the enormous map of our Motherland. Indicate with fat black dots all provincial capitals, all railroad junctions, all transfer points where the railroad line ends in a river route, and where rivers bend and trails begin. What is this? Has the entire map been speckled by infectious flies? What it is, in fact, is precisely the majestic map of the ports of the Archipelago.

It is a rare zek who has not known from three to five transit prisons and camps; many remember a dozen or so, and *the sons of Gulag* can count up to fifty of them without the slightest difficulty. However, in memory they get all mixed up together because they are so similar: in the illiteracy of their convoys, in their inept roll calls based on *case files;* the long waiting under the beating sun or autumn drizzle; the still longer *body searches* that involve undressing completely; their haircuts with unsanitary clippers; their cold, slippery baths; their foul-smelling toilets; their damp and moldy corridors; their perpetually crowded, nearly always dark, wet cells; the warmth of human flesh flanking you on the floor or on the board bunks; the bumpy ridges of bunk heads knocked together from boards; the wet, almost liquid, bread; the gruel cooked from what seems to be silage.

And whoever has a good sharp memory and can recollect precisely what distinguishes one from another has no need to travel about the country because he knows its geography full well on the basis of transit prisons. Novosibirsk? I know it. I was there. Very strong barracks there, made from thick beams. Irkutsk? That was where the windows had been bricked over in several stages, you could see how they had been in Tsarist times, and each course had been laid separately, and only small slits had been left between them. Vologda? Yes, an ancient building with towers. The toilets right on top of one another, the wooden partitions rotten, and the ones above leaking down into the ones underneath. Usman? Of course. A lice-ridden stinking hole of a jail, an ancient vaulted structure. And they used to pack it so full that whenever they took prisoners out for a transport you couldn't imagine where they'd put them all—a line strung out halfway through the city.

You had better not tell such a connoisseur that you know some city without a transit prison. He will prove to you conclusively that there are no such cities, and he will be right. You must realize, dear sir, that every town has to have its own transit prison. After all, the courts operate everywhere. And how are prisoners to be delivered to camp? By air?

The transit prison at Kotlas was tenser and more aboveboard than many. Tenser because it opened the way to the whole Northeast of European Russia, and more aboveboard because it was already deep in the Archipelago, and there was no need to pretend to anybody. It was simply a piece of land divided into cages by fencing and the cages were all kept locked. Although it had been thickly settled by peasants when they were exiled in 1930 (one must realize that they had no roofs over their heads, but nobody is left to tell about it), even in 1938 there simply wasn't room for everyone in the frail one-story wooden barracks made of discarded end-pieces of lumber and covered with . . . tarpaulin. Under the wet autumn snow and in freezing temperatures people simply lived there on the ground, beneath the heavens. True, they weren't allowed to grow numb from inactivity. They were being counted endlessly; they were invigorated by check-ups (twenty thousand people were there at a time) or by sudden night searches. Later on tents were pitched in these cages, and log houses two stories high were built in some of them, but to reduce the construction costs sensibly, no floor was laid between the stories—six-story bunks with stepladders were simply built into the sides, up and down which prisoners on their last legs, on the verge of dying, had to clamber like sailors. In the winter of 1944–1945, when everyone had a roof over his head, there was room for only 7,500 prisoners, and fifty of them died every day, and the stretchers on which they were carried to the morgue were never idle.

The Knyazh-Pogost transit point (latitude 63 degrees north) consisted of shacks built on a swamp. Their pole frames were covered with torn tarpaulin tenting that didn't quite reach the ground. The double bunks inside them were also made of poles (from which, incidentally, the branches had been only partially removed), and the aisle was floored with poles also. During the day, the wet mud squelched through the flooring, and at night it froze. In various parts of the area, the walkways were laid on frail and shaky poles and here and there people whom weakness had made clumsy fell into the water and ooze. In 1938 they fed the prisoners in Knyazh-Pogost the same thing every day: a mash made of crushed grits and fish bones. This was convenient because there were no bowls, spoons, or forks at the transit prison and the prisoners had none of their own either. They were herded to the boiler by the dozens and the mash was ladled into their caps or the flaps of their jackets.

The imagination of writers is poverty-stricken in regard to the native life and customs of the Archipelago. When they want to write about the most reprehensible and disgraceful aspect of prison, they always accuse the *latrine bucket.* In literature the latrine bucket has become the symbol of prison, a symbol of humiliation, of stink. Oh, how frivolous can you be? Now was the latrine bucket really an evil for the prisoner? On the contrary, it was the most merciful device of the prison administration. The actual horror began the moment there was *no* latrine bucket in the cell.

In 1937 *there were no latrine buckets* in certain Siberian prisons, or there weren't enough. Not enough of them had been made ahead of time—Siberian industry hadn't caught up with the full scope of arrests. There were no latrine barrels in the warehouses for the newly created cells. There were old latrine buckets in the cells, but they were antiquated and small, and the only reasonable thing to do at that point was to remove them, since they amounted to nothing at all for the new reinforcements of prisoners. So if long ago the Minusinsk Prison had been built for five hundred people (Vladimir Ilyich Lenin was never inside it; he moved about freely), and there were now ten thousand in it, it meant that each latrine bucket ought to have become twenty times bigger. But it had not.

Our Russian pens write only in large letters. We have lived through so very much, and almost none of it has been described and called by its right name. But, for Western authors, peering through a microscope at the living cells of everyday life, shaking a test tube in the beam of a strong light, this is after all a whole epic, another ten volumes of *Remembrance of Things Past:* to describe the perturbation of a human soul placed in a cell filled to twenty times its capacity and with no latrine bucket, where prisoners are taken out to the toilet only once a day! Of course, much of the texture of this life is bound to be quite unknown to Western writers; they wouldn't realize that in this situation one solution was to urinate in your canvas hood, nor would they at all understand one prisoner's advice to another to urinate in his boot! And yet that advice was the fruit of wisdom derived from vast experience, and it didn't involve spoiling the boot and it didn't reduce the boot to the status of a pail. It meant that the boot had to be taken off, turned upside down, the boot tops turned inside out and up—and thus a cylindrical vessel was formed that constituted the much-needed container. But, at the same time, with what psychological twists and turns Western writers could enrich their literature (without in the least risking any banal repetition of the famous masters) if they only knew about the scheme of things in that same Minusinsk Prison: there was only one food bowl for every four prisoners; and one mug of drinking water per day was issued to each (there were enough mugs to go around). And it could happen that one of the four contrived to use the bowl allotted to him and three others to relieve his internal pressure and then refuse to hand over his daily water ration to wash it out before lunch. What a conflict! What a clash of four personalities! What nuances! (And I am not joking. That is when the rock bottom of a human being is revealed. It is only that Russian pens are too busy to write about it, and Russian eyes don't have time to read about it. I am not joking—because only doctors can tell us how months in such a cell will ruin a human being's health for his entire life, even if he wasn't shot under Yezhov and was rehabilitated under Khrushchev.)

■

No one would have believed the story of Erik Arvid Andersen had it not been for his unshorn locks—a miracle unique in all Gulag. And that foreign bearing of his. And his fluent English, German, and Swedish speech. According to him he was the son of a rich Swede—not merely a millionaire but a billionaire. (Well, let's assume he embellished a little.) On his mother's side he was a nephew of the British General Robertson, who commanded the British Zone in occupied Germany. A Swedish subject, he had served as a volunteer in the British Army and had actually landed in Normandy, and after the war he had become a Swedish career officer. However, the investigation of social systems remained one of his principal interests. His thirst for socialism was stronger than his attachment to his father's capital. He looked upon Soviet socialism with feelings of profound sympathy, and he had even had the chance to become convinced of its flourishing state with his own eyes when he had come to Moscow as a member of a Swedish military delegation. They had been given banquets and taken to country homes and there they had encountered no obstacles at all to establishing contact with ordinary Soviet citizens—with pretty actresses who for some reason never had to rush off to work and who willingly spent time with them, even tête-à-tête. And thus convinced once and for all of the triumph of our social system, Erik on his return to the West wrote articles in the press defending and praising Soviet socialism. And this proved to be his undoing. In those very years, in 1947 and 1948, they were roping in from all sorts of nooks and crannies progressive young Westerners prepared to renounce the West publicly (and it appeared that if they could only have collected another dozen or so the West would shudder and collapse). Erik's newspaper articles caused him to be regarded as suitable for this category. At the time he was serving in West Berlin, and he had left his wife in Sweden. And out of pardonable male weakness he used to visit an unmarried German girl in East Berlin. And it was there that he was bound and gagged one night (and is not this the significance of the proverb which says: "He went to see his cousin, and he ended up in prison"? This had probably been going on for a long time, and he wasn't the first). They took him to Moscow, where Gromyko, who had once dined at his father's home in Stockholm and who knew the son also, not only returned the hospitality but proposed to the young man that he renounce publicly both capitalism and his own father. And in return he was promised full and complete capitalist maintenance to the end of his days here in our country. But to Gromyko's surprise, although Erik would not have suffered any material loss, he became indignant and uttered some very insulting words. Since they didn't believe in his strength of mind, they locked him up in a dacha outside Moscow, fed him like a prince in a fairy tale (sometimes they used "awful methods of repression" on him: they refused to accept his orders for the following day's menu and instead of the spring chicken he ordered they simply brought him a steak, just like that), surrounded him with the works of Marx-Engels-Lenin-Stalin, and waited a year for him to be re-educated. To their surprise it didn't happen. At that point they quartered with him a former lieutenant general who had already served two years in Norilsk. They probably calculated that by relating the horrors of camp the lieutenant general would persuade Erik to surrender. But either he carried out that assignment badly or else he didn't want to carry it out. After ten months of their being imprisoned together, the only thing he had taught Erik was broken Russian, and he had bolstered Erik's growing repugnance for the bluecaps. In the summer of 1950 they once more summoned Erik to Vyshinsky and he once more refused (in so doing, he made existence contingent on consciousness, thereby violating all the Marxist-Leninist rules!). And then Abakumov himself read Erik the decree: twenty years in prison (what for???). They themselves already regretted having gotten mixed up with this ignoramus, but at the same time they couldn't release him and let him go back to the West. And so they transported him in a separate compartment, and it was there that he had heard the story of the Moscow girl through the partition and seen through the train window in the dawn light the rotting straw-thatched roofs of the age-old Russia of Ryazan.

Those two years had very strongly confirmed him in his loyalty to the West. He believed blindly in the West. He did not want to recognize its weaknesses. He considered Western armies unbeatable and Western political leaders faultless. He refused to believe us when we told him that during the period of his imprisonment Stalin had begun a blockade of Berlin and had gotten away with it perfectly well. Erik's milky neck and creamy cheeks blushed with indignation whenever we ridiculed Churchill and Roosevelt. And he was also certain that the West would not countenance his, Erik's, imprisonment; that on the basis of information from the Kuibyshev Transit Prison the Western intelligence services would immediately learn that Erik had not drowned in the Spree River but had been imprisoned in the Soviet Union—and either he would be ransomed or someone would be exchanged for him. (This faith of his in the individual importance of *his own* fate among other prisoners' fates was reminiscent of our own well-intentioned orthodox Soviet Communists.) Notwithstanding our heated arguments, he invited my friend and me to Stockholm whenever we could come. ("Everyone knows us there," he said with a tired smile. "My father virtually maintains the Swedish King's whole court.") For the time being, however, the son of the billionaire had nothing to dry himself with, and I presented him with an extra tattered towel as a gift. And soon they took him away on a prisoner transport.

■

Human nature, if it changes at all, changes not much faster than the geological face of the earth. And the very same sensations of curiosity, relish, and sizing up which slave-traders felt at the slave-girl markets twenty-five centuries ago of course possessed the Gulag bigwigs in the Usman Prison in 1947, when they, a couple of dozen men in MVD uniform, sat at several desks covered with sheets (this was for their self-importance, since it would have seemed awkward otherwise), and all the women prisoners were made to undress in the box next door and to walk in front of them bare-footed and bare-skinned, turn around, stop, and answer questions. "Drop your hands," they ordered those who had adopted the defensive pose of classic sculpture. (After all, these officers were very seriously selecting bedmates for themselves and their colleagues.)

And so it was that for the new prisoner various manifestations foreshadowed the camp battle of the morrow and cast their pall over the innocent spiritual joys of the transit prison.

Chapter 3

■

The Slave Caravans

It was painful to travel in a Stolypin, unbearable in a Black Maria, and the transit prison would soon wear you down—and it might just be better to skip the whole lot and go straight to camp in the red cattle cars.

As always, the interests of the state and the interests of the individual coincided here. It was also to the state's advantage to dispatch sentenced prisoners straight to the camps by direct routing and thus avoid overloading the city trunk-line railroads, automotive transport, and transit-camp personnel. They had long since grasped this fact in Gulag, and it had been taken to heart: witness the caravans of *red cows* (red cattle cars), the caravans of barges, and, where there were no rails and no water, the caravans on foot (after all, prisoners could not be allowed to exploit the labor of horses and camels).

The red trains were always a help when the courts in some particular place were working swiftly or the transit facilities were overcrowded. It was possible in this way to dispatch a large number of prisoners in one batch. That is how the millions of peasants were transported in 1929–1931. That is how they exiled Leningrad from Leningrad. That is how they populated the Kolyma in the thirties: every day Moscow, the capital of our country, belched out one such train to Sovetskaya Gavan, to Vanino Port. And each provincial capital also sent off red trainloads, but not on a daily schedule. That is how they removed the Volga German Republic to Kazakhstan in 1941, and later all the rest of the exiled nations were sent off in the same way. In 1945 Russia's prodigal sons and daughters were sent from Germany, from Czechoslovakia, from Austria, and simply from western border areas—whoever had gotten there on his own—in such trains as these. In 1949 that is how they collected the 58's in Special Camps.

The Stolypins follow routine railroad schedules. And the red trains travel on imposing waybills, signed by important Gulag generals. The Stolypins cannot go to an empty site, to "nowhere"; their destination must always be a station, even if it's in some nasty little two-bit town with some preliminary detention cells in an attic. But the red trains can go into emptiness: and wherever one does go, there immediately rises right next to it, out of the sea of the steppe or the sea of the taiga, a new island of the Archipelago.

Not every red cattle car is ready as is to transport prisoners. First it has to be prepared. But not in the sense some of our readers might expect: that the coal or lime it carried before it was assigned to carry people has to be swept out and the car cleaned—that isn't always done. Nor in the sense that it needs to be calked and have a stove installed if it is winter. Here is what was involved in preparing a red cattle car for prisoners: The floors, walls, and ceilings had to be tested for strength and checked for holes or faults. Their small windows had to be barred. A hole had to be cut in the floor to serve as a drain, and specially protected by sheet iron firmly nailed down all around it. The necessary number of platforms on which convoy guards would stand with machine guns had to be evenly distributed throughout the train, and if there were too few, more had to be built. Access to the roofs of the cars had to be provided. Sites for searchlights had to be selected and supplied with uninterrupted electric power. Long-handled wooden mallets had to be procured. A passenger car had to be hooked on for the staff, and if there wasn't one, then instead heated freight cars had to be prepared for the chief of convoy, the Security officer, and the convoy. Kitchens had to be built—for the convoy and for the prisoners. And only after all this had been done was it all right to walk along the cattle cars and chalk on the sides: "Special Equipment" or "Perishable Goods." (In her chapter, "The Seventh Car," Yevgeniya Ginzburg described a transport of red cars very vividly, and her description largely obviates the necessity of presenting details here.)

The preparation of the train has been completed—and ahead lies the complicated combat operation of *loading* the prisoners into the cars. At this point there are two important and obligatory *objectives:*

- to conceal the loading from ordinary citizens
- to terrorize the prisoners

To conceal the loading from the local population was necessary because approximately a thousand people were being loaded on the train simultaneously (at least twenty-five cars), and this wasn't your little group from a Stolypin that could be led right past the townspeople. Everyone knew, of course, that arrests were being made every day and every hour, but no one was to be horrified by the sight of large numbers of them *together*. In Orel in 1938 you could hardly hide the fact that there was no home in the city where there hadn't been arrests, and weeping women in their peasant carts blocked the square in front of the Orel Prison just as in Surikov's painting *The Execution of the Streltsy*. (Oh, who one day will paint this latter-day tragedy for us? But no one will. It's not fashionable, not fashionable. . . .) But you don't need to show our Soviet people an entire trainload of them collected in one day. (And in Orel that year there were.) And young people mustn't see it either—for young people are our future. Therefore it was done only at night—and every night, too, each and every night, and that was the way it went for several months. The black line of prisoners to be transported was driven from the prison to the station on foot. (Meanwhile the Black Marias were busy making new arrests.) True, the women realized, the women somehow found out, and at night they came to the station from all over the city and kept watch over the trains on the siding. They ran along the cars, tripping over the ties and rails, and shouting at every car: "Is So-and-so in there?" "Is So-and-so in there?" And they ran on to the next one, and others ran up to this one: "Is So-and-so in there?" And suddenly an answer would come from the sealed car: "I'm in here. I'm here!" Or else: "Keep looking for him. He's in another car." Or else: "Women! Listen! My wife is somewhere out there, near the station. Run and tell her."

These scenes, unworthy of our contemporary world, testify only to the then inept organization of train embarkations. The mistakes were noted, and after a certain night the trains were surrounded in depth by cordons of snarling and barking police dogs.

However, although the convoy had no use for the superfluous light of the sun by day, on the other hand they made use of suns by night —the searchlights. They were more efficient since they could be concentrated on the necessary area, where the prisoners were seated on the earth in a frightened pack awaiting the command: "Next unit of five —stand up! To the car—on the run!" (Only on the run, so as not to have time to look around, to think things over, to run as though chased by the dogs, afraid of nothing so much as falling down.) On that uneven path. Up the loading ramp, scrambling. And clear, hostile searchlight beams not only provided light but were an important theatrical element in terrorizing the prisoners, along with yells, threats, gunstock blows on those who fell behind, and the order: "Sit down." (And sometimes, as in the station square of that same Orel: "Down on your knees." And like some new breed of believers at prayer, the whole thousand would get down on their knees.) Along with that running to the car, quite unnecessary except for intimidation—for which it was very important. Along with the enraged barking of the dogs. Along with the leveled gun barrels (rifles or automatic pistols, depending on the decade). And the main thing was to undermine, to crush the prisoner's will power so he wouldn't think of trying to escape, so that for a long time he wouldn't notice his new advantage: the fact that he had exchanged a stone-walled prison for a railroad car with thin plank walls.

But in order to load one thousand prisoners into railroad cars at night so precisely, the prison had to start jerking them out of their cells and processing them for transport the morning before, and the convoy had to spend the entire day on a long-drawn-out and strict procedure of checking them in while still in prison and then holding those who'd been checked in for long hours, not, of course, in the cells by now, but in the courtyard, on the ground, so as not to mix them up with the prisoners still belonging in the prison. Thus for the prisoner the loading at night was only a relief after a whole day of torment.

Besides the ordinary counts, verifications, hair clipping, clothing roasting, and baths, the core of the preparation for the prisoner transport was general *frisking*. This search was carried out not by the prison but by the convoy receiving the prisoners. The convoy was expected, in accordance with the directives regarding the red transports and in accordance with their own operational requirements, to carry out this search so that the prisoners would not be left in possession of anything that might help them to escape; to take away: everything that could saw or cut; all powders (tooth powder, sugar, salt, tobacco, tea) so they could not be used to blind the convoy; all string, cord, twine, belts, and straps because they could all be used in escaping (and that meant all kinds of straps! and so they cut off the straps which held up the artificial limb of a one-legged man—and the cripple had to carry his artificial leg on his shoulder and hop with the help of those on either side of him).

A search begins. (Kuibyshev, summer of 1949.) Naked prisoners approach, carrying their possessions and the clothes they've taken off. A mass of armed soldiers surrounds them. It doesn't look as though they are going to be led to a prisoner transport but as though they are going to be shot immediately or put to death in a gas chamber—and in that mood a human being ceases to concern himself with his possessions. The convoy does everything with intentional brusqueness, rudely, sharply, not speaking one word in an ordinary human voice. After all, the purpose is to terrify and dishearten. Suitcases are shaken apart, and things fall all over the floor and are then stacked up in separate piles. Cigarette cases, billfolds, and other pitiful "valuables" are all taken away and thrown without any identifying marks into a *barrel* that is standing nearby. (And, for some reason, the fact that this particular receptacle isn't a safe, or a trunk, or a box, but a barrel particularly depresses the naked prisoners there, and it seems so terribly futile to protest.) The naked prisoner has all he can do simply to snatch up his well-searched rags from the floor and knot them together or tie them up in a blanket. Felt boots? You can check them, throw them over there, sign for them on the list! (You aren't the one who gets the receipt, but *you* are the one who signs for having surrendered them, certifying that you threw them onto the pile!) And when at dusk the last truck leaves the prison yard with the prisoners, they see the convoy guards rushing to grab the best leather suitcases from the pile and select the best cigarette cases from the barrel. And after them, the jailers scurry for their booty, too, and last of all the transit prison *trusties.* . . .

They don't heat the car, they don't protect the other prisoners from the thieves, they don't give you enough to drink, and they don't give you enough to eat—but on the other hand they don't let you sleep either. During the day the convoy can see the whole train very clearly and the tracks behind them, and can be sure that no one has jumped out the side or slipped down on the rails. But at night vigilance possesses them. With long-handled wooden mallets (the standard Gulag equipment) they knock resoundingly on every board of the car at every stop: maybe someone has sawed through it. And at certain stops the door of the car is thrown open. The light of the lantern or the beam of the searchlight: "Checkup!" And this means: Get on your feet and be ready to go where they tell you—everyone run to the left or to the right. The convoy guards jump inside with their mallets (others have ranged themselves in a semicircle outside with automatic pistols), and they point: to the left! That means that those on the left are in place and those on the right must get over there on the jump like fleas hopping over each other and landing where they can. And whoever isn't nimble, whoever gets caught daydreaming, gets whacked on the ribs and back with the mallets to give him more energy. And by this time the convoy jackboots are already trampling your pauper's pallet and all your lousy *duds* are being thrown in every direction and everywhere there are lights and hammering: Have you sawed through any place? No. Then the convoy guards stand in the middle and begin to shift you from left to right, counting: "First . . . second . . . third." It would be quite enough to count simply with a wave of the finger, but if that were done, it wouldn't be terrifying, and so it is more vivid, less subject to error, more energetic and faster, to beat out that count with the same mallet on your ribs, shoulders, heads, wherever it happens to land. They have counted up to forty. So now they will go about their tossing, lighting up, and hammering at the other end of the car. It's all over finally and the car is locked up.

The red train differs from other long-distance trains in that those who have embarked on it do not know whether or not they will disembark. When they unloaded a trainload from the Leningrad prisons (1942) in Solikamsk, the entire embankment was covered with corpses, and only a few got there alive. In the winters of 1944–1945 and 1945–1946 in the village of Zheleznodorozhny (Knyazh-Pogost), as in all the main rail junctions in the North, the prisoner trains from liberated territories (the Baltic states, Poland, Germany) arrived with one or two carloads of corpses tacked on behind. That meant that en route they had carefully taken the corpses out of the cars that contained the living passengers and put them in the dead cars. But not always. There were many occasions when they found out who was still alive and who was dead only when they opened up the car after arriving at the Sukhobezvodnaya (Unzhlag) Station. Those who didn't come out were dead.

It was terrifying and deadly to travel this way in winter because the convoy, with all its bother about security, wasn't able to haul coal for twenty-five stoves. But it wasn't so cushy to travel this way in hot weather either. Two of the four tiny windows were tightly sealed and the car roof would overheat and the convoy wasn't about to exert itself in hauling water for a thousand prisoners—after all, they couldn't even manage to give just one Stolypin car enough to drink. The prisoners considered April and September the best months for transports. But even the best of seasons was too short if the train was en route for *three months*. (Leningrad to Vladivostok in 1935.)

No, damn that red cattle car train too, even though it did carry the prisoners straight to their destination without changing trains. Anyone who has ever been in one will never forget it. Just as well get to camp sooner! Just as well arrive sooner.

But it was not at all unusual for the red trains to arrive nowhere, and the end of the journey often marked the opening day of a *new* camp. They might simply stop somewhere in the taiga under the northern lights and nail to a fir tree a sign reading: "FIRST SEPARATE CAMP." And there they would chew on dried fish for a week and try to mix their flour with snow.

■

But here I note that I am again beginning to repeat myself. And this will be boring to write, and boring to read, because the reader already knows everything that is going to happen ahead of time.

■

Shut your eyes, reader. Do you hear the thundering of wheels? Those are the Stolypin cars rolling on and on. Those are the red cows rolling. Every minute of the day. And every day of the year. And you can hear the water gurgling—those are prisoners' barges moving on and on. And the motors of the Black Marias roar. They are arresting someone all the time, cramming him in somewhere, moving him about. And what is that hum you hear? The overcrowded cells of the transit prisons. And that cry? The complaints of those who have been plundered, raped, beaten to within an inch of their lives.

We have reviewed and considered all the methods of delivering prisoners, and we have found that they are all . . . *worse.* We have examined the transit prisons, but we have not found any that were good. And even the last human hope that there is something better ahead, that it will be better in camp, is a false hope.

In camp it will be . . . worse.

Chapter 4

■

From Island to Island

As the title suggests, this chapter charts the transporting of lone prisoners from one camp to another. It includes the following brief autobiographical vignette.

While we—I, my codefendant, and others of our age—had been fighting for four years at the front, a whole new generation had grown up here in the rear. And had it been very long since we ourselves had tramped the parquet floors of the university corridors, considering ourselves the youngest and most intelligent in the whole country and, for that matter, on earth? And then suddenly pale youths crossed the tile floors of the prison cells to approach us haughtily, and we learned with astonishment that we were no longer the youngest and most intelligent—they were. But I didn't take offense at this; at that point I was already happy to move over a bit to make room. I knew so very well their passion for arguing with everyone, for finding out everything, I understood their pride in having chosen a worthy lot and in not regretting it. It gave me gooseflesh to hear the rustle of the prison halos hovering over those self-enamored and intelligent little faces.

One month earlier, in another Butyrki cell, a semihospital cell, I had just stepped into the aisle and had still not seen any empty place for myself—when, approaching in a way that hinted at a verbal dispute, even at an entreaty to enter into one, came a pale, yellowish youth, with a Jewish tenderness of face, wrapped, despite the summer, in a threadbare soldier's overcoat shot full of holes: he was chilled. His name was Boris Gammerov. He began to question me; the conversation rolled along: on one hand, our biographies, on the other, politics. I don't remember why, but I recalled one of the prayers of the late President Roosevelt, which had been published in our newspapers, and I expressed what seemed to me a self-evident evaluation of it:

"Well, that's hypocrisy, of course."

And suddenly the young man's yellowish brows trembled, his pale lips pursed, he seemed to draw himself up, and he asked me: "Why? Why do you not admit the possibility that a political leader might sincerely believe in God?"

And that is all that was said! But what a direction the attack *had* come from! To hear such words from someone born in 1923? I could have replied to him very firmly, but prison had already undermined my certainty, and the principal thing was that some kind of clean, pure feeling does live within us, existing apart from all our convictions, and right then it dawned upon me that I had not spoken out of conviction but because the idea had been implanted in me from outside. And because of this I was unable to reply to him, and I merely asked him: "Do you believe in God?"

"Of course," he answered tranquilly.

PART III

The Destructive-Labor Camps

■

Chapter 1

■

The Fingers of Aurora

Rosy-fingered Eos, so often mentioned in Homer and called Aurora by the Romans, caressed, too, with those fingers the first early morning of the Archipelago.

When our compatriots heard via the BBC that M. Mihajlov claimed to have discovered that concentration camps had existed in our country as far back as 1921, many of us (and many in the West too) were astonished: That early really? Even in 1921?

Of course not! Of course Mihajlov was in error. In 1921, in fact, concentration camps were already in full flower (already even *coming to an end*).

And how could it have been otherwise? Let us pause to ponder.

Didn't Marx and Engels teach that the old bourgeois machinery of compulsion had to be broken up, and *a new one created* immediately in its place? And included in the machinery of compulsion were: the army (we are not surprised that the Red Army was created at the beginning of 1918); the police (the militia was inaugurated even sooner than the army); the courts (from November 22, 1917); and the prisons. How, in establishing the dictatorship of the proletariat, could they delay with a new type of prison?

That is to say that it was altogether impermissible to delay in the matter of prisons, whether old or new. In the first months after the October Revolution Lenin was already demanding "the most decisive, draconic measures to tighten up discipline." And are draconic measures possible—without prison?

What new could the proletarian state contribute here? Lenin was feeling out new paths. In December, 1917, he suggested for consideration the following assortment of punishments: "confiscation of all property . . . confinement in prison, dispatch to the front and forced labor for all who disobey the existing law." Thus we can observe that the leading idea of the Archipelago—*forced labor*—had been advanced in the first month after the October Revolution.

And even while sitting peacefully among the fragrant hay mowings of Razliv and listening to the buzzing bumblebees, Lenin could not help but ponder the future penal system. Even then he had worked things out and reassured us: "The suppression of the minority of exploiters by the majority of the hired slaves of yesterday is a matter so comparatively easy, simple and natural, that it is going to cost much less in blood . . . will be much cheaper for humanity" than the preceding suppression of the majority by the minority.

According to the estimates of émigré Professor of Statistics Kurganov, this "comparatively easy" internal repression cost us, from the beginning of the October Revolution up to 1959, a total of . . . sixty-six million—66,000,000—lives. We, of course, cannot vouch for his figure, but we have none other that is official. And just as soon as the official figure is issued the specialists can make the necessary critical comparisons.

It is interesting to compare other figures. How large was the total staff of the *central* apparatus of the terrifying Tsarist Third Department, which runs like a strand through all the great Russian literature? At the time of its creation it had sixteen persons, and at its height it had forty-five. A ridiculously small number for even the remotest Cheka provincial headquarters in the country. Or, how many political prisoners did the February Revolution find in the Tsarist "Prison of the Peoples"? All these figures do exist somewhere. In all probability there were more than a hundred such prisoners in the Kresty Prison alone, and several hundred returned from Siberian exile and hard labor, and how many more were languishing in the prison of every provincial capital! But it is interesting to know—exactly how many. Here is a figure for Tambov, taken from the fiery local papers. The February Revolution, which opened wide the doors of the Tambov Prison, found there political prisoners in the number of . . . seven (7) persons. And there were more than forty provinces. (It is superfluous to recall that from February to July, 1917, there were no political arrests, and after July the number imprisoned could be counted on one's fingers.)

Here, however, was the trouble: The first Soviet government was a coalition government, and a portion of the people's commissariats had to be allotted, like it or not, to the Left SR's, including, unhappily, the

People's Commissariat of Justice, which fell to them. Guided by rotten petty bourgeois concepts of freedom, this People's Commissariat of Justice brought the penal system to the verge of ruin. The sentences turned out to be too light, and they made hardly any use at all of the progressive principle of forced labor. In February, 1918, the Chairman of the Council of People's Commissars, Comrade Lenin, demanded that the number of places of imprisonment be increased and that repression of criminals be intensified, and in May, already going over to concrete guidance, he gave instructions that the sentence for bribery must be *not less than* ten years of prison and ten years of forced labor *in addition,* i.e., a total of twenty years. This scale might seem pessimistic at first: would forced labor really still be necessary after twenty years? But we know that forced labor turned out to be a very long-lived measure, and that even after fifty years it would still be extremely popular.

The reader has already read the term concentration camp—"kont-slager"—several times in the sentences of the tribunals and concluded, perhaps, that we were guilty of an error, of making careless use of terminology subsequently developed? No, this is not the case.

In August, 1918, Vladimir Ilyich Lenin wrote in a telegram to Yevgeniya Bosh and to the Penza Provincial Executive Committee (they were unable to cope with a peasant revolt): "Lock up all the doubtful ones [not "guilty," mind you, but *doubtful*—A.S.] in *a concentration camp* outside the city." (And in addition "carry out merciless mass terror"—this was before the decree.)

Only on September 5, 1918, ten days after this telegram, was the Decree on the Red Terror published. In addition to the instructions on mass executions, it stated in particular: "Secure the Soviet Republic against its class enemies by isolating them in *concentration camps.*"

So that is *where* this term—*concentration camps*—was discovered and immediately seized upon and confirmed—one of the principal terms of the twentieth century, and it was to have a big international future! And this is *when* it was born—in August and September, 1918. The word itself had already been used during World War I, but in relation to POW's and undesirable foreigners. But here in 1918 it was for the first time applied to the citizens of one's own country.

There is no one now to tell us about most of those first concentration camps. And only from the last testimony of those few surviving first concentration camp inmates can we glean and preserve a little bit.

At that time the authorities used to love to set up their concentration camps in former monasteries: they were enclosed by strong walls, had good solid buildings, and they were empty. (After all, monks are not human beings and could be tossed out at will.)

Here is how they fed them in a camp in 1921: half a pound of bread (plus another half-pound for those who fulfilled the norm), hot water for tea morning and evening, and, during the day, a ladle of gruel (with several dozen grains and some potato peelings in it).

Camp life was embellished on the one hand by the denunciations of provocateurs (and arrests on the basis of the denunciations), and on the other by a dramatics and glee club. They gave concerts for the people of Ryazan in the hall of the former noblemen's assembly, and the deprivees' brass band played in the city park. The deprivees got better and better acquainted with and more friendly with the inhabitants, and this became intolerable—and at that point they began to send the so-called "war prisoners" to the Northern Special Purpose Camps.

The lesson of the instability and laxity in these concentration camps lay in their being surrounded by civilian life. And that was why the special northern camps were required. (Concentration camps were abolished in 1922.)

This whole dawn of the camps deserves to have its spectrum examined much more closely. And glory to him who can—for all I have in my own hands is crumbs.

At the end of the Civil War the two labor armies created by Trotsky had to be dissolved because of the grumbling of the soldiers kept in them. And by this token, the role of camps in the structure of the R.S.F.S.R. not only did not diminish but intensified. By the end of 1920 in the R.S.F.S.R. there were eighty-four camps in forty-three provinces. If one believes the official statistics (even though classified), 25,336 persons and in addition 24,400 "prisoners of war of the Civil War" were held in them at this time. Both figures, particularly the second, seem to be understated. However, if one takes into consideration that by *unloading prisons,* sinking barges, and other types of mass annihilation the figure had often begun with zero and been reduced to zero over and over, then perhaps these figures are accurate.

On the threshold of the "reconstruction period" (meaning from 1927) "the role of camps *was growing* [Now just what was one to think? Now after all the victories?]—against the most dangerous, hostile elements, wreckers, the kulaks, counterrevolutionary propaganda."

And so it was that the Archipelago was not about to disappear into the depths of the sea! The Archipelago would live!

Chapter 2

■

The Archipelago Rises from the Sea

On the White Sea, where the nights are white for half a year at a time, Bolshoi Solovetsky Island lifts its white churches from the water within the ring of its bouldered kremlin walls, rusty-red from the lichens which have struck root there—and the grayish-white Solovetsky seagulls hover continually over the kremlin and screech.

"In all this brightness it is as if there were no sin present. . . . It is as if nature here had not yet matured to the point of sin" is how the writer Prishvin perceived the Solovetsky Islands.

Without us these isles rose from the sea; without us they acquired a couple of hundred lakes replete with fish; without our help they were settled by capercaillies, hares, and deer, while foxes, wolves, and other beasts of prey never ever appeared there.

The glaciers came and went, the granite boulders littered the shores of the lakes; the lakes froze during the Solovetsky winter nights, the sea howled under the wind and was covered with an icy sludge and in places froze; the northern lights blazed across half the sky; and it grew bright once again and warm once again, and the fir trees grew and thickened, and the birds cackled and called, and the young deer trumpeted—and the planet circled through all world history, and kingdoms fell and rose, and here there were still no beasts of prey and no human being.

Sometimes the men of Novgorod landed there and they counted the islands as belonging to their Obonezhskaya "pyatina." Karelians lived there too. Half a hundred years after the Battle of Kulikovo Field and half a thousand years before the GPU, the monks Savvaty and German crossed the mother-of-pearl sea in a tiny boat and came to look on this island without a beast of prey as sacred. The Solovetsky Monastery began with them. . . .

Military thought: It was impermissible for some sort of feckless monks just to live on just an island. The island was on the borders of the Great Empire, and, consequently, it was required to fight with the Swedes, the Danes, the English, and, consequently, it was required to build a fortress with walls eight yards thick and to raise up eight towers on the walls, and to make narrow embrasures in them, and to provide for a vigilant watch from the cathedral bell tower.

Prison thought: How glorious—good stone walls standing on a separate island! What a good place to confine important criminals—and with someone already there to provide guard. We won't interfere with their saving their souls: just guard our prisoners!

Had Savvaty thought about that when he landed on the holy island?

They imprisoned church heretics here and political heretics as well. . . .

But when power passed into the hands of the workers, what was to be done with these malevolent parasitical monks? They sent Commissars, socially tried-and-true leaders, and they proclaimed the monastery a state farm, and ordered the monks to pray less and to work harder for the benefit of the workers and peasants. The monks worked, and their herring, which was astonishing in its flavor, and which they had been able to catch because of their special knowledge of where and when to cast nets, was shipped off to Moscow to be used for the Kremlin tables.

However, the abundance of valuables concentrated in the monastery, especially in the sacristy, troubled some of the leaders and overseers who had arrived there: instead of passing into the workers' hands (i.e., *their own*), these valuables lay there as a dead religious burden. And at that point, contradicting to a certain degree the Criminal Code but corresponding in a very genuine way with the general spirit of expropriation of the property of nonworkers, the monastery was set on fire (on May 25, 1923); the buildings were damaged, and many valuables disappeared from the sacristy; and, the principal thing, all the inventory records burned up, and it was quite impossible to determine how much and exactly what had disappeared.

But without even conducting an investigation, what is our revolutionary sense of justice (sense of smell) going to hint to us? Who if not

The Solovetsky kremlin

The Herring Gates

the black gang of monks themselves could have been to blame for the arson of the monastery wealth? So throw them out onto the mainland, and concentrate all the Northern Special Purpose Camps on the Solovetsky Islands! The eighty-year-old and even hundred-year-old monks begged on their knees to be allowed to die on the "holy soil," but they were all thrown out with proletarian ruthlessness except for the most necessary among them: the artels of fishermen, the cattle specialists on Muksalma; Father Methodius, the cabbage salter; Father Samson, the foundry specialist; yes, and other such useful fathers as well. (They were allotted a corner of the kremlin separate from the camp, with their own exit—the Herring Gates. They were christened *a Workers' Commune,* but out of condescension for their total stupefaction they were left for their prayers the Onufriyev Church at the cemetery.)

And that is how one of the favorite sayings constantly repeated by the prisoners came true: A holy place is never empty. The chimes of bells fell silent, the icon lamps and the candle stands fell dark, the liturgies and the vespers resounded no longer; psalms were no longer chanted around the clock, the iconostases were wrecked (though they left the one in the Cathedral of the Transfiguration)—but on the other hand courageous Chekists, in overcoats with superlong flaps which reached all the way down to the heels, and particularly distinctive black Solovetsky cuffs and lapels and black-edged service caps without stars, arrived there in June, 1923, to set up a model camp, a model of severity, the pride of the workers' and peasants' Republic.

■

The magazine *The Solovetsky Islands* (1930, No. 1) declared it was the "dream of many prisoners" to receive standard clothing.

Only the children's colony was completely dressed. And the women, for example, were given neither underwear nor stockings nor even kerchiefs to cover their heads. They had grabbed the old biddy in a summer dress; she just had to go on wearing it the whole Arctic winter. Because of this many prisoners remained in their company quarters in nothing but their underwear, and no one chased them out to work.

Government-issue clothing was so precious that no one on Solovki found the following scene either astonishing or weird: In the middle of winter a prisoner undressed and took his shoes off near the kremlin, then carefully handed in his uniform and ran naked for two hundred yards to another group of people, where he was given clothes to put on. This meant that he was being transferred from the kremlin administration to the administration of the Filimonovo Branch Railroad—but if he had been transferred wearing clothes, those taking him over might not have returned the clothes or have cheated by switching them.

And here is another winter scene—the same customs, though the reason is different. The Medical Section infirmary is found to be infectious, and orders are issued to scald it down and wash it out with boiling water. But where are the sick prisoners to be put in the meanwhile? All the kremlin accommodations are overcrowded, the density of the population of the Solovetsky Archipelago is greater than that of Belgium—so what must it be like in the Solovetsky kremlin? And therefore all the sick prisoners are carried out on blankets and laid out on the snow for three hours. When they have washed out the infirmary, they haul the patients in again.

Our newcomer knows nothing of the Second World War or of Buchenwald! What he sees is this: *The squad leaders* drive workers out with long clubs—with *staves*. He sees that sledges and carts are drawn not by horses but by men (several harnessed into one rig)—and there is also another word, *VRIDLO* (an acronym meaning a "Temporary Replacement for a Horse").

And from other Solovetsky inhabitants he learns things more awful than his eyes perceive. People pronounce the fatal word *"Sekirka"* to him. This means Sekirnaya Hill. Punishment cells were set up in the two-story cathedral there. And here is how they kept prisoners in the punishment cells: Poles the thickness of an arm were set from wall to wall and prisoners were ordered to sit on these poles all day. (At night they lay on the floor, one on top of another, because it was overcrowded.) The height of the poles was set so that one's feet could not reach the ground. And it was not so easy to keep balance. In fact, the prisoner spent the entire day just trying to maintain his perch. If he fell, the jailers jumped in and beat him. Or else they took him outside to a flight of stairs consisting of 365 steep steps (from the cathedral to the lake, just as the monks had built it). They tied the person lengthwise to a "balan" (a beam), for the added weight, and rolled him down (and there wasn't even one landing, and the steps were so steep that the log with the human being on it would go all the way down without stopping).

Well, after all, for *poles* you didn't have to go to Sekirka. They were right there in the kremlin punishment block, which was always overcrowded. Or they might put the prisoners on a sharp-edged boulder on which one could not stay long either. Or, in summer, "on the

Church of the Beheading on Sekirnaya Hill

stump," which meant naked among the mosquitoes. But in that event one had to keep an eye on the culprit; whereas if he was bound naked to a tree, the mosquitoes would look after things themselves. And then they could put whole companies out in the snow for disobedience. Or they might drive a person into the marsh muck up to his neck and keep him there. And then there was another way: to hitch up a horse in empty shafts and fasten the culprit's legs to the shafts; then the guard mounted the horse and kept on driving the horse through a forest cut until the groans and the cries from behind simply came to an end.

■

So all the scares were just a joke! But a shout comes in broad daylight in the kremlin yard where prisoners are crowded as thick as on Nevsky Prospekt: "Make way! Make way!" And three foppish young men with the faces of junkies (the lead man drives back the crowd of prisoners not with a club but with a riding crop) drag along swiftly by the shoulders a prisoner with limp arms and legs dressed only in his underwear. His face is horrible—*flowing* like liquid! They drag him off *beneath the bell tower.* They squeeze him through that little door and shoot him in the back of the head—steep stairs lead down inside, and he tumbles down them, and they can pile up as many as seven or eight men in there, and then send men to drag out the corpses and detail women (mothers and wives of men who have emigrated to Constantinople and religious believers who refuse to recant their faith and to allow their children to be torn from it) to wash down the steps.

But why like this? Couldn't they have done it at night—quietly? But why do it quietly? In that case a bullet would be wasted. In the daytime crowd the bullet had an educational function. It, so to speak, struck down ten with one shot.

They shot them in a different way too—right at the Onufriyev cemetery, behind the women's barracks (the former guest house for women pilgrims). And in fact that road past the women's barracks was christened *execution* road. In winter one could see a man being led barefoot along it, in only his underwear, through the snow (no, it was not for torture! it was just so his footgear and clothes should not go to waste), his hands bound behind his back with wire, and the condemned man would bear himself proudly and erectly, and with his lips alone, without the help of his hands, smoke the last cigarette of his life.

In the thirties a new camp era began, when Solovki even ceased to be Solovki—and became a mere run-of-the-mill "Corrective Labor Camp." And the black star of the ideologist of that new era, Naftaly Frenkel, rose in the heavens while his formula became the supreme law of the Archipelago:

"We have to squeeze everything out of a prisoner in the first three months—after that we don't need him any more."

The Golgotha-Crucifixion Monastery on Anzer was a penalty work site, where they cured patients . . . by murdering them. There in the Golgotha Church prisoners lay dying from lack of food and from cruelty, enfeebled priests next to syphilitics, and aged invalids next to young thieves. At the request of the dying, and in order to ease his own problem, the Golgotha doctor gave terminal cases strychnine; and in the winter the bearded corpses in their underwear were kept in the church for a long time. Then they were put in the vestibule, stacked standing up since that way they took up less space. And when they carried them out, they gave them a shove and let them roll on down Golgotha Hill.

They say that in December, 1928, on Krasnaya Gorka in Karelia, the prisoners were left to spend the night in the woods as punishment for failure to fulfill the assigned norm of work—and 150 men froze to death there. This was a standard Solovetsky trick. Hard to doubt the story.

And so, imperceptibly—via work parties—the former concept of the Special Purpose Camp, totally isolated on its islands, dissolved. And the Archipelago, born and come to maturity on Solovki, began its malignant advance through the nation.

A problem arose: The territory of this country had to be spread out in front of the Archipelago—but without allowing the Archipelago to conquer it, to distract it, to take it over or assimilate it to itself. Every little island and every little hillock of the Archipelago had to be encircled by a hostile, stormy Soviet seascape. It was permissible for the two worlds to interlock in separate strata—but not to intermingle!

Now, with the spread of the Archipelago, escapes multiplied. There was the hopelessness of the logging and road-building work parties—yet at the same time there was a whole continent beneath the feet of the escapees. So there was hope in spite of all.

But how could they escape from Solovki? For half a year the sea was frozen over, but not solidly, and in places there was open water, and the snowstorms raged, and the frost bit hard, and things were enveloped in mists and darkness. And in the spring and for a large part of the summer there were the long white nights with clear visibility over long distances for the patrolling cutters. And it was only when the nights began to lengthen, in the late summer and the autumn, that the time was right. Not for prisoners in the kremlin, of course, but for those who were out in work parties, where a prisoner might have freedom of movement and time to build a boat or a raft near the shore—and to cast off at night (even just riding off on a log for that matter) and strike out at random, hoping above all to encounter a foreign ship. The bustle among the guards and the embarkation of the cutters would reveal to the islanders the fact of an escape—and there would be a tremor of rejoicing among the prisoners, as if they were themselves escaping. They would ask in a whisper: Had he been caught yet? Had he been found yet? Many must have drowned without ever getting anywhere. One or another of them reached the Karelian shore perhaps—and if he did was more silent than the grave.

And there was a famous escape from Kem to England. This particular daredevil (his name is unknown to us—that's the breadth of our horizon!) knew English and concealed it. He managed to get assigned to loading timber in Kem, and he told his story to the Englishmen. The convoy discovered he was missing and delayed the ship for nearly a whole week and searched it several times without finding the fugitive. (What happened was that whenever a search party started from the shore, they lowered him overboard on the opposite side on the anchor chain, where he clung under water with a breathing pipe held in his teeth.) An enormous fine had to be paid for delaying the ship, so they finally decided to take a chance and let the ship go, thinking that perhaps the prisoner had drowned.

Then a book came out in England, even it would seem, in more than one printing. Evidently *An Island Hell* by S. A. Malsagoff.

This book astounded Europe (and no doubt they accused its fugitive author of exaggerating, for, after all, the friends of the New Society could not permit themselves to believe this slanderous volume) because it contradicted what was already well known; the newspaper *Rote Fahne* had described Solovki as a paradise. (And we hope that the paper's correspondent spent time in the Archipelago later on.) And it also contradicted those albums about Solovki disseminated by Soviet diplomatic missions in Europe: fine-quality paper and true-to-life photographs of the cozy monks' cells.

Slander or not, the breach had been a misfortune! And so a commission of VTsIK, under the chairmanship of the "conscience of the Party," Comrade Solts was sent off to find out what was going on there on those Solovetsky Islands (for, of course, they didn't have the least

Aron Solts

idea!). But in fact the commission merely rode along the Murmansk Railroad, and they didn't do much of anything even there. And they thought it right to send to the islands—no, to implore to go there!— none less than the great proletarian writer Maxim Gorky, who had recently returned to live in the proletarian Fatherland. His testimony would be the very best refutation of that repulsive foreign forgery.

The rumor reached Solovki before Gorky himself—and the prisoners' hearts beat faster and the guards hustled and bustled. One has to know prisoners in order to imagine their anticipation! The falcon, the stormy petrel, was about to swoop down upon the nest of injustice, violence, and secrecy. The leading Russian writer! He will give them hell! He will show them! He, the father, will defend! They awaited Gorky almost like a universal amnesty.

The chiefs were alarmed too: as best they could, they hid the monstrosities and polished things up for show. Transports of prisoners were sent from the kremlin to distant work parties so that fewer would remain there; many patients were discharged from the Medical Section and the whole thing was cleaned up. And they set up a "boulevard" of fir trees without roots, which were simply pushed down into the ground. (They only had to last a few days before withering.) It led to the Children's Colony, opened just three months previously and the pride of USLON, where everyone had clothes and where there were no socially hostile children, and where, of course, Gorky would be very interested in seeing how juveniles were being re-educated and saved for a future life under socialism.

Only in Kem was there an oversight. On Popov Island the ship *Gleb Boky* was being loaded by prisoners in underwear and sacks, when Gorky's retinue appeared out of nowhere to embark on that steamer! You inventors and thinkers! Here is a worthy problem for you, given that, as the saying goes, every wise man has enough of the fool in him: a barren island, not one bush, no possible cover—and right there, at a distance of three hundred yards, Gorky's retinue has shown up. Your solution? Where can this disgraceful spectacle—these men dressed in sacks—be hidden? The entire journey of the great Humanist will have been for naught if he sees them now. Well, of course, he will try hard not to notice them, but help him! Drown them in the sea? They will wallow and flounder. Bury them in the earth? There's no time. No, only a worthy son of the Archipelago could find a way out of this one. The work assigner ordered: "Stop work! Close ranks! Still closer! Sit down on the ground! Sit still!" And a tarpaulin was thrown over them. "Anyone who moves will be shot!" And the former stevedore Maxim

Gorky ascended the ship's ladder and admired the landscape from the steamer for a full hour till sailing time—and *he didn't notice!*

That was June 20, 1929. The famous writer disembarked from the steamer in Prosperity Gulf. Surrounded by the commanding officer corps of the GPU, Gorky marched with long swift strides through the corridors of several barracks. The room doors were all wide open, but he entered hardly any. In the Medical Section doctors and nurses in clean robes formed up for him in two rows, but he didn't even look around and went on out. From there the Chekists of USLON fearlessly took him to Sekirka. And what was there to see there? It turned out that there was no overcrowding in the punishment cells, and—the main point—no *poles.* None at all. Thieves sat on benches (there was already a multitude of thieves in Solovki), and they were all . . . reading newspapers. None of them was so bold as to get up and complain, but they did think up one trick: they held the newspapers upside down! And Gorky went up to one of them and in silence turned the newspaper right side up! He had noticed it! He had understood! He would not abandon them. He would defend them!

They went to the Children's Colony. How decent everything was there. Each was on a separate cot, with a mattress. They all crowded around in a group and all of them were happy. And all of a sudden a fourteen-year-old boy said: "Listen here, Gorky! Everything you see here is false. Do you want to know the truth? Shall I tell you?" Yes, nodded the writer. Yes, he wanted to know the truth. (Oh, you bad boy, why do you want to spoil the just recently arranged prosperity of the literary patriarch? A palace in Moscow, an estate outside Moscow . . .) And so everyone was ordered to leave—and the boy spent an hour and a half telling the whole story to the lanky old man. Gorky left the barracks, streaming tears. He was given a carriage to go to dinner at the villa of the camp chief. And the boys rushed back into the barracks. "Did you tell him about the *mosquito treatment?*" "Yes." "Did you tell him about the *pole torture?*" "Yes." "Did you tell him about *the prisoners hitched up instead of horses?*" "Yes." "And how they roll them down the stairs? And about the sacks? And about being made to spend the night in the snow?" And it turned out that the truth-loving boy had told all . . . all . . . all!!!

But we don't even know his name.

On June 22, in other words after his chat with the boy, Gorky left the following inscription in the "Visitors' Book," which had been specially made for this visit:

"I am not in a state of mind to express my impressions in just a few words. I wouldn't want, yes, and I would likewise be ashamed [!], to permit myself banal praise of the remarkable energy of people who, while remaining vigilant and tireless sentinels of the Revolution, are able, at the same time, to be remarkably bold creators of culture."

On June 23 Gorky left Solovki. Hardly had his steamer pulled away from the pier than they shot the boy. (Oh, great interpreter of the human heart! Great connoisseur of human beings! How could he have failed to take the boy along with him?!)

And that is how faith in justice was instilled in the new generation.

They try to tell us that up there on the summit the chief of literature made excuses, that he didn't want to publish praise of USLON. But how can that be, Aleksei Maximovich? With bourgeois Europe looking on?! But *right now, right at this very moment,* which is so dangerous and so complicated! And the camp regimen there? We'll change it, we'll change the camp regimen.

And he did publish his statement, and it was republished over and over in the big free press, both our own and that of the West, claiming it was nonsense to frighten people with Solovki, and that prisoners lived remarkably well there and were being well reformed.

Chapter 3

■

The Archipelago Metastasizes

Well, the Archipelago did not develop on its own but side by side with the whole country. As long as there was unemployment in the nation there was no feverish demand for prisoner manpower, and arrests took place not as a means of mobilizing labor but as a means of sweeping clean the road. But when the concept arose of stirring up the whole 180 million with an enormous mixing paddle, when the plan for superindustrialization was rejected in favor of the plan for supersuper-superindustrialization, when the liquidation of the kulaks was already foreseen along with the massive public works of the First Five-Year Plan—on the eve of the Year of the Great Fracture the view of the Archipelago and everything in the Archipelago changed too.

On March 26, 1928, the Council of People's Commissars conducted a review of the status of penal policy in the nation and of conditions in places of imprisonment. In regard to penal policy, it was admitted that it was inadequate. And it was decreed that harsh measures of repression should be applied to class enemies and hostile-class elements, that the camp regimen should be made more severe (and that *socially unstable elements* should not be given terms at all). And in addition: forced labor should be set up in such a way that the prisoner should not earn anything from his work but that the state should derive economic profit from it. "And to consider it necessary from now on *to expand the capacity* of labor colonies." In other words, putting it simply, it was proposed that more camps be prepared in anticipation of the abundant arrests planned. Throughout the nation unemployment was abolished, and *the economic rationale* for expansion of the camps appeared.

Back in 1923 no more than three thousand persons had been imprisoned on Solovki. And by 1930 there were already about fifty thousand, yes, and another thirty thousand in Kem. In 1928 the Solovetsky cancer began to creep outward, first through Karelia, on road-building projects and in logging for export. Just as willingly SLON began to "sell" its engineers: they went off without convoy to work in any northern locality and their wages were credited to the camp. By 1929 SLON camp sites had already appeared at all points on the Murmansk Railroad from Lodeinoye Pole to Taibola. From there the movement continued along the Vologda Railroad—and so active was it that at Zvanka Station it proved necessary to open up a SLON transport control center. By 1930 Svirlag had already grown strong in Lodeinoye Pole and stood on its own legs, and in Kotlas Kotlag had already been formed. In 1931 BelBaltlag had been born, with its center in Medvezhyegorsk, which was destined over the next two years to bring glory to the Archipelago for eternity and on five continents.

And the malignant cells kept on creeping and creeping. They were blocked on one side by the sea and on the other by the Finnish border, but there was nothing to hinder the founding of a camp near Krasnaya Vishera in 1929. And the main thing was that all the paths to the east through the Russian North lay open and unobstructed. Very soon the Soroka-Kotlas road was reaching out. Creeping on to the Northern Dvina River, the camp cells formed SevDvinlag. Crossing it, they fearlessly marched on the Urals. By 1931 the Northern Urals department of SLON was founded, which soon gave rise to the independent Solikamlag and SevUrallag. The Berezniki Camp began the construction of a big chemical combine which in its time was much publicized. In the summer of 1929 an expedition of unconvoyed prisoners was sent to the Chibyu River from Solovki, under the leadership of the geologist M. V. Rushchinsky, in order to prospect for petroleum, which had been discovered there as far back as the eighties of the nineteenth century. The expedition was successful—and a camp was set up on the Ukhta, Ukhtlag. But it, too, did not stand still on its own spot, but quickly metastasized to the northeast, annexed the Pechora, and was transformed into UkhtPechlag. Soon afterward it had its Ukhta, Inta, Pechora, and Vorkuta sections—all of them the bases of great independent future camps.

The opening up of so expansive a roadless northern region as this required the building of a railroad: from Kotlas via Knyazh-Pogost and Ropcha to Vorkuta. This called forth the need for two more independent camps which were railroad-building camps: SevZhelDorlag—on the sector from Kotlas to the Pechora River—and Pechorlag (not to be confused with the industrial UkhtPechlag!)—on the sector from the Pechora River to Vorkuta.

And thus from the depths of the tundra and the taiga rose hundreds of new medium-sized and small islands. And on the march, in battle order, a new system of organization of the Archipelago was created: Camp Administrations, Camp Divisions, Camps (OLP's—Separate Camps; KOLP's—Commandant's Camps; GOLP's—Head Camps), Camp Sectors (and these were the same as "work parties" and "work subparties"). And in the Administrations there were Departments, and in the Divisions there were Sections: I. Production (P.); II. Records and Classification (URCh); III. Security Operations (again the *third!*).

And so all the northern portion of the Archipelago sprang from Solovki. But not from there alone. In response to the great appeal, Corrective Labor Camps (ITL's) and Corrective Labor Colonies (ITK's) burst out in a rash throughout our whole great country. Every province acquired its own ITL's and ITK's. Millions of miles of barbed wire ran on and on, the strands crisscrossing one another and interweaving, their barbs twinkling gaily along railroads, highways, and around the outskirts of cities. And the peaked roofs of ugly camp watchtowers became the most dependable landmarks in our landscape, and it was only by a surprising concatenation of circumstances that they were not seen in either the canvases of our artists or in scenes in our films.

As had been happening from the Civil War on, monastery buildings were intensively *mobilized* for camp needs, were ideally adapted for isolation by their very locations. The Boris and Gleb Monastery in Torzhok was put to use as a transit camp (still there today), while the Valdai Monastery was put to use for a colony of juveniles (across the lake from the future country house of Zhdanov). Nilova Hermitage on Stolbny Island in Lake Seliger became a camp. Sarovskaya Hermitage was used for the nest of Potma camps, and there is no end to this enumeration. Camps arose in the Donbas, on the upper, middle, and lower Volga, in the central and southern Urals, in Transcaucasia, in central Kazakhstan, in Central Asia, in Siberia, and in the Far East. It is officially reported that in 1932 the area devoted to Agricultural Corrective Labor Colonies in the Russian Republic alone—was 625,000 acres, and in the Ukrainian Republic 138,000.

Estimating the average colony at 2,500 acres, we learn that at this time, without counting the other Soviet republics, there were already more than three hundred such *Selkhozy* alone, in other words the lowest grade and most privileged form of camp.

■

A stubborn legend persists in the Archipelago to the effect that *"The camps were thought up by Frenkel."*

It seems to me that this fanciful idea, both unpatriotic and even insulting to the authorities, is quite sufficiently refuted by the preceding chapters. Even with the meager means at our disposal we succeeded, I hope, in showing the birth of camps for repression and labor back in 1918. Without any Frenkel whatsoever they arrived at the conclusion that prisoners must not waste their time in moral contemplation ("The purpose of Soviet corrective labor policy is not at all individual correction in its traditional meaning") but must labor, and at the same time must be given very severe, almost unbearable work norms to achieve. Long before Frenkel they already used to say: "correction through labor" (and as far back as Eichmans they already understood this to mean "destruction through labor").

Nonetheless, Frenkel really did become the nerve of the Archipelago. He was one of those successful men of action whom History hungrily awaits and summons to itself. It would seem that there had been camps even before Frenkel, but they had not taken on that final and unified form which savors of perfection. Every genuine prophet arrives when he is most acutely needed. Frenkel arrived in the Archipelago just at the beginning of the metastases.

Naftaly Aronovich Frenkel, a Turkish Jew, was born in Constantinople. He graduated from the commercial institute there and took up the timber trade. He founded a firm in Mariupol and soon became a millionaire, "the timber king of the Black Sea." He had his own steamers, and he even published his own newspaper in Mariupol called *The Kopeck,* whose function was to slander and persecute his competitors. During World War I Frenkel conducted some speculative arms deals through Gallipoli. In 1916, sensing the pending storm in Russia, he transferred his capital to Turkey even *before* the February Revolution, and in 1917 he himself went to Constantinople in pursuit of it.

And he could have gone on living the sweetly exciting life of a merchant, and he would have known no bitter grief and would not have turned into a legend. But some fateful force beckoned him to the Red power.

The rumor is unverified that in those years in Constantinople he became the resident Soviet intelligence agent (perhaps for ideological reasons, for it is otherwise difficult to see why he needed it). But it is a fact that in the NEP years he came to the U.S.S.R., and here, on secret instructions from the GPU, created, as if in his own name, a black market for the purchase of valuables and gold in return for Soviet paper rubles. Business operators and manipulators remembered him very well indeed from the old days; they trusted him—and the gold flowed into the coffers of the GPU. The purchasing operation came to an end, and, in gratitude, the GPU arrested him. Every wise man has enough of the simpleton in him.

However, inexhaustible and holding no grudges, Frenkel, while still in the Lubyanka or on the way to Solovki, sent some sort of declaration to the top. Finding himself in a trap, he evidently decided to make a business analysis of this life too. He was brought to Solovki in 1927, but was immediately separated from the prisoner transport, settled into a stone booth outside the bounds of the monastery itself, provided with an orderly to look after him, and permitted free movement about the island. We have already recalled that he became the Chief of the Economic Section (the privilege of a free man) and expressed his famous thesis about using up the prisoner in the first three months. In 1928 he was already in Kem. There he created a profitable auxiliary enterprise. He brought to Kem the leather which had been accumulated by the monks for decades and had been lying uselessly in the monastery warehouses. He recruited furriers and shoemakers from among the prisoners and supplied fashionable high-quality footwear and leather goods directly to a special shop on Kuznetsky Most in Moscow.

One day in 1929 an airplane flew from Moscow to get Frenkel and brought him to an appointment with Stalin. The Best Friend of prisoners (and the Best Friend of the Chekists) talked interestedly with Frenkel for three hours. The stenographic report of this conversation will never become public. There simply was none. But it is clear that Frenkel unfolded before the Father of the Peoples dazzling prospects for constructing socialism through the use of prisoner labor. Much of the geography of the Archipelago being described in the aftermath by my obedient pen, he sketched in bold strokes on the map of the Soviet Union to the accompaniment of the puffing of his interlocutor's pipe. It was Frenkel in person, and in this very conversation, who proposed renouncing the reactionary system of equality in feeding prisoners and who outlined a unified system of redistribution of the meager food supplies for the whole Archipelago—*a scale for bread rations and a scale for hot-food rations* which was adapted by him from the Eskimos:

a fish on a pole held out in front of the running dog team. In addition, he proposed *time off sentence* and release ahead of term as rewards for good work (but in this respect he was hardly original—for in 1890, in Sakhalin hard labor, Chekhov discovered both the one and the other). In all probability the first experimental field was set up here too—the great Belomorstroi, the White Sea–Baltic Canal Construction Project, to which the enterprising foreign-exchange and gold speculator would soon be appointed—not as chief of construction nor as chief of a camp either, but to the post especially dreamed up for him of "works chief" —the chief overseer of the labor battle.

And here he is himself (see pages 206 and 209). It is evident from his face how he brimmed with a vicious human-hating animus. In the book on the Belomor Canal—the White Sea–Baltic Canal—wishing to laud Frenkel, one Soviet writer would soon describe him thus: " . . . the eyes of an interrogator and a prosecutor, the lips of a skeptic and a satirist . . . A man with enormous love of power and pride, for whom the main thing is unlimited power. If it is necessary for him to be feared, then let him be feared. He spoke harshly to the engineers, attempting to humiliate them."

This last phrase seems to us a keystone—to both the character and biography of Frenkel.

By the start of Belomorstroi Frenkel had been freed.

■

The whole long history of the Archipelago, about which it has fallen to me to write this home-grown, homemade book, has, in the course of half a century, found in the Soviet Union almost no expression whatever in the printed word. In this a role was played by that same unfortunate happenstance by which camp watchtowers never got into scenes in films nor into landscapes painted by our artists.

But this was not true of the White Sea–Baltic Canal nor of the Moscow-Volga Canal. There is a book about each at our disposal, and we can write this chapter at least on the basis of documentary and responsible source material.

In diligently researched studies, before making use of a particular source, it is considered proper to characterize it. We shall do so.

Here before us lies the volume, in format almost equal to the Holy Gospels, with the portrait of the Demigod engraved in bas-relief on the cardboard covers. The book, entitled *The White Sea–Baltic Stalin Canal,* was issued by the State Publishing House in 1934 and dedicated by the authors to the Seventeenth Congress of the Soviet Communist

Party, and it was evidently published for the Congress. It is an extension of the Gorky project of "Histories of Factories and Plants." Its editors were Maxim Gorky, Ida Leonidovna Averbakh, and S. G. Firin. This last name is little known in literary circles, and we shall explain why: Semyon Firin, notwithstanding his youth, was Deputy Chief of Gulag.

Nobody was more famous in Soviet literature at that time than Leopold Leonidovich Averbakh, brother of Ida Leonidovna. He was the director of the magazine *Na literaturnom postu ("On Literary Guard"),* he played a prominent role among those in charge of clubbing writers, and he was also Sverdlov's nephew.

That charming Sverdlov family somehow remained in the shadows of revolutionary history, thanks to the early death of Yakov, who had managed, however, to take an active part in the executions, including that of the tsar and his family. Here we have his amiable nephews; there was also his son Andrew, quite a remarkable prosecutor and hangman, whose hobby was to pretend that he was a prisoner in order to denounce his fellow inmates. Sverdlov's wife, Claudia Novgorodtseva, kept at home a fortune in diamonds and brilliants, the so-called "party fund": a product of Bolshevik robberies during the revolution. The Politburo gang had put aside this reserve in case they lost power and had to flee the government buildings in a hurry.

The history of the above-mentioned book is as follows: On August 17, 1933, an *outing* of 120 writers took place aboard a steamer on the just completed canal. D. P. Vitkovsky, a prisoner who was a construction superintendent on the canal, witnessed the way these people in white suits crowded on the deck during the steamer's passage through the locks, summoned prisoners from the area of the locks (where by this time they were more operational workers than construction workers), and, in the presence of the canal chiefs, asked a prisoner whether he loved his canal and his work, and did he think that he had managed to reform here, and did the chiefs take enough interest in the welfare of the prisoners? There were many questions, all in this general vein, and all asked from shipboard to shore in the presence of the chiefs and only while the steamer was passing through the locks. And after this outing eighty-four of these writers somehow or other managed nonetheless to worm their way out of participating in Gorky's collective work (though perhaps they wrote their own admiring verses and essays), and the remaining thirty-six constituted an authors' collective. By virtue of intensive work in the fall and winter of 1933 they created this unique book.

This book was published to last for all eternity, so that future generations would read it and be astounded. But by a fateful coincidence, most of the leaders depicted in its photographs and glorified in its text were exposed as enemies of the people within two or three years. Naturally all copies of the book were thereupon removed from libraries and destroyed. Private owners also destroyed it in 1937, not wishing to earn themselves *a term* for owning it. And that is why very few copies have remained intact to the present; and there is no hope that it may be reissued—and therefore all the heavier is the obligation to my fellow countrymen I feel on my shoulders not to permit the principal ideas and facts described in this book to perish.

Their general enthusiasm for the camp way of life led the authors of the collective work to this panegyric: "No matter to what corner of the Soviet Union fate should take us, even if it be the most remote wilderness and backwoods, the imprint of order . . . of precision and of conscientiousness . . . marks each OGPU organization." And what OGPU organization exists in the Russian backwoods? Only the camps. *The camp as a torch of progress*—that is the level of this historical source of ours.

The editor in chief has something to say about this himself. Addressing the last rally of Belomorstroi officials on August 25, 1933, in the city of Dmitrov (they had already moved over to the Moscow-Volga Canal project), Gorky said: "Ever since 1928 I have watched how the GPU re-educates people." (And what this means is that even before his visit to Solovki, even before that boy was shot, ever since, in fact, he first returned to the Soviet Union, he had been watching them.) And by then hardly able to restrain his tears, he addressed the Chekists present: "You devils in woolen overcoats, you yourselves don't know what you have done." And the authors note: the Chekists there merely *smiled*. (They knew *what* they had done. . . .) And Gorky noted *the extraordinary modesty* of the Chekists in the book itself. (This dislike of theirs for publicity was truly a touching trait.)

The collective authors do not simply keep silent about the deaths on the Belomor Canal during construction. They do not follow the cowardly recipe of *half-truths*. Instead, they write directly that *no one* died during construction. (Probably they calculated it this way: One hundred thousand started the canal and one hundred thousand finished. And that meant they were all alive. They simply forgot about the prisoner transports devoured by the construction in the course of two fierce winters. But this is already on the level of the cosine of the cheating engineering profession.)

The authors see nothing more inspiring than this camp labor. They find in forced labor one of the highest forms of blazing, conscientious creativity. Here is the theoretical basis of re-education: "Criminals are the result of the repulsive conditions of former times, and our country is beautiful, powerful and *generous,* and it needs to be *beautified.*" In their opinion all those driven to work on the canal would never have found their paths in life if the employers had not assigned them to unite the White Sea with the Baltic. Because, after all, *"Human raw material is immeasurably more difficult to work than wood."* What language! What profundity! Who said that? Gorky said it in his book, disputing the "verbal trumpery of humanism." And Zoshchenko, with profound insight, wrote: "Reforging—this is not the desire to serve out one's term and be freed [So such suspicions did exist?—A.S.], but is in actual fact a restructuring of the consciousness and the pride of a builder." What a student of man! Did you ever push a canal wheelbarrow—and on a penalty ration too?

We were in such a rush that trainloads of zeks kept on arriving and arriving at the canal site before there were any barracks there, or supplies, or tools, or a precise plan. And what was to be done?

And our authors rave. In the jolly tone of inveterate merrymakers they tell us: Women came in silk dresses and were handed a wheelbarrow on the spot! Almost choking with laughter, they tell us: From the Krasnovodsk camps in Central Asia, from Stalinabad, from Samarkand, they brought Turkmenians and Tadzhiks in their Bukhara robes and turbans—here to the Karelian subzero winter cold! Now that was something the *Basmachi* rebels never expected! The norm here was *to break up two and a half cubic yards of granite and to move it a distance of a hundred yards in a wheelbarrow.* And the snow kept falling and covering everything up, and the wheelbarrows somersaulted off the gangways into the snow.

The basic transportation at Belomorstroi consisted of *grabarki,* dray carts, with boxes mounted on them for carrying earth, as we learn from the book. And in addition there were also *Belomor Fords!* And here is what they were: heavy wooden platforms placed on four wooden logs (rollers), and two horses dragged this *Ford* along and carried stones away on it. And a wheelbarrow was handled by a team of two men—on slopes it was caught and pulled upward by a *hookman*—a worker using a hook. And how were trees to be felled if there were neither saws nor axes? Our inventiveness could find the answer to that one: ropes were tied around the trees, and they were rocked back and forth by brigades pulling in different directions—*they rocked the trees out.* Our inventiveness can solve any problem at all—and why? Because the *canal was being built on the initiative and instructions of Comrade Stalin!*

The very grandeur of this construction project consisted in the fact that it was carried out without contemporary technology and equipment and without any supplies from the nation as a whole! "These are not the tempos of noxious European-American capitalism, these are socialist tempos!" the authors brag.

No, it would be unjust, most unjust, unfair, to compare this most savage construction project of the twentieth century, this continental canal built "with wheelbarrow and pick," with the Egyptian pyramids; after all, the pyramids were built with the *contemporary* technology!! And we used the technology of forty centuries earlier!

That's what our gas execution van consisted of. We didn't have any gas for the gas chamber.

And meanwhile it is incessantly dinned into our ears: *"The canal is being built on the initiative and orders of Comrade Stalin!"* "The radio in the barracks, on the canal site, by the stream, in a Karelian hut, on a truck, the radio which *sleeps* neither day *nor night* [just imagine it!], those innumerable black mouths, those black masks without eyes [imagery!] cry out incessantly: what do the Chekists of the whole country think about the canal project, what does the Party have to say about it?" And you, too, better think the same! You, too, better think the same! *"Nature we will teach—and freedom we will reach."* Hail socialist competition and the shock-worker movement. Competition between work brigades! Competition between phalanxes (from 250 to 300 persons)! Competition between labor collectives! Competition between locks! And then, finally, the *Vokhrovtsy*—the Militarized Camp Guards —entered into competition with the zeks. (And the obligation of the Vokhrovtsy? To guard you better.)

But the main reliance was, of course, on the *socially friendly elements—in other words,* the thieves! These concepts had already merged at the canal. Deeply touched, Gorky shouted to them from the rostrum: "After all, any capitalist steals more than all of you combined!" The thieves roared with approval, flattered.

The food norm was not provided, it was cold in the barracks, there was an infestation of lice, and people were ill—never mind, we'll *manage!* An atmosphere of constant battle alert was created. All of a sudden *a night of storm assault* was proclaimed. It was decided: *to double the work norms!* So a universal *day of records* is proclaimed! A

The wooden cranes

The earliest machinery

Women's shock brigade

"Volunteers"

blow against *tempo interrupters*. *Bonus pirozhki* are distributed to a brigade. Why such haggard faces? The longed-for moment—but no gladness . . .

In January came *the storm of the watershed!* All the phalanxes, with their kitchens and property, were to be thrown into one single sector! There were not enough tents for everyone. They slept out on the snow—never mind. *We'll manage!*

In April there was an incessant forty-eight-hour storm assault— hurrah! *Thirty thousand people did not sleep!*

And by May 1, 1933, People's Commissar Yagoda reported to his beloved Teacher that the canal had been completed on time.

D. P. Vitkovsky, a Solovetsky Islands veteran, who worked on the White Sea Canal as a work supervisor and saved the lives of many prisoners with that very same "tukhta," the falsification of work reports, draws a picture of the evenings:

At the end of the workday there were corpses left on the work site. The snow powdered their faces. One of them was hunched over beneath an overturned wheelbarrow, he had hidden his hands in his sleeves and frozen to death in that position. Someone had frozen with his head bent down between his knees. Two were frozen back to back leaning against each other. They were peasant lads and the best workers one could possibly imagine. They were sent to the canal in tens of thousands at a time, and the authorities tried to work things out so no one got to the same subcamp as his father; they tried to break up families. And right off they gave them norms of shingle and boulders that you'd be unable to fulfill even in summer. No one was able to teach them anything, to warn them; and in their village simplicity they gave all their strength to their work and weakened very swiftly and then froze to death, embracing in pairs. At night the sledges went out and collected them. The drivers threw the corpses onto the sledges with a dull clonk.

And in the summer bones remained from corpses which had not been removed in time, and together with the shingle they got into the concrete mixer. And in this way they got into the concrete of the last lock at the city of Belomorsk and will be preserved there forever.

The Belomorstroi newspaper choked with enthusiasm in describing how many Canal Army Men, who had been "aesthetically carried away" by their great task, had in their own free time (and, obviously, without any payment in bread) decorated the canal banks with stones —simply for the sake of beauty.

Yes, and it was quite right for them to set forth on the banks of the canal the names of the six principal lieutenants of Stalin and

Frenkel, Firin, and Uspensky

Distribution of the food bonus

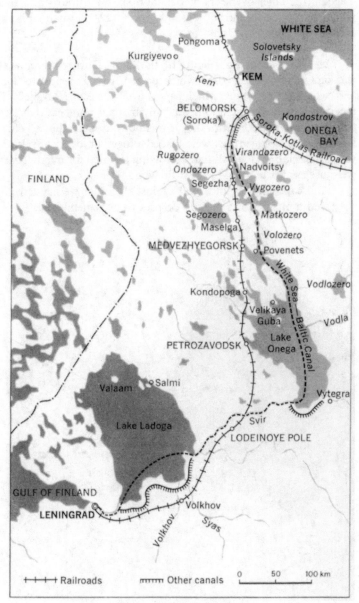

Map of the Belomor Canal

Yagoda, the chief overseers of Belomor, six hired murderers each of whom accounted for thirty thousand lives: Firin—Berman—Frenkel—Kogan—Rappoport—Zhuk.

In 1966 I spent eight hours by the canal. During this time there was one self-propelled barge which passed from Povenets to Soroka, and one, identical in type, which passed from Soroka to Povenets. Their numbers were different, and it was only by their numbers that I could tell them apart and be sure that it was not the same one as before on its way back. Because they were loaded altogether identically: with the very same pine logs which had been lying exposed for a long time and were useless for anything except firewood.

And canceling the one load against the other we get zero.

And a quarter of a million corpses to be remembered.

Sergei Zhuk

Naftaly Frenkel

Yakov Rappoport

Matvei Berman

Lazar Kogan

Genrikh Yagoda

Chapter 4

■

The Archipelago Hardens

And the clock of history was striking.

In 1933, at the January session of the Central Committee of the Communist Party, the Great Leader, who at that time was already computing the number of two-legged beings who had yet to be exterminated in this country, declared that the dying off of the state, so firmly promised by Lenin and fervently expected by humanists, would not come about through weakening the state, but on the contrary through strengthening it to the utmost, which was necessary in order to kill off the moribund classes. . . .

This was so unexpectedly brilliant that it was not given to every little mind to grasp it, but Vyshinsky, ever the loyal apprentice, immediately picked it up: "And this means the maximum *strengthening* of corrective-labor institutions."

Entry into socialism via the maximum strengthening of prison! And this was not some satirical magazine cracking a joke either, but was said by the Prosecutor General of the Soviet Union!

And an iron curtain descended around the Archipelago. No one other than the officers and sergeants of the NKVD could enter and leave via the camp gatehouse. That harmonious order of things was established which the zeks themselves would soon come to consider the only conceivable one. . . .

And that is when the wolf's fangs were bared! And that is when the bottomless pit of the Archipelago gaped wide!

"I'll shoe you in tin cans, but you're going to go out to work!"

"If there aren't enough railroad ties, I'll make one out of you!"

They say that in February–March, 1938, a secret instruction was circulated in the NKVD: *Reduce the number of prisoners.* (And not by releasing them, of course.) I do not see anything in the least impossible here: this was a logical instruction because there was simply not enough housing, clothing, or food. Gulag was grinding to a halt from exhaustion.

And this was when the pellagra victims lay down and died en masse. This was when the chiefs of convoy began to test the accuracy of machine-gun fire by shooting at the stumbling zeks. And this was when every morning the orderlies hauled the corpses to the gatehouse, stacking them there.

In the Kolyma, that pole of cold and cruelty in the Archipelago, that very same about-face took place with a sharpness worthy of a pole.

According to the recollections of Ivan Semyonovich Karpunich-Braven (former commander of the 40th Division and of the XII Corps, who recently died with his notes incomplete and scattered), a most dreadfully cruel system of food, work, and punishment was established in the Kolyma. The prisoners were so famished that at Zarosshy Spring they ate the corpse of a horse which had been lying dead for more than a week and which not only stank but was covered with flies and maggots. At Utiny Goldfields the zeks ate half a barrel of lubricating grease, brought there to grease the wheelbarrows. At Mylga they ate Iceland moss, like the deer. And when the passes were shut by snowdrifts, they used to issue three and a half ounces of bread a day at the distant goldfields, without ever making up for previous deficiencies. Multitudes of "goners," unable to walk by themselves, were dragged to work on sledges by other "goners" who had not yet become quite so weak. Those who lagged behind were beaten with clubs and torn by dogs. Working in 50 degrees below zero Fahrenheit, they were forbidden to build fires and warm themselves.

■

The beginning of the war shook the Archipelago chieftains: the course of the war at the very start was such that it might very likely have led to the breakdown of the entire Archipelago, and perhaps even to the employers having to answer to the workers. As far as one can judge from the impressions of the zeks from various camps, the course of events gave rise to two different kinds of conduct among the bosses. Some of them, those who were either more reasonable or perhaps more cowardly, relaxed their regime and began to talk with the prisoners almost gently, particularly during the weeks of military defeats. They were unable, of course, to improve the food or the maintenance. Others who were more stubborn and more vicious began, on the contrary, to be even stricter and more threatening with the 58's, as if to promise them death before liberation. And in some camps (sensing intuitively the direction of future policy) they began to isolate the 58's from the nonpolitical offenders in compounds guarded with particular strictness, put machine guns up on the watchtowers, and even spoke thus to the zeks who had formed up: "You are hostages!"

From the first days of the war, everywhere in the Archipelago (on opening the packages of mobilization instructions) they halted all releases of 58's. There were even cases of released prisoners being sent back to camp while on their way home. In Ukhta on June 23 a group released was already outside the perimeter waiting for a train when the convoy chased them back and even cursed them: "It's because of you the war began!" Karpunich received his release papers on the morning of June 23 but had not yet succeeded in getting through the gatehouse when they coaxed them out of him by fraud: "Show them to us!" He *showed* them and was kept in camp for another five years. This was considered to mean "until *special* orders." (When the war had already come to an end, in many camps they were forbidden even to go to the Classification and Records Section and ask when they would be freed. The point was that after the war there were not enough people for a while, and many local administrations, even if Moscow allowed them to release prisoners, issued their own "special orders" so as to hold on to manpower.

Here's what the wartime camp was: more work and less food and less heat and worse clothes and ferocious discipline and more severe punishment—and that still wasn't all.

For the 58's the wartime camps were particularly unbearable because of their pasting on *second terms,* which hung over the prisoners' heads worse than any ax. The Security officers, busily engaged in saving themselves from the front, discovered in well-set-up backwaters and backwoods, in logging expeditions, plots involving the participation of the world bourgeoisie, plans for armed revolts and mass escapes.

■

Such are the forms into which the islands of the Archipelago hardened, but one need not think that as it hardened it ceased to exude more metastases from itself.

In 1939, before the Finnish War, Gulag's alma mater, Solovki, which had come too close to the West, was moved via the Northern Sea Route to the mouth of the Yenisei River and there merged into the already created Norillag, which soon reached 75,000 in size. So malignant was Solovki that even in dying it threw off one last metastasis— and what a metastasis!

The Archipelago's conquest of the unpeopled deserts of Kazakhstan belongs to the prewar years. That was where the nest of Karaganda camps swelled like an octopus; and fertile metastases were propagated in Dzhezkazgan with its poisoned cuprous water, in Mointy and in Balkhash. And camps spread out over the north of Kazakhstan also.

New growths swelled in Novosibirsk Province (the Mariinsk Camps), in the Krasnoyarsk region (the Kansk Camps and Kraslag), in Khakassiya, in Buryat-Mongolia, in Uzbekistan, even in Gornaya Shoriya.

Nor did the Russian North, so beloved by the Archipelago, end its own growth.

Chapter 5

■

What the Archipelago Stands On

The camps are not merely the "dark side" of our postrevolutionary life but very nearly the very liver of events.

Just as every point is formed by the intersection of at least two lines, every event is formed by the intersection of at least two necessities —and so although on one hand our economic requirements led us to the system of camps, this by itself might have led us to labor armies, but it intersected with the theoretical justification for the camps, fortunately already formulated.

And so they met and grew together. And that is how the Archipelago was born.

The economic need manifested itself, as always, openly and greedily; for the state which had decided to strengthen itself in a very short period of time and which did not require anything from outside, the need was manpower:

a. Cheap in the extreme, and better still—for free.
b. Undemanding, capable of being shifted about from place to place any day of the week, free of family ties, not requiring either established housing, or schools, or hospitals, or even, for a certain length of time, kitchens and baths.

It was possible to obtain such manpower only by swallowing up one's own sons.

The theoretical justification could not have been formulated with such conviction in the haste of those years had it not had its beginnings in the past century. Engels discovered that the human being had arisen not through the perception of a moral idea and not through the process of thought, but out of happenstance and meaningless work (an ape picked up a stone—and with this everything began). Marx, concerning himself with a less remote time ("Critique of the Gotha Program"), declared with equal conviction that the *one and only* means of correcting offenders (true, he referred here to criminals; he never even conceived that his pupils might consider politicals offenders) was not solitary contemplation, not moral soul-searching, not repentance, and not languishing (for all that was superstructure!)—but productive labor. He himself had never in his life taken a pick in hand. To the end of his days he never pushed a wheelbarrow, mined coal, felled timber, and we don't even know how his firewood was split—but he wrote that down on paper, and the paper did not resist.

And for his followers everything now fell into place: To compel a prisoner to labor every day (sometimes fourteen hours at a time, as at the Kolyma mine faces) was humane and would lead to his correction. On the contrary, to limit his confinement to a prison cell, courtyard, and vegetable garden, to give him the chance to read books, write, think, and argue during these years meant to treat him "like cattle." (This is from that same "Critique of the Gotha Program.")

True, in the heated times immediately following the October Revolution they paid little heed to these subtleties, and it seemed even more humane simply to shoot them.

Oh, "what an intelligent, farsighted humane administration from top to bottom," as Supreme Court Judge Leibowitz of New York State wrote in *Life* magazine, after having visited Gulag. "In serving out his term of punishment the prisoner retains a feeling of dignity." That is what he comprehended and saw.

And oh, you well-fed, devil-may-care, nearsighted, irresponsible foreigners with your notebooks and your ball-point pens—beginning with those correspondents who back in Kem asked the zeks questions in the presence of the camp chiefs—how much you have harmed us in your vain passion to shine with understanding in areas where you did not grasp a lousy thing!

Human dignity! Of persons condemned without trial? Who are made to sit down beside Stolypin cars at stations with their rear ends in the mud? Who, at the whistle of the citizen jailer's lash, scrape up with their hands the urine-soaked earth and carry it away, so as not to be sentenced to the punishment block? Of those educated women who, as a great honor, have been found worthy of laundering the linen of the citizen chief of the camp and of feeding his privately owned pigs? And who, at his first drunken gesture, have to make themselves available, so as not to perish *on general work* the next day?

■

Serfs! This comparison occurred to many when they had the time to think about it, and not accidentally either. Not just individual features, but the whole central meaning of their existence was identical for serfdom and the Archipelago; they were forms of social organization for the forced and pitiless exploitation of the unpaid labor of millions of slaves.

But there are some who will object that nonetheless there are really not so many similarities between serfs and prisoners. There are more differences.

And we agree with that: there are more differences. But what is surprising is that all the differences are to the credit of serfdom! All the differences are to the discredit of the Gulag Archipelago!

The serfs did not work longer than from sunrise to sunset. The zeks started work in darkness and ended in darkness (and they didn't always end either). For the serfs Sundays were sacred; and the twelve sacred Orthodox holidays as well, and local saints' days, and a certain number of the twelve days of Christmas (they went about in mummers' costumes). The prisoner was fearful on the eve of every Sunday: he didn't know whether they would get it off. And he never got holidays at all; those firsts of May and those sevenths of November involved more miseries, with searches and special regimen, than the holidays were worth (and a certain number were put into punishment blocks every year precisely on those very days). For the serfs Christmas and Easter were genuine holidays; and as for a *body search* either after work or in the morning or at night, the serfs knew not of these! The serfs lived in permanent huts, regarding them as their own, and when at night they lay down on top of their stoves, or on their sleeping platform between ceiling and stove—or else on a bench, they knew: This is my own place, I have slept here forever and ever, and I always will. The prisoner did not know what barracks he would be in on the morrow (and even when he returned from work he could not be certain that he would sleep in that place that night). He did not have his "own" sleeping shelf or his "own" multiple bunk. He went wherever they drove him.

Old Russia, which experienced Asiatic slavery for seven whole centuries, did not for the most part know *famine*. "In Russia no one has ever died of starvation," said the proverb. And a proverb is not made up out of lies and nonsense. The serfs were slaves, but they had full bellies. The Archipelago lived for decades in the grip of cruel famine. The zeks would scuffle over a herring tail from the garbage pail. For Christmas and Easter even the thinnest serf peasant broke his fast with fat bacon. But even the best worker in camp could get fat bacon only in parcels from home.

The serfs lived in families. The sale or exchange of a serf away from his family was a universally recognized and proclaimed barbarism. Popular Russian literature waxed indignant over this. Hundreds of serfs, perhaps thousands (but this is unlikely), were torn from their families. But not millions. The zek was separated from his family on the first day of his arrest and, in 50 percent of all cases—forever. If a son was arrested with his father (as we heard from Vitkovsky) or a wife together with her husband, the greatest care was taken to see that they did not meet at the same camp. And if by some chance they did meet, they were separated as quickly as possible. Similarly, every time a man and a woman zek came together in camp for fleeting or real love, they hastened to penalize them with the punishment cell, to separate them and send them away from one another. . . .

■

There was a famous incantation repeated over and over again: "In the new social structure there can be no place for the discipline of the stick on which serfdom was based, nor the discipline of starvation on which capitalism is based."

And there you are—the Archipelago managed miraculously to combine the one and the other.

All in all, the particular techniques required for this totaled three: (1) *the differentiated ration pot;* (2) *the brigade;* and (3) *two sets of bosses.*

We have already explained about the differentiated ration pot. This was a redistribution of bread and cereals aimed at making our zek beat his head against the wall and break his back for the average prisoner's ration, which in parasitical societies is issued to an inactive prisoner. To fix it so that our zek could get his own lawful ration only in extra dollops of three and a half ounces and by being considered a shock worker. Percentages of output above 100 conferred the right to supplementary spoonfuls of kasha (those previously taken away). What a merciless knowledge of human nature! Neither those pieces of bread nor those cereal patties were comparable with the expenditure of strength that went into earning them. But as one of his eternal, disastrous traits the human being is incapable of grasping the ratio of an object to its price. For a cheap glass of vodka a soldier is roused to attack in a war not his own and lays down his life; in the same way the zek, for those pauper's handouts, slips off a log, gets dunked in the icy freshet of a northern river, or kneads clay for mud huts barefoot in icy water, and because of this those feet are never going to reach the land of *freedom*.

Also, the *brigade* was invented. Slave-driving the prisoners with club and ration, the brigadier has to cope with the brigade in the absence of the higher-ups, the supervisors, and the convoy. Shalamov cites examples in which the whole membership of the brigade died several times over in the course of one gold-washing season on the Kolyma but the brigadier remained the same.

And the *two sets of bosses* were also convenient—in just the same way that pliers need both a right and left jaw. Two bosses—these were the hammer and the anvil, and they hammered out of the zek what the state required, and when he broke, they brushed him into the garbage bin. Two bosses—this was two tormenters instead of one, in shifts too, and placed in a situation of competition to see: who could squeeze more out of the prisoner and give him less.

Just as always in our well-thought-out social system, two different *plans* collided head on here too: the production plan, whose objective was to have the lowest possible expenditures for wages, and the MVD plan, whose objective was to extract the largest possible earnings from camp production. To an observer on the sidelines it seems strange: why set one's own plans in conflict with one another? Oh, but there is a profound meaning in it! Conflicting plans flatten the human being. This is a principle which far transcends the barbed wire of the Archipelago.

Chapter 6

■

"They've Brought the Fascists!"

"The Fascists"—as a nickname for the 58's [political prisoners]— was introduced by the sharp-eyed thieves and very much approved by the chiefs. This chapter tells of zeks arriving at camps in 1945, shortly after the Japanese surrender.

Chapter 7

■

The Way of Life and Customs of the Natives

To describe the native life in all its outward monotony would seem to be both very easy and very readily attainable. Yet it is very difficult at the same time. As with every different way of life, one has to describe the round of living from one morning until the next, from one winter to the next, from birth (arrival in one's first camp) until death (death). And simultaneously describe everything about all the many islands and islets that exist.

No one is capable of encompassing all this, of course, and it would merely be a bore to read whole volumes.

And the life of the natives consists of work, work, work; of starvation, cold, and cunning. This work, for those who are unable to push others out of the way and set themselves up in a soft spot, is that selfsame *general work* which raises socialism up out of the earth, and drives us down into the earth.

One cannot enumerate nor cover all the different aspects of this work, nor wrap your tongue about them. To push a wheelbarrow. ("Oh, the machine of the OSO, two handles and one wheel, so!") To carry hand barrows. To unload bricks barehanded (the skin quickly wears off the fingers). To haul bricks on one's own body by "goat" (in a shoulder barrow). To break up stone and coal in quarry and mine, to dig clay and sand. To hack out eight cubic yards of gold-bearing ore with a pick and haul them to the screening apparatus. Yes, and just to dig in the earth, just to "chew" up earth (flinty soil and in winter). To cut coal underground. And there are ores there too—lead and copper. Yes, and one can also . . . pulverize copper ore (a sweet taste in the mouth, and one waters at the nose). One can impregnate ties with creosote (and one's whole body at the same time too). One can carve out tunnels for railroads. And build roadbeds. One can dig peat in the bog, up to one's waist in the mud. One can smelt ores. One can cast metal. One can cut hay on hummocks in swampy meadows (sinking up to one's ankles in water). One can be a stableman or a drayman (yes, and steal oats from the horse's bag for one's own pot, but the horse is government-issue, the old grass-bag, and she'll last it out, most likely, but you can drop dead). Yes, and generally at the *"selkhozy"*—the Agricultural Camps —you can do every kind of peasant work (and there is no work better than that: you'll grab something from the ground for yourself).

But the father of all is our Russian forest with its genuinely golden tree trunks (gold is mined from them). And the oldest of all the kinds of work in the Archipelago is logging. It summons everyone to itself and has room for everyone, and it is not even out of bounds for cripples (they will send out a three-man gang of armless men to stamp down the foot-and-a-half snow). Snow comes up to your chest. You are a lumberjack. First you yourself stamp it down next to the tree trunk. You cut down the tree. Then, hardly able to make your way through the snow, you cut off all the branches (and you have to feel them out in the snow and get to them with your ax). Still dragging your way through the same loose snow, you have to carry off all the branches and make piles of them and burn them. (They smoke. They don't burn.) And now you have to saw up the wood to size and stack it. And the work norm for you and your brother for the day is six and a half cubic yards each, or thirteen cubic yards for two men working together. (In Burepolom the norm was nine cubic yards, but the thick pieces also had to be split into blocks.) By then your arms would not be capable of lifting an ax nor your feet of moving.

During the war years (on war rations), the camp inmates called three weeks at logging *"dry execution."*

You come to hate this forest, this beauty of the earth, whose praises have been sung in verse and prose. You come to walk beneath the arches of pine and birch with a shudder of revulsion! For decades in the future, you only have to shut your eyes to see those same fir and aspen trunks which you have hauled on your back to the freight car, sinking into the snow and falling down and hanging on to them tight, afraid to let go lest you prove unable to lift them out of the snowy mash.

Work at hard labor in Tsarist Russia was limited for decades by the Normative Statutes of 1869, which were actually issued for free persons. In assigning work, the physical strength of the worker and the degree to which he was accustomed to it were taken into consideration. (Can one nowadays really believe this?) The workday was set at seven hours (!) in winter and at twelve and a half hours in summer. As for Dostoyevsky's hard labor in Omsk, it is clear that in general they simply loafed about, as any reader can establish. The work there was agreeable and went with a swing, and the prison administration there even dressed them up in *white* linen jackets and trousers! After work the hard-labor convicts of the "House of the Dead" used to spend a long time *strolling* around the prison courtyard. That means that they were *not* totally fagged out! Indeed, the Tsarist censor did not want to pass the manuscript of *The House of the Dead* for fear that the *easiness* of the life depicted by Dostoyevsky would fail to deter people from crime. And so Dostoyevsky added new pages for the censor which demonstrated that life in hard labor was *nonetheless* hard! In our camps only the trusties went strolling around on Sundays, yes, and even they hesitated to. And Shalamov remarks with respect to the *Notes of Mariya Volkonskaya* that the Decembrist prisoners in Nerchinsk had a norm of 118 pounds of ore to mine and load each day. (One hundred and eighteen pounds! One could lift that all at once!) Whereas Shalamov on the Kolyma had a work norm per day of 28,800 pounds. And Shalamov writes that in addition their summer workday was sometimes sixteen hours long! I don't know how it was with sixteen, but for many it was thirteen hours long—on earth-moving work in Karlag and at the northern logging operations—and these were hours on the job itself, over and above the three miles' walk to the forest and three back. And anyway, why should we argue about the length of the day? After all, the *work norm* was senior in rank to the length of the workday, and when the brigade didn't fulfill the norm, the only thing that was changed at the end of the shift was the convoy, and the work sloggers were left in the woods by the light of searchlights until midnight—so that they got back to the camp just before morning in time to eat their dinner along with their breakfast and go out into the woods again.

There is no one to tell about it either. They all died.

And then here's another way they raised the norms and proved it was possible to fulfill them: In cold lower than 60 degrees below zero, workdays were written off; in other words, on such days the records showed that the workers had not gone out to work; but they chased them out anyway, and whatever they squeezed out of them on those days was added to the other days, thereby raising the percentages. (And the servile Medical Section wrote off those who froze to death on such cold days on some other basis. And the ones who were left who could no longer walk and were straining every sinew to crawl along on all fours on the way back to camp, the convoy simply shot, so that they wouldn't escape before they could come back to get them.)

And how did they feed them in return? They poured water into a pot, and the best one might expect was that they would drop unscrubbed small potatoes into it, but otherwise black cabbage, beet tops, all kinds of trash. Or else vetch or bran, they didn't begrudge these. (And wherever there was a water shortage, as there was at the Samarka Camp near Karaganda, only one bowl of gruel was cooked a day, and they also gave out a ration of two cups of turbid salty water.) Everything any good was always and without fail stolen for the chiefs, for the trusties, and for the thieves—the cooks were all terrorized, and it was only by submissiveness that they kept their jobs. Certain amounts of fat and meat "subproducts" (in other words, not real food) were signed out from the warehouses, as were fish, peas, and cereals. But not much of that ever found its way into the mouth of the pot. The worse the food, the more of it they gave the zeks. They used to give them horse meat from exhausted horses driven to death at work, and, even though it was quite impossible to chew it, it was a feast.

It was impossible to try to keep nourished on Gulag norms anyone who worked out in the bitter cold for thirteen or even ten hours. And it was completely impossible once the basic ration had been plundered. . . .

And how were our natives dressed and shod?

All archipelagoes are like all archipelagoes: the blue ocean rolls about them, coconut palms grow on them, and the administration of the islands does not assume the expense of clothing the natives—they go about barefoot and almost naked. But as for our cursed Archipelago, it would have been quite impossible to picture it beneath the hot sun; it was eternally covered with snow and the blizzards eternally raged over it. And in addition to everything else it was necessary to clothe and to shoe all that horde of ten to fifteen million prisoners.

Fortunately, born outside the bounds of the Archipelago, the zeks arrived here not altogether naked. They wore what they came in—more accurately, what the *socially friendly* elements might leave of it—except that as a brand of the Archipelago, a piece had to be torn off, just as they clip one ear of the ram. . . . But alas, the clothing of free men is not eternal, and footgear can be in shreds in a week from the stumps and hummocks of the Archipelago. And therefore it is necessary to clothe the natives, even though they have nothing with which to pay for the clothing.

Someday the Russian stage will yet see this sight! And the Russian cinema screen! The pea jackets one color and their sleeves another. Or so many patches on the pea jacket that its original cloth is totally invisible. Or a *flaming* pea jacket—with tatters on it like tongues of flame. Or patches on britches made from the wrappings of someone's food parcel from home, and for a long while to come one can still read the address written in the corner with an indelible pencil.

And on their feet the tried and true Russian "lapti"—bast sandals —except that they had no decent "onuchi"—footcloths—to go with them. Or else they might have a piece of old automobile tire, tied right on the bare foot with a wire, an electric cord. (Grief has its own inventiveness. . . .)

And then, in addition, bronze-gray camp faces will appear on the screen. Eyes oozing with tears, red eyelids. White cracked lips, covered with sores. Skewbald, unshaven bristles on the faces. In winter . . . a summer cap with earflaps sewn on.

I recognize you! It is you, the inhabitants of my Archipelago!

But no matter how many hours there are in the working day— sooner or later sloggers will return to the barracks.

Their barracks? Sometimes it is a dugout, dug into the ground. And in the North more often . . . *a tent*—true, with earth banked and reinforced hit or miss with boards. Often there are kerosene lamps in place of electricity, but sometimes there are the ancient Russian "splinter lamps" or else cotton-wool wicks. It is by this pitiful light that we will survey this ruined world.

Sleeping shelves in two stories, sleeping shelves in three stories, or, as a sign of luxury, "vagonki"—multiple bunks—the boards most often bare and nothing at all on them; on some of the work parties they steal so thoroughly (and then sell the spoils through the free employees) that nothing government-issue is given out and no one keeps anything of his own in the barracks; they take both their mess tins and their mugs to work with them (and even tote the bags containing their belongings— and thus laden they dig in the earth); those who have them put their blankets around their necks (a film scene!), or else lug their things to trusty friends in a guarded barracks. During the day the barracks are as empty as if uninhabited. At night they might turn over their wet work clothes to be dried in the drier (if there is a drier!)—but undressed like that you are going to freeze on the bare boards! And so they dry their clothes on themselves. At night their caps may freeze to the wall of the tent—or, in a woman's case, her hair. They even hide their bast sandals under their heads so they won't be stolen off their feet. In the middle of the barracks there is an oil drum with holes in it which has been converted into a stove, and it is good when it gets red-hot—then the steamy odor of drying footcloths permeates the entire barracks—but it sometimes happens that the wet firewood in it doesn't burn. Some of the barracks are so infested with insects that even four days' fumigation with burning sulphur doesn't help and when in the summer the zeks go out to sleep on the ground in the camp compound the bedbugs crawl after them and find them even there. And the zeks boil the lice off their underwear in their mess tins after dining from them.

All this became possible only in the twentieth century, and comparison here with the prison chroniclers of the past century is to no avail; they didn't write of anything like this.

And one must remember as well that everything that has been said refers to the established camp in operation for some time. But that camp had to be *started* at some time and by someone (and by whom if not by our unhappy brother zeks, of course?): they came to a cold, snowy woods, they stretched wire on the trees, and whoever managed to survive until the first barracks knew those barracks would be for the guard anyway.

Now that is the way of life of my Archipelago.

■

Philosophers, psychologists, medical men, and writers could have observed in our camps, as nowhere else, in detail and on a large scale the special process of the narrowing of the intellectual and spiritual horizons of a human being, the reduction of the human being to an animal and the process of dying alive. But the psychologists who got into our camps were for the most part not up to observing; they themselves had fallen into that very same stream that was dissolving the personality into feces and ash.

Just as nothing that contains life can exist without getting rid of its wastes, so the Archipelago could not keep swirling about without precipitating to the bottom its principal form of waste—the *last-leggers*. And everything built by the Archipelago had been squeezed out of the muscles of the last-leggers (before they became last-leggers). And those who survived, who reproach *the last-leggers with being themselves to blame,* must take upon themselves the disgrace of their own preserved lives.

And among the surviving, the orthodox Communists now write me lofty protests: How base are the thoughts and feelings of the heroes of your story *One Day in the Life of Ivan Denisovich!* Where are their anguished cogitations about the course of history? Everything is about bread rations and gruel, and yet there are sufferings much more unbearable than hunger.

Oh—so there are! Oh—so there are indeed much more unbearable sufferings (the sufferings of orthodox thought)? You in your medical sections and your storerooms, you never knew hunger there, orthodox loyalist gentlemen!

It has been known for centuries that Hunger . . . rules the world! (And all your Progressive Doctrine is, incidentally, built on Hunger, on the thesis that hungry people will inevitably revolt against the well-fed.) Hunger rules every hungry human being, unless he has himself consciously decided to die. Hunger, which forces an honest person to reach out and steal ("When the belly rumbles, conscience flees"). Hunger, which compels the most unselfish person to look with envy into someone else's bowl, and to try painfully to estimate what weight of ration his neighbor is receiving. Hunger, which darkens the brain and refuses to allow it to be distracted by anything else at all, or to think about anything else at all, or to speak about anything else at all except food, food, and food. Hunger, from which it is impossible to escape even in dreams—dreams are about food, and insomnia is over food. And soon —just insomnia. Hunger, after which one cannot even eat up; the man has by then turned into a one-way pipe and everything emerges from him in exactly the same state in which it was swallowed.

And this, too, the Russian cinema screen must see: how the last-leggers, jealously watching their competitors out of the corners of their eyes, stand duty at the kitchen porch waiting for them to bring out the slops in the dishwater. How they throw themselves on it, and fight with one another, seeking a fish head, a bone, vegetable parings. And how one last-legger dies, killed in that scrimmage. And how immediately afterward they wash off this waste and boil it and eat it. (And inquisitive cameramen can continue with their shooting and show us how, in 1947 in Dolinka, Bessarabian peasant women who had been brought in from *freedom* hurled themselves with that very same intent on slops which the last-leggers had *already checked over.*) The screen will show bags of bones which are still joined together lying under blankets at the hospital, dying almost without movement—and then being carried out. And on the whole . . . how simply a human being dies: he was speaking —and he fell silent; he was walking along the road—and he fell down. "Shudder and it's over." How the fat-faced, socially friendly work assigner jerks a zek by the legs to get him out to line-up—and he turns out to be dead, and the corpse falls on its head on the floor. "Croaked, the scum!" And he gaily gives him a kick for good measure. (At those camps during the war there was no doctor's aide, not even an orderly, and as a result there were no sick, and anyone who pretended to be sick was taken out to the woods in his comrades' arms, and they also took a board and rope along so they could drag the corpse back the more easily. At work they laid the sick person down next to the bonfire, and it was to the interest of both the zeks and the convoy to have him die the sooner.)

What the screen cannot catch will be described to us in slow, meticulous prose, which will distinguish between the nuances of the various paths to death, which are sometimes called scurvy, sometimes pellagra, sometimes alimentary dystrophy. For instance, if there is blood on your bread after you have taken a bite—that is scurvy. From then on your teeth begin to fall out, your gums rot, ulcers appear on your legs, your flesh will begin to fall off in whole chunks, and you will begin to smell like a corpse. Your bloated legs collapse. They refuse to take such cases into the hospital, and they crawl on all fours around the camp compound. But if your face grows dark and your skin begins to peel and your entire organism is racked by diarrhea, this is pellagra. It is necessary to halt the diarrhea somehow—so they take three spoons of chalk a day, and they say that in this case if you can get and eat a lot of herring the food will begin to hold. But where are you going to get herring? The man grows weaker, weaker, and the bigger he is, the faster it goes. He has already become so weak that he cannot climb to the top bunks, he cannot step across a log in his path; he has to lift his leg with his two hands or else crawl on all fours. The diarrhea takes out of a man both strength and all interest—in other people, in life, in himself. He grows deaf and stupid, and he loses all capacity to weep, even when he is being dragged along the ground behind a sledge. He is no longer afraid of death; he is wrapped in a submissive, rosy glow. He has crossed all boundaries and has forgotten the name of his wife, of his children, and finally his own name too. Sometimes the entire body of a man dying of starvation is covered with blue-black pimples like peas, with pus-filled heads smaller than a pinhead—his face, arms, legs, his trunk, even his scrotum. It is so painful he cannot be touched. The tiny boils come to a head and burst and a thick wormlike string of pus is forced out of them. The man is rotting alive.

If black astonished head lice are crawling on the face of your neighbor on the bunks, it is a sure sign of death.

Fie! What naturalism. Why keep talking about all that?

And that is what they usually say today, those who did not themselves suffer, who were themselves the executioners, or who have washed their hands of it, or who put on an innocent expression: Why remember all that? Why rake over old wounds? (*Their* wounds!!)

∎

In our glorious Fatherland, which was capable *for more than a hundred years* of not publishing the work of Chaadayev because of his reactionary views, you see, you are not likely to surprise anyone with the fact that the most important and boldest books are never read by contemporaries, never exercise an influence on popular thought in good time. And thus it is that I am writing this book solely from a sense of obligation—because too many stories and recollections have accumulated in my hands and I cannot allow them to perish. I do not expect to see it in print anywhere with my own eyes; and I have little hope that those who managed to drag their bones out of the Archipelago will ever read it; and I do not at all believe that it will explain the truth of our history in time for anything to be corrected.

∎

But there is one form of *early release* that no bluecap can take away from the prisoner. This release is—death.

And this is the most basic, the steadiest form of Archipelago output there is—with no norms.

From the fall of 1938 to February, 1939, at one of the Ust-Vym camps, 385 out of 550 prisoners died. Certain work brigades died off totally, including the brigadiers. In the autumn of 1941, Pechorlag (the railroad camp) had a listed population of fifty thousand prisoners, and in the spring of 1942, ten thousand. *During this period not one prisoner transport was sent out of Pechorlag anywhere—so where* did the *forty thousand* prisoners go? I have written *thousand* here in italics—why? Because I learned these figures accidentally from a zek who had access to them. But you would not be able to get them for all camps in all periods nor to total them up.

Corpses withered from pellagra (no buttocks, and women with no breasts), or rotting from scurvy, were checked out in the morgue cabin and sometimes in the open air. This was seldom like an autopsy—a long vertical cut from neck to crotch, breaking leg bones, pulling the skull apart at its seam. Mostly it was not a surgeon but a convoy guard who *verified* the corpse—to be certain the zek was really dead and not pretending. And for this they ran the corpse through with a bayonet or smashed the skull with a big mallet. And right there they tied to the big toe of the corpse's right foot a tag with his prison file number, under which he was identified in the prison lists.

At one time they used to bury them in their underwear but later on in the very worst, lowest-grade, which was dirty gray. And then came an across-the-board regulation not to waste any underwear on them at all (it could still be used for the living) but to bury them naked.

At one time in Old Russia it was thought that a corpse could not get along without a coffin. Even the lowliest serfs, beggars, and tramps were buried in coffins. Even the Sakhalin and the Akatui hard-labor prisoners were buried in coffins. But in the Archipelago this would have amounted to the unproductive expenditure of millions on labor and lumber. When at Inta after the war one honored foreman of the woodworking plant was actually buried in a coffin, the Cultural and Educational Section was instructed to make propaganda: Work well *and you, too, will be buried in a wooden coffin.*

The corpses were hauled away on sledges or on carts, depending on the time of year. Sometimes, for convenience, they used one box for six corpses, and if there were no boxes, then they tied the hands and legs with cord so they didn't flop about. After this they piled them up like logs and covered them with bast matting. If there was ammonal available, a special brigade of gravediggers would dynamite pits for them. Otherwise they had to dig the graves, always common graves, in the ground: either big ones for a large number or shallow ones for four at a time. (In the springtime, a stink used to waft into the camp from the shallower graves, and they would then send last-leggers to deepen them.)

On the other hand, no one can accuse us of gas chambers.

Where there was more time to spare on such things—as, for example, in Kengir—they would set out little posts on the hillocks, and a representative of the Records and Classification Section, no less, would personally inscribe on them the inventory numbers of those buried there. However, in Kengir someone also did some wrecking: Mothers and wives who came there were shown the cemetery and they went there to mourn and weep. Thereupon the chief of Steplag, Comrade

Colonel Chechev, ordered the bulldozers to bulldoze down the little grave posts and level off the hillocks—because of this lack of gratitude.

Now that, fair reader, is how your father, your husband, your brother, was buried.

Chapter 8

■

Women in Camp

And how could one not think of them, even back during interrogation? One day, one of the Butyrki jailers was fussing with a lock, and left our men's cell to stand half a minute at the windows in the well-lit upper corridor, and, peering underneath the "muzzle" of a corridor window, we suddenly saw down below, in the little green garden on a corner of asphalt, standing in line in pairs like us—and also waiting for a door to be opened—women's shoes and ankles! All we could see were just ankles and shoes, but on high heels! And it was like a Wagnerian blast from *Tristan and Isolde.* We could see no higher than that and the jailer was already driving us into the cell, and once inside we raved there, illumined and at the same time beclouded, and we pictured all the rest to ourselves, imagining them as heavenly beings dying of despondency. What were they like? What were they like!

But it seems that things were no harder for them and maybe even easier. I have so far found nothing in women's recollections of interrogation which could lead me to conclude that they were any more disheartened than we were or that they became any more deeply depressed. The gynecologist N. I. Zubov, who served ten years himself and who in camp was constantly engaged in treating and observing women, says, to be sure, that statistically women react more swiftly and more sharply to arrest than men and to its principal effect—the loss of the family. The woman arrested is spiritually wounded and this expresses itself most often in the cessation of the vulnerable female functions.

But of course for all of us, and for women in particular, prison was just the flower. The berries came later—camp. And it was precisely in camp that the women would either be broken or else, by bending and degenerating, adapt themselves.

In camp it was the opposite—everything was harder for the women than for us men. Beginning with the camp filth. Having already suffered from the dirt in the transit prisons and on the prisoner transports themselves, the woman would then find no cleanliness in camp either. In the average camp, in the women's work brigades, and also, it goes without saying, in the common barracks, it was almost never possible for her to feel really clean, to get warm water (and sometimes there was no water at all). There was no lawful way a woman could lay hands on either cheesecloth or rags. No place there, of course, to do laundry!

A bath? Well! The initial arrival in camp began with a bath—if one doesn't take into account the unloading of the zeks from the cattle car onto the snow, and the march across with one's things on one's back surrounded by convoy and dogs. In the camp bath the naked women were examined like merchandise. Whether there was water in the bath or not, the inspection for lice, the shaving of armpits and pubic hair, gave the barbers, by no means the lowest-ranking aristocrats in the camp, the opportunity to look over the new women. And immediately after that they would be inspected by the other trusties. This was a tradition going right back to the Solovetsky Islands. Except that there, at the dawn of the Archipelago, a shyness still existed, not typical of the natives—and they were inspected clothed, during auxiliary work. But the Archipelago hardened, and the procedure became more brazen. Fedot S. and his wife (it was their fate to be united!) now recollect with amusement how the male trusties stood on either side of a narrow corridor and passed the newly arrived women through the corridor naked, not all at once, but one at a time. And then the trusties decided among themselves who got whom. (According to the statistics of the twenties there was one woman serving time for every six or seven men. After the decrees of the thirties and forties the proportion of women to men rose substantially—but still not sufficiently for women not to be valued, particularly the attractive ones.) In certain camps a polite procedure was preserved: The women were conducted to their barracks —and then the well-fed, self-confident, and impudent trusties entered the barracks, dressed in new padded jackets (any clothing in camp which was not in tatters and soiled seemed mad foppery). Slowly and deliberately they strolled between the bunks and made their choices. They sat down and chatted. They invited their choices to "visit" them. And they were living, too, not in a common-barracks situation, but in cabins occupied by several men. And there they had hot plates and frying pans. And they had fried potatoes too! An unbelievable dream! The first time, the chosen women were simply feasted and given the chance to make comparisons and to discover the whole spectrum of camp life. Impatient trusties demanded "payment" right after the potatoes, while those more restrained escorted their dates home and explained the future. You'd better make your arrangements, make your arrangements, inside *the camp compound,* darling, while it is being proposed in a gentlemanly way. There's cleanliness here, and laundry facilities, and decent clothes and unfatiguing work—and it's all yours.

And it is true there are women who by their own nature, out in freedom too, by and large, get together with men easily, without being choosy. Such women, of course, always had open to them easy ways out. Personal characteristics do not get distributed simply on the basis of the *articles* of the Criminal Code, yet we are not likely to be in error if we say that the majority of women among the 58's were not of this kind. For some of them, from the beginning to the end, this step was less bearable than death. Others would bridle, hesitate, be embarrassed (and they were held back by shame before their girl friends too), and when they had finally decided, when they had reconciled themselves— it might be too late, they might not find a camp taker any longer.

Because not every one was lucky enough *to get propositioned.*

Thus many of them gave in during the first few days. The future looked too cruel—and there was no hope at all. And this choice was made by those who were almost little girls, along with solidly married women and mothers of families. And it was the little girls in particular, stifled by the crudity of camp life, who quickly became the most reckless of all.

What if you said . . . no? All right, that's your lookout! Put on britches and pea jacket. And go marching off to the woods, with your formless, fat exterior, and your frail inner being. You'll come crawling yet. You'll go down on bended knees.

And what of it if you loved someone out in freedom and wanted to remain true to him? What profit is there in the fidelity of a female corpse? *"When you get back to freedom—who is going to need you?"* Those were the words which kept ringing eternally through the women's barracks. You grow coarse and old and your last years as a woman are cheerless and empty. Isn't it smarter to hurry up and grab something too, even from this savage life?

And it was all made easier by the fact that no one here condemned anyone else. "Everyone lives like that here."

And hands were also untied by the fact that there was no meaning, no purpose, left in life.

A multiple bunk curtained off with rags from the neighboring women was a classic camp scene. But things could be a great deal simpler than that too. This again refers to the Krivoshchekovo Camp No. 1, 1947–1949. (We know of this No. 1, but how many were there?) At this camp there were thieves, nonpolitical offenders, juveniles, invalids, women and nursing mothers, all mixed up together. There was just one women's barracks—but it held five hundred people. It was indescribably filthy, incomparably filthy and rundown, and there was an oppressive smell in it and the bunks were without bedding. There was an official prohibition against men entering it, but this prohibition was ignored and no one enforced it. Not only men went there, but juveniles too, boys from twelve to thirteen, who flocked in to learn. First they began with simple observation of what was going on; there was no false modesty there, whether because there were no rags or perhaps not enough time; at any rate *the bunks were not curtained off.* And, of course, the light was never doused either. Everything took place very naturally as in nature in full view, and in several places at once. Obvious old age and obvious ugliness were the only defenses for a woman there —nothing else. Attractiveness was a curse. Such a woman had a constant stream of visitors on her bunk and was constantly surrounded. They propositioned her and threatened her with beatings and knives— and she had no hope of being able to stand up against it but only to be smart about whom she gave in to—to pick the kind of man to defend her with his name and his knife from all the rest, from the next in line, from the whole greedy queue, from those crazy juveniles gone berserk, aroused by everything they could see and breathe in there. And it wasn't only men that she had to be defended against either. Nor only the juveniles who were aroused. What about the women next to them, who day after day had to see all that but were not themselves invited by the men? In the end those women, too, would explode in an uncontrollable rage and hurl themselves on their successful neighbors and beat them up.

And then, too, venereal diseases were nearly epidemic at Krivoshchekovo. There was a rumor that nearly half the women were infected, but there was no way out, and on and on both the sovereigns and the suppliants kept crossing the same threshold. And only those who were very foresighted, like the accordionist K., who had his own connections in the Medical Section, could each time check the secret list of the venereal-disease patients for himself and his friends in order not to get caught.

Here is what *women's* work was like in Krivoshchekovo. At the brickyard, when they had completed working one section of the clay pit, they used to take down the overhead shelter (before they had mined there, it had been laid out on the surface of the earth). And now it was necessary to hoist wet beams ten to twelve yards up out of a big pit. How was it done? The reader will say: with machines. Of course. A women's brigade looped a cable around each end of a beam, and in two rows like barge haulers, keeping even so as not to let the beam drop and then have to begin over again, pulled one side of each cable and . . . out came the beam. And then a score of them would hoist up one beam on their shoulders to the accompaniment of command oaths from their out-and-out slave driver of a woman brigadier and would carry the beam to its new place and dump it there. A tractor, did you say? But, for pity's sakes, where would you get a tractor in 1948? A crane, you say? But you have forgotten Vyshinsky: "work, the miracle worker which transforms people from nonexistence and insignificance into heroes"? If there were a crane . . . then what about the miracle worker? If there were a crane . . . then these women would simply wallow in insignificance!

The body becomes worn out at that kind of work, and everything that is feminine in a woman, whether it be constant or whether it be monthly, ceases to be. If she manages to last to the next "commissioning," the person who undresses before the physicians will be not at all like the one whom the trusties smacked their lips over in the bath corridor: she has become ageless; her shoulders stick out at sharp angles, her breasts hang down in little dried-out sacs; superfluous folds of skin form wrinkles on her flat buttocks; there is so little flesh above her knees that a big enough gap has opened up for a sheep's head to stick through or even a soccer ball; her voice has become hoarse and rough and her face is tanned by pellagra. (And, as a gynecologist will tell you, several months of logging will suffice for the prolapse and falling out of a more important organ.)

Work—*the miracle worker!*

External legislation (for outside Gulag) seemingly abetted camp love. An All-Union Decree of July 8, 1944, on the strengthening of marriage ties was accompanied by an unpublished decree of the Council of People's Commissars and an instruction of the People's Commissariat of Justice dated November 27, 1944, in which it was stated that the court was required to dissolve unconditionally a marriage with a spouse in prison (or in an insane asylum) at the first indication of desire on the part of a free Soviet person, and even to encourage this by freeing such a person from the fee for issuance of a divorce decree. (And at the same time no one was obliged legally to inform the other spouse of the accomplished divorce!) By this token, citizenesses and citizens were called on to abandon their imprisoned wives and husbands all the more speedily in misfortune. And prisoners were correspondingly invited . . . to forget about their marriages all the more thoroughly.

Yes, the zeks were to forget about their marriages, but Gulag instructions also forbade indulgence in love affairs as a diversionary action against the production plan. After all, these unscrupulous women who wandered about the work sites, forgetting their obligations to the state and the Archipelago, were ready to lie down on their backs anywhere at all—on the damp ground, on wood chips, on road stone, on slag, on iron shavings—and the plan would collapse! And the Five-Year Plan would mark time! And there would be no prize money for the Gulag chiefs! And besides some of those *zechkas* secretly nurtured a desire to get pregnant and, on the strength of this pregnancy, exploiting the humanitarianism of our laws, to snatch several months off their terms, which were often a short three or five years anyway, and not work at all those months. That was why Gulag instructions required that any prisoners caught cohabiting should be immediately separated, and that the less useful of the two should be sent off on a prisoner transport.

Plundered of everything that fulfills female life and indeed human life in general—of family, motherhood, the company of friends, familiar and perhaps even interesting work, in some cases perhaps in art or among books, and crushed by fear, hunger, abandonment, and savagery—what else could the women camp inmates turn to except love? With God's blessing the love which came might also be almost not of the flesh, because to do it in the bushes was shameful, to do it in the barracks in everyone's presence was impossible, and the man was not always up to it, and then the jailers would drag the culprits out of every *hideout* (seclusion) and put them in the punishment block. But from its unfleshly character, as the women remember today, the spirituality of camp love became even more profound. And it was particularly because of the absence of the flesh that this love became more poignant than out in freedom! Women who were already elderly could not sleep nights because of a chance smile, because of some fleeting mark of attention they had received. So sharply did the light of love stand out against the dirty, murky camp existence!

But it was not only the custodial staff and camp chiefs who would break up camp marriages. The Archipelago was such an upside-down land that in it a man and a woman could be split up by what ought to have united them even more firmly: the birth of a child. A month before giving birth a pregnant woman was transported to another camp, where there was a camp hospital with a maternity ward and where husky little voices shouted that they did not want to be zeks because of the sins of their parents.

And these issues of whether to give birth or not, which were difficult enough for any woman at all, were still more confused for a woman camp inmate. And what would happen to the child subsequently? And if such a fickle camp fate gave one the chance to become pregnant by one's loved one, then how could one go ahead and have an abortion? Should you have the child? That meant certain separation immediately, and when you left would he not pair off with some other woman in the same camp? And what kind of child would it be? (Because of the malnutrition of the parents it was often defective.) And when you stopped nursing the child and were sent away (you still had many years left to serve), would they keep an eye out so as not to do him in? And would you be able to take the child into your own family? (For some this was excluded.) And if you didn't take him, would your conscience then torment you all your life?

But why rake up all that past? Why reopen the old wounds of those who were living in Moscow and in country houses at the time, writing for the newspapers, speaking from rostrums, going off to resorts and abroad?

Why recall all that when it is still the same even today? After all, you can only write about whatever "will not be repeated."

Chapter 9

■

The Trusties

"Trusties" were prisoners who got themselves what were by camp standards soft jobs. They were despised by other prisoners.

Chapter 10

■

In Place of Politicals

But in that grim world where everyone gnawed up whomever he could, where a human's life and conscience were bought for a ration of soggy bread—in that world who and where were *the politicals,* bearers of the honor and the torch of all the prison populations of history?

We have already traced how the original "politicals" were divided, stifled, and exterminated.

And in their place?

Well—what did take their place? Since then we have had no politicals. And we could not possibly have any. What kind of "politicals" could we have if universal justice had been established? They simply . . . abolished the politicals. There are none, and there won't be any.

The village club manager went with his watchman to buy a bust of Comrade Stalin. They bought it. The bust was big and heavy. They ought to have carried it in a hand barrow, both of them together, but the manager's status did not allow him to. "All right, you'll manage it if you take it slowly." And he went off ahead. The old watchman couldn't work out how to do it for a long time. If he tried to carry it at his side, he couldn't get his arm around it. If he tried to carry it in front of him, his back hurt and he was thrown off balance backward. Finally he figured out how to do it. He took off his belt, made a noose for Comrade Stalin, put it around his neck, and in this way carried it over his shoulder through the village. Well, there was nothing here to argue about. It was an open-and-shut case. Article 58-8, terrorism, ten years.

A shepherd in a fit of anger swore at a cow for not obeying: "You collective-farm wh—!" And he got 58, and a term.

A *deaf and dumb* carpenter got a term for counterrevolutionary *agitation!* How? He was laying floors in a club. Everything had been removed from a big hall, and there was no nail or hook anywhere. While he was working, he hung his jacket and his service cap on a bust of Lenin. Someone came in and saw it. 58, ten years.

The children in a collective farm club got out of hand, had a fight, and accidentally knocked some poster or other off the wall with their backs. The two eldest were sentenced under Article 58. (On the basis of the Decree of 1935, children from the age of twelve on had full criminal responsibility for all crimes!) They also sentenced the parents for having allegedly told them to and sent them to do it.

A sixteen-year-old Chuvash schoolboy made a mistake in Russian in a slogan in the wall newspaper; it was not his native language. Article 58, five years.

And in a state farm bookkeeping office the slogan hung: "Life has become better; life has become more gay. (Stalin)" And someone added a letter in red pencil to Stalin's name, making the slogan read as though life had become more gay *for* Stalin. They didn't look for the guilty party—but sentenced the entire bookkeeping office.

Boris Mikhailovich Vinogradov, with whom I served time in prison, had in his youth been a locomotive engineer. After the workers' school and an institute, he became a railway transport engineer (and was not put immediately on Party work, as often happens too), and he was a good engineer (in the sharashka he carried out complex calculations in gas dynamics for jet turbines). But by 1941, it's true, he had become the Party organizer of the Moscow Institute for Railroad Engineering. In the bitter Moscow days of October 16 and 17, 1941, seeking instructions, he telephoned but no one replied. He went to the District Party Committee, the City Party Committee, the Provincial Party Committee, and found no one there; everyone had scattered to the winds; their chambers were empty. And it seems he didn't go any higher than that. He returned to his own people in the Institute and declared: "Comrades! All the leaders have run away. But we are Communists, we will join the defense." And they did just that. But for that remark of his, "They have all run away," those who had run away sent him who had not run away to prison for eight years—for Anti-Soviet Propaganda. He was a quiet worker, a dedicated friend, and only in heart-to-heart conversation would he disclose that he believed, believes, and will go on believing. And he never wore it on his sleeve.

Irina Tuchinskaya was arrested while leaving church. (The intention was to arrest their whole family.) And she was charged with having "prayed in church for the death of Stalin." (Who could have heard that prayer?!) Terrorism! Twenty-five years!

However, for the most part fantastic accusations were not really required. There existed a very simple standardized collection of charges from which it was enough for the interrogator to pick one or two and stick them like postage stamps on an envelope:

- Discrediting the Leader
- A negative attitude toward the collective-farm structure
- A negative attitude toward state loans (and what normal person could have had a positive attitude!)
- A negative attitude toward the Stalinist constitution
- A negative attitude toward whatever was the immediate, particular measure being carried out by the Party
- Sympathy for Trotsky
- Friendliness toward the United States
- Etc., etc., etc.

The pasting on of these stamps of varying value was monotonous work requiring no artistry whatsoever. All the interrogator needed was the next victim in line, so as not to lose time. Such victims were selected on the basis of arrest quotas by Security chiefs of local administrative districts, military units, transportation departments, and educational institutions. And so that the Security chiefs did not have to strain their brains, denunciations from informers came in very handy.

In the conflicts between people in freedom, denunciations were the superweapon, the X-rays: it was sufficient to direct an invisible little ray at your enemy—and he fell. And it always worked. I can affirm that I heard *many* stories in imprisonment about the use of denunciations in lovers' quarrels: a man would remove an unwanted husband; a wife would dispose of a mistress, or a mistress would dispose of a wife; or a mistress would take revenge on her lover because she had failed to separate him from his wife.

Europe, of course, won't believe it. Not until Europe itself *serves time* will she believe it. Europe has believed our glossy magazines and can't get anything else into her head.

Chapter 11

■

The Loyalists

Here we shall concern ourselves particularly with those orthodox Communists who made a display of their ideological orthodoxy first to the interrogator, then in the prison cells, and then in camp to all and everyone, and now recall their camp past in this light.

By a strange selective process none of them will be sloggers. Such people ordinarily had held big jobs before their arrest, and had had an enviable situation; and in camp they found it hardest of all to reconcile themselves to extinction, and they fought fiercest of all to rise above the universal zero. In this category were all the interrogators, prosecutors, judges, and camp officials who had landed behind bars. And all the theoreticians, dogmatists, and loud-mouths.

We have to understand them, and we won't scoff at them. It was painful for them to fall. "When you cut down trees, the chips will fly!" was the cheerful proverb of justification. And then suddenly they themselves were chopped off with all the other chips.

To say that things were *painful* for them is to say almost nothing. They were incapable of assimilating such a blow, such a downfall, and from their *own people* too, from their dear Party, and, from all appearances, for nothing at all. After all, they had been guilty of nothing as far as the Party was concerned—nothing at all.

It was painful for them to such a degree that it was considered taboo among them, uncomradely, to ask: "What were you imprisoned for?" The only squeamish generation of prisoners! The rest of us, in 1945, with tongues hanging out, used to recount our arrests, couldn't wait to tell the story to every chance newcomer we met and to the whole cell, as if it were an anecdote.

Here's the sort of people they were. Olga Sliozberg's husband had already been arrested, and they had come to carry out a search and arrest her too. The search lasted four hours—and she spent those four hours sorting out the minutes of the congress of Stakhanovites of the bristle and brush industry, of which she had been the secretary until the previous day. The incomplete state of the minutes troubled her more than her children, whom she was leaving forever! Even the interrogator conducting the search could not resist telling her: "Come on now, say farewell to your children!"

Here's the sort of people they were. A letter from her fifteen-year-old daughter came to Yelizaveta Tsvetkova in the Kazan Prison for long-term prisoners: "Mama! Tell me, write to me—are you guilty or not? I hope you weren't guilty, because then I won't join the Komsomol, and I won't forgive them because of you. But if you are guilty —I won't write you any more and will hate you." And the mother was stricken by remorse in her damp gravelike cell with its dim little lamp: How could her daughter live without the Komsomol? How could she be permitted to hate Soviet power? Better that she should hate me. And she wrote: "I am guilty. . . . Enter the Komsomol!"

How could it be anything but hard! It was more than the human heart could bear: to fall beneath the beloved ax—then to have to justify its wisdom.

But that is the price a man pays for entrusting his God-given soul to human dogma.

Even today any orthodox Communist will affirm that Tsvetkova acted correctly. Even today they cannot be convinced that this is precisely the "perversion of small forces," that the mother perverted her daughter and harmed her soul.

Here's the sort of people they were: Y.T. gave sincere testimony against her husband—anything to aid the Party!

Oh, how one could pity them if at least now they had come to comprehend their former wretchedness!

This whole chapter could have been written quite differently if today at least they had forsaken their earlier views!

Loyalty? And in our view it is just plain pigheadedness. These devotees of the theory of development construed loyalty to that development to mean renunciation of any personal development whatsoever. As Nikolai Adamovich Vilenchik said, after serving seventeen years: "We believed in the Party—and we were *not mistaken!*" Is this loyalty or pigheadedness?

No, it was not for show and not out of hypocrisy that they argued in the cells in defense of all the government's actions. They needed ideological arguments in order to hold on to a sense of their own rightness—otherwise insanity was not far off.

How easily one could sympathize with them all! But they all see so clearly what their sufferings were—and they don't see wherein lies their own guilt.

This sort of person was not arrested before 1937. And after 1938 very few such people were arrested. And that is why they were named the "call-up of 1937," and this would be permissible but shouldn't be allowed to blur the overall picture: even at the peak they were not the only ones being arrested, and those same peasants, and workers, and young people, and engineers, and technicians, and agronomists, and economists, and ordinary believers continued to stream in as well.

The "call-up of 1937" was very loquacious, and having access to the press and radio created the "legend of 1937," a legend consisting of two points:

1. If they arrested people at all under the Soviet government, it was only in 1937, and it is necessary to speak out and be indignant only about 1937.
2. In 1937 they were . . . the only ones arrested.

At the very beginning of our book we gave a conspectus of the *waves* pouring into the Archipelago during the two decades up to 1937. How long all that dragged on! And how many millions there were! But the future call-up of 1937 didn't bat an eyelid and found it all normal. They remained calm while *society* was being imprisoned. Their "outraged reason boiled" when *their own fellowship* began to be imprisoned.

Of course, they did not remember how very recently they themselves had helped Stalin destroy the opposition, yes, and even themselves too. After all, Stalin gave his own weak-willed victims the opportunity of taking a chance and rebelling, for this game was not without its satisfactions for him. To arrest each member of the Central Committee required the sanction of all the others! That is something the playful tiger thought up. And while the sham plenums and conferences proceeded, a paper was passed along the rows which stated impersonally that materials had been received compromising a certain individual; and it was requested that consent be given (or refused!) to his expulsion from the Central Committee. (And someone else watched to see whether the person reading this paper held it for a long time.) And they all . . . signed their names. And that was how the Central Committee of the All-Union Communist Party (Bolsheviks) shot itself. (Stalin had calculated and verified their weakness even earlier than that: once the top level of the Party had accepted as their due high wages, secret provisioning facilities, private sanatoriums, it was already in the trap and there was no way to backtrack.)

And they had forgotten even more (yes, and had never read it anyway) such ancient history as the message of the Patriarch Tikhon to the Council of People's Commissars on October 26, 1918. Appealing for mercy and for the release of the innocent, the staunch Patriarch warned them: "That the blood of all the prophets which was shed from the foundation of the world may be required of this generation." (Luke 11:50.) And: ". . . for all they that take the sword shall perish with the sword." (Matthew 26:52.) But at that time it seemed absurd, impossible! How could they imagine at that time that History sometimes does know revenge, a sort of voluptuous and delayed justice, but chooses strange forms for it and unexpected executors of its will.

And though the curses of the women and children shot in the Crimean spring of 1921, as Voloshin has told us, were incapable of piercing the breast of Bela Kun, this was done by his own comrade in the Third International.

Here is their inevitable moral: I have been imprisoned for nothing and that means I am good, and that all these people around me are enemies and have been imprisoned for good cause.

And here is how their energy is spent: Six and twelve times a year they send off complaints, declarations, and petitions. And what do they write about? What do they scrawl in them? Of course, they swear loyalty to the Great Genius (and without that they won't be released). Of course, they dissociate themselves from those already shot in their case. Of course, they beg to be forgiven and permitted to return to their old jobs at the top. And tomorrow they will gladly accept any Party assignment whatever—even to run this camp! (And the fact that all the complaints and petitions were met with just as thick a shoal of rejections—well, that was because they didn't reach Stalin! He would have understood! He would have forgiven, the benefactor!) Fine "politicals" they were if they begged the government for . . . forgiveness.

Here was the level of their consciousness: V. P. Golitsyn, son of a district physician, a road engineer, was imprisoned for 140 (one hundred forty!) days in a death cell (plenty of time to think!). And then he got fifteen years, and after that external exile. "In my mind nothing changed. I was the same non-Party Bolshevik as before. My faith in the

Party helped me, the fact that the evil was being done not by the Party and government but by the evil will of *certain people* [what an analysis!] who came and went [but somehow they never seemed to go . . .], but all the rest [!!] remained. . . .

However—why this whole chapter? Why this whole lengthy survey and analysis of the loyalists? Instead we shall just write in letters a yard high:

JANOS KADAR, WLADYSLAW GOMULKA, and GUSTAV HUSAK

All three of them underwent unjust arrest and interrogation with torture, and all three served time so-and-so many years.

And the whole world sees how much they learned. The whole world has learned what they are worth.

Chapter 12

■

Knock, Knock, Knock . . .

In our technological years cameras and photoelectric elements often work in place of eyes, and microphones, tape recorders, and laser listening devices often replace ears. But for the entire epoch covered by this book almost the only eyes and almost the only ears of the Cheka-GB were *stool pigeons*.

Without having the experience and without having thought the matter over sufficiently, it is difficult to evaluate the extent to which we are permeated and enveloped by stool-pigeoning. Just as, without a transistor in hand, we do not sense in a field, in a forest, or on a lake that multitudes of radio waves are constantly pouring through us.

It is difficult to school oneself to ask that constant question: *Who is the stool pigeon among us?* In our apartment, in our courtyard, in our watch-repair shop, in our school, in our editorial office, in our workshop, in our design bureau, and even in our police. It is difficult to school oneself, and it is repulsive to become schooled—but for safety one must. It is impossible to expel the stoolies or to fire them—they will recruit new ones. But you have to *know* them—sometimes in order to beware of them; sometimes to put on an act in their presence, to pretend to be something you aren't; sometimes in order to quarrel openly with the informer and by this means devalue his testimony against you.

The poetry of recruitment of stool pigeons still awaits its artist. There is a visible life and there is an invisible life. The spiderwebs are stretched everywhere, and as we move we do not notice how they wind about us.

Selecting tools available for recruitment is like selecting master keys: No. 1, No. 2, No. 3. No. 1: "Are you a Soviet person?" No. 2 is to promise that which the person being recruited has fruitlessly sought by lawful means for many years. No. 3 is to bring pressure to bear on some weak point, to threaten a person with what he fears most of all. No. 4 . . .

You see, it only takes a tiny bit of pressure. A certain A.G. is called in, and it is well known that he is a nincompoop. And so to start he is instructed: "Write down a list of the people you know who have anti-Soviet attitudes." He is distressed and hesitates: "I'm not sure." He didn't jump up and didn't thump the table: "How dare you!" (Who does in our country? Why deal in fantasies!) "Aha, so you are not sure? Then write a list of people you can guarantee are one hundred percent Soviet people! But *you are guaranteeing,* you understand? If you provide even one of them with false references, you yourself will *go to prison* immediately. So why aren't you writing?" "Well, I . . . can't guarantee." "Aha, you can't? That means you know they are anti-Soviet. So write down immediately the ones you know about!" And so the good and honest rabbit A.G. sweats and fidgets and worries. He has too soft a soul, formed before the Revolution. He has sincerely accepted this pressure which is bearing down on him: Write either that they are Soviet or that they are anti-Soviet. He sees no third way out.

A stone is not a human being, and even stones get crushed.

■

Though he was an enlightened and irreligious person, U. discovered that the only defense against the security officers was to hide behind Christ. This was not very honest, but it was a sure thing. He lied: "I must tell you frankly that I had a Christian upbringing, and therefore it is quite impossible for me to work with you!"

And that ended it! And all the lieutenant's chatter, which had by then lasted many hours, simply stopped! The lieutenant understood he had drawn a bad number. "We need you like a dog needs five legs," he exclaimed petulantly. "Give me a written refusal." (Once again "written"!) "And write just that, explaining about your damned god!"

Apparently they have to close the case of every informer with a separate piece of paper, just as they open it with one. The reference to Christ satisfied the lieutenant completely: none of the security officers would accuse him subsequently of failing to use every effort he could.

And does the impartial reader not find that they flee from Christ like devils from the sign of the cross, from the bells calling to matins?

And that is why our Soviet regime can never come to terms with Christianity!

Chapter 13

■

Hand Over Your Second Skin Too!

Can you behead a man whose head has already been cut off? You can. Can you skin the hide off a man when he has already been skinned? You can!

This was all invented in our camps. This was all devised in the Archipelago! So let it not be said that the *brigade* was our only Soviet contribution to world penal science. Is not *the second camp term* a contribution too? The waves which surge into the Archipelago from outside do not die down there and do not subside freely, but are pumped through the pipes of the second interrogation.

Oh, blessed are those pitiless tyrannies, those despotisms, those savage countries, where a person once arrested cannot be arrested a second time! Where once in prison he cannot be reimprisoned. Where a person who has been tried cannot be tried again! Where a sentenced person cannot be sentenced again!

But in our country everything is permissible. When a man is flat on his back, irrevocably doomed and in the depths of despair, how convenient it is to poleax him again! The ethics of our prison chiefs are: "Beat the man who's down." And the ethics of our Security officers are: "Use corpses as steppingstones!"

We may take it that camp interrogations and camp court were born on Solovki, although what they did there was simply to push them into the bell-tower basement and finish them off. During the period of the Five-Year Plans and of the metastases, they began to employ the second camp term instead of the bullet.

For how otherwise, without second (or third or fourth) terms, could they secrete in the bosom of the Archipelago, and destroy, all those marked down for destruction?

The generation of new prison terms, like the growing of a snake's rings, is a form of Archipelago life. As long as our camps thrived and our exile lasted, this black threat hovered over the heads of the convicted: to be given a new term before they had finished the first one. Second camp terms were handed out every year, but most intensively in 1937 and 1938 and during the war years. (In 1948–1949 the burden of second terms was transferred outside: they *overlooked,* they missed, prisoners who should have been resentenced in camp—and then had to haul them back into camp from outside. These were even called *repeaters,* whereas those resentenced inside didn't get a special name.)

And it was a mercy—an automated mercy—when, in 1938, second camp terms were given out without any second arrest, without a camp interrogation, without a camp court, when the prisoners were simply called up in brigades to the Records and Classification Section and told to sign for their second terms. (For refusing to sign—you were simply put in punishment block, as for smoking where it wasn't allowed.) And they also had it all explained to them in a very human way: "We aren't telling you that you are guilty of anything, but just sign that you have been informed." And it was useless to try to get out of it as if, in the dark infinity of the Archipelago, eight was in any way distinct from eighteen, or a tenner at the start from a tenner at the end of a sentence. The only important thing was that they did not claw and tear your body today.

Now we can understand: The epidemic of camp sentences in 1938 was the result of a directive from above. It was there at the top that they suddenly came to their senses and realized that they had been handing out too little, that they had to pile it on (and shoot some too)—and thus frighten the rest.

But the epidemic of camp cases during the war was stimulated by a happy spark from *below* too, by the features of popular initiative. In all likelihood there was an order from above that during the war the most colorful and notable individuals in each camp, who might become centers of rebellion, had to be suppressed and isolated. The bloody local boys immediately sensed the riches in this vein—their own *deliverance from the front.* This was evidently guessed in more than one camp and rapidly taken up as useful, ingenious, and a salvation. The camp Chekists also helped fill up the machine-gun embrasures—but with other people's bodies.

Let the historian picture to himself the pulse of those years: The front was moving east, the Germans were around Leningrad, outside Moscow, in Voronezh, on the Volga, and in the foothills of the Caucasus. In the rear there were ever fewer men. Every healthy male figure aroused reproachful glances. Everything for the front! There was no price too big for the government to pay to stop Hitler. And only the camp officers (and their confreres in State Security) were well fed, white, soft-skinned, idle—all in their places in the rear. And the farther into Siberia and the North they were, the quieter things were. But we must soberly understand: theirs was a shaky prosperity. Due to end at the first outcry: Bring out those rosy-cheeked, smart camp fellows! No battle experience? So they had ideology. And they would be lucky to end up in the police, or in the behind-the-lines "obstacle" detachments, but it could happen otherwise; otherwise it was into officer battalions and be thrown into the Battle of Stalingrad! In the summer of 1942 they picked up whole officer-training schools and hurled them into the front, uncertified, their courses unfinished. All the young and healthy convoy guards had already been scraped up for the front. And the camps hadn't fallen apart. It was all right. And they wouldn't fall apart if the security officers were called up either! (There were already rumors.)

Draft deferment—that was life. Draft deferment—that was happiness. How could you keep your draft deferment? Easy—you simply had to prove your *importance!* You had to prove that if it were not for Chekist vigilance the camps would blow apart, that they were a caldron of seething tar! And then our whole glorious front would collapse! It was right here in the camps in the tundra and the taiga that the white-chested security chiefs were holding back the Fifth Column, holding back Hitler! This was their contribution to victory! Not sparing themselves, they conducted interrogation after interrogation, exposing plot after plot.

Until now only the unhappy, worn-out camp inmates, tearing the bread from each other's mouths, had been fighting for their lives! But now the omnipotent Chekist security officers shamelessly entered the fray. "You croak today, me tomorrow." Better you should perish and put off my death, you dirty animal.

And what was this? Plots were discovered in every camp! More plots! Still more! Ever larger in scale! And ever broader! Oh, those perfidious last-leggers! They were just feigning that they could be blown over by the wind—their paper-thin, pellagra-stricken hands were secretly reaching for the machine guns! Oh, thank you, Security Section! Oh, savior of the Motherland—the Third Section!

And—you? You thought that in camp at least you could unburden your soul? That here you could at least complain aloud: "My sentence is too long! They fed me badly! I have too much work!" Or you thought that here you could at least *repeat* what you got your term for? But if you say any of this aloud—you are done for! You are doomed to get a new "tenner." (True, once a new camp tenner begins, at least the first is erased, so that as it works out you serve not twenty, but some thirteen or fifteen or the like. . . . Which will be more than you can survive.)

But you are sure you have been silent as a fish? And then you are grabbed anyway? Quite right! They couldn't help grabbing you no matter how you behaved. After all, they don't grab *for something* but *because.* It's the same principle according to which they clip the wool off *freedom* too. When the Third Section gang goes hunting, it picks a list of the most noticeable people in the camp. And that is the list they then dictate to Babich. . . .

In camp, after all, it is even more difficult to hide, everything is out in the open. And there is only one salvation for a person: to be a zero! A total zero. A zero from the very beginning.

To stick you with a charge presents no problem. When the "plots" came to an end after 1943 (the Germans began to retreat), a multitude of cases of "propaganda" appeared. (Those "godfathers" still didn't want to go to the front!) In the Burepolom Camp, for example, the following selection was available:

- Hostile activity against the policy of the Soviet Communist Party and the Soviet government (and what it was you can guess for yourself!)
- Expression of defeatist fabrications
- Expression of slanderous opinions about the material situation of the workers of the Soviet Union (Telling the truth was slander.)
- Expression of a desire (!) for the restoration of the capitalist system
- Expression of a grudge against the Soviet government (This was particularly impudent! Who are you, you bastard, to nurse grudges! So you got a "tenner" and you should have kept your mouth shut!)

A seventy-year-old former Tsarist diplomat was charged with making the following propaganda:

- That the working class in the U.S.S.R. lives badly
- That Gorky was a bad writer (!!)

To say that they had gone too far in bringing these charges against him is out of the question. They always handed out sentences for Gorky; that's how he had set himself up. Skvortsov, for example, in Lokchimlag (near Ust-Vym), harvested fifteen years, and among the charges against him was the following:

- He had unfavorably contrasted the proletarian poet Mayakovsky with *a certain* bourgeois poet.

That's what it said in the formal charges against him, and it was enough to get him convicted. And from the minutes of the interrogation we can establish who that *certain* bourgeois poet was. It was Pushkin! To get a sentence for Pushkin—that, in truth, was a rarity!

After that, therefore, Martinson, who really did say in the tin shop that "the U.S.S.R. was one big *camp*," ought to have sung praise to God that he got off with a "tenner."

As ought those refusing to work who got a "tenner" instead of execution.

But it was not the number of years, not the empty and fantastic length of years, that made these second terms so awful—but *how* you got them. How you had to crawl through that iron pipe in the ice and snow to get them.

It would seem that arrest would be a nothing for a camp inmate. For a person who had once been arrested from his warm domestic bed —what did it matter to be arrested again from an uncomfortable barracks with bare bunks? But it certainly did! In the barracks the stove was warm and a full bread ration was given. But here came the jailer and jerked you by the foot at night. "Gather up your things!" Oh, how you didn't want to go! People, people, I love you. . . .

Chapter 14

■

Changing One's Fate!

To defend yourself in that savage world was impossible. To go on strike was suicide. To go on hunger strikes was useless.

And as for dying, there would always be time.

So what was left for the prisoner? To break out! To go *change one's fate!*

Chekhov used to say that if a prisoner was not a philosopher who could get along equally well in all possible circumstances (or let us put it this way: who could retire into himself) then he could *not* but wish to escape and he *ought to* wish to.

He could not but wish to! That was the imperative of a free soul. True, the natives of the Archipelago were far from being like that. They were much more submissive than that. But even among them there were always those who thought about escape or who were just about to. The continual escapes in one or another place, even those that did not succeed, were a true proof that the energy of the zeks had not yet been lost.

Here is a camp compound. It is well guarded; the fence is strong and the inner cordon area is reliable and the watchtowers are set out correctly—every spot is open to view and open to fire. But all of a sudden you grow sick to death of the thought that you are condemned to die right here on this bit of fenced-off land. So why not try your luck? Why not burst out and change your fate? This impulse is particularly strong at the beginning of your term of imprisonment, in the first year, and it is not even deliberate. In that first year when, generally speaking, the prisoner's entire future and whole prison personality are being decided. Later on this impulse weakens somehow; there is no longer the conviction that it is more important for you to be *out there,* and all the threads binding you to the outer world weaken, and the cauterizing of the soul is transformed into decay, and the human being settles into camp harness.

During all the years of the camps, there were evidently quite a few escapes. Here are some statistics accidentally come by: In the month of March, 1930, alone, 1,328 persons escaped from imprisonment in the R.S.F.S.R. (And how inaudible and soundless this was in our society.)

With the enormous development of the Archipelago after 1937, particularly during the war years, when battle-fit infantrymen were rounded up and sent to the front, it became even more difficult to provide proper convoy, and not even the evil notion of self-guarding could solve all the problems of the chiefs. So they relied on certain invisible chains which kept the natives reliably in their place.

The strongest of these chains was the prisoners' universal submission and total surrender to their situation as slaves. Almost to a man, both the 58's and the nonpolitical offenders were hardworking family people capable of manifesting valor only in lawful ways, on the orders of and the approval of the higher-ups. Even when they had been imprisoned for five and ten years they could not imagine that singly—or, God forbid, collectively—they might rise up for their liberty since they saw arrayed against them the state (*their own* state), the NKVD, the police, the guards, and the police dogs. And even if you were fortunate enough to escape unscathed, how could you live afterward on a false passport, with a false name, when documents were checked at every intersection, when suspicious eyes followed passers-by from behind every gateway?

Another chain was *the death factory*—camp starvation. Although it was precisely this starvation that at times drove the despairing to wander through the taiga in the hope of finding more food than there was in camp, yet it was this starvation that also weakened them so that they had no strength for a long flight, and because of it it was impossible to save up a stock of food for the journey.

And there was another chain too—the threat of a new term. A political prisoner was given a new tenner for an escape attempt under that same Article 58 (and gradually it proved best to give Article 58-14 —Counter-Revolutionary Sabotage).

Another thing restraining the zeks was not the compound but the privilege of going without convoy. The ones guarded the least, who enjoyed the small privilege of going to work and back without a bayonet at their backs, or once in a while dropping into the free settlement, highly prized their advantages. And after an escape these were taken away.

The geography of the Archipelago was also a solid obstacle to escape attempts—those endless expanses of snow or sandy desert, tundra, taiga.

The hostility of the surrounding population, encouraged by the authorities, became the principal hindrance to escapes. The authorities were not stingy about rewarding the captors. (This was an additional form of political indoctrination.) And the nationalities inhabiting the areas around Gulag gradually came to assume that the capture of a fugitive meant a holiday, enrichment, that it was like a good hunt or like finding a small gold nugget.

But the desperate heart sometimes did not weigh things. It saw: the river was flowing and a log was floating down it—and one jump! We'll float on down. Vyacheslav Bezrodny from the Olchan Camp, barely released from the hospital, still utterly weak, escaped down the Indigirka River on two logs fastened together—to the Arctic Ocean! Where was he going? And what was he hoping for? In the end he was not so much caught as picked up on the open sea and returned over the winter road to Olchan to that very same hospital.

It is not possible to say of everyone who didn't return to camp on his own, who was not brought in half alive, or who was not brought in dead, that he had escaped. Perhaps he had only exchanged an involuntary and long-drawn-out death in camp for the free death of a beast in the taiga.

The *quiet* escapes were usually more fortunate in their results. Some of them were surprisingly successful. But we rarely hear of these happy stories; *those who broke out* do not give interviews; they have changed their names, and they are in hiding. Kuzikov-Skachinsky, who escaped successfully in 1942, tells the story now only because he was caught in 1959—after seventeen years.

And we have learned of the successful escape of Zinaida Yakovlevna Povalyayeva because in the end it fell through. She got her term because she had stayed on as a teacher in her school during the German occupation. But she was not immediately arrested when the Soviet armies arrived, and before her arrest she was married to a pilot. Then she was arrested and sent to Mine No. 8 at Vorkuta. Through some Chinese working in the kitchen she established communication with freedom and with her husband. He was employed in civil aviation and arranged a trip to Vorkuta for himself. On an appointed day Zina went to the bath in the work zone, where she shed her camp clothing and released her hair, which had been curled the night before, from under her head scarf. Her husband was waiting for her in the work sector. There were security officers on duty at the river ferry, but they paid no attention to a girl with curly hair who was arm in arm with a flier. They flew out on a plane. Zina spent one year living on false papers. But she couldn't resist the desire to see her mother again—and her mother was under surveillance. At her new interrogation she managed to convince them she had escaped in a coal car. And they never did find out about her husband's participation.

We have not yet described the group escapes, and there were many of them too. They say that in 1956 a whole small camp escaped near Monchegorsk.

The history of all the escapes from the Archipelago would be a list too long to be read, too long to be leafed through. And any one person who wrote a book solely about escapes, to spare his reader and himself, would be forced to omit hundreds of cases.

Chapter 15

■

Punishments

Among the many joyous renunciations brought us by the new world were the renunciation of exploitation, the renunciation of colonies, the renunciation of obligatory military service, the renunciation of secret diplomacy, secret assignments and transfers, the renunciation of secret police, the renunciation of "divine law," and many, many other fairy-tale renunciations in addition. But not, to be sure, a renunciation of prisons.

So it must have seemed ridiculous not only to the prison keepers but to the prisoners themselves that for some reason or other there was no punishment cell, that it should have been banned. For if you didn't intimidate the prisoner, if there was no further punishment you could apply—how could he be compelled to submit to the regimen?

And where could you put the captured fugitives?

What was the ShIzo given for? For whatever they felt like: You didn't please your chief; you didn't say hello the way you should have; you didn't get up on time; you didn't go to bed on time; you were late for roll call; you took the wrong path; you were wrongly dressed; you smoked where it was forbidden; you kept extra things in your barracks —take a day, three, five. You failed to fulfill the work norm, you were caught with a broad—take five, seven, or ten. And for *work shirkers* there was even fifteen days. And even though, according to the law (what law?), fifteen days was the maximum in penalty cells (though according to the Corrective Labor Code of 1933 even that was impermissible!), this accordion could be stretched out to a whole year. In 1932 in Dmitlag (this is something Averbakh writes about—black on white!) they used to give *one year* of ShIzo for *self-mutilation!* And if one bears in mind that they used to refuse treatment in such cases, then this meant they used to put a sick, wounded person in a punishment cell to rot—for a whole year!

What was required of a ShIzo? It had to be: (a) cold; (b) damp; (c) dark; (d) for starvation. Therefore there was no heat: not even when the temperature outside was 22 degrees below zero Fahrenheit. They did not replace missing glass panes in the winter. They allowed the walls to get damp. (Or else they put the penalty-block cellar in moist ground.) The windows were microscopic or else there were none at all (more usual). They fed a *"Stalin" ration* of ten and a half ounces a day and issued a "hot" meal, consisting of thin gruel, only on the third, sixth, and ninth days of your imprisonment there. But at Vorkuta-Vom they gave only seven ounces of bread, and a piece of *raw* fish in place of a *hot* dish on the third day. This is the framework in which one has to imagine all the penalty cells.

It is very naïve to think that a penalty cell has of necessity to be like a cell—with a roof, door, and lock. Not at all! At Kuranakh-Sala, at a temperature of minus 58 degrees Fahrenheit, the punishment cell was a sodden frame of logs. (The free physician Andreyev said: "I, as a *physician,* declare it *possible* to put a prisoner in that kind of punishment cell.") Let us leap the entire Archipelago: at Vorkuta-Vom in 1937, the punishment cell for work shirkers was a log frame *without a roof.* And in addition there was *a plain hole in the ground.* Arnold Rappoport lived in a hole like that (to get shelter from the rain they used to pull some kind of rag over themselves), like Diogenes in a barrel.

In the Mariinsk Camp (as in many others, of course) there was snow on the walls of the punishment cell—and in such-and-such a punishment cell the prisoners were not allowed to keep their camp clothes on but were forced *to undress to their underwear.*

The BUR—the Strict Regimen Barracks—could be the most ordinary kind of barracks, set apart and fenced off by barbed wire, with the prisoners in it being taken out daily to the hardest and most unpleasant work in the camp. It could also be a masonry prison inside the camp with a full prison system—with beatings of prisoners summoned one by one to the jailers' quarters (a favorite method that didn't leave marks was to beat with a felt boot with a brick inside it); with bolts, bars, locks, and peepholes on every door; with concrete floors to the cells and an additional separate punishment cell for BUR inmates.

The favored candidates for the penalty compounds were: religious believers, stubborn prisoners, and thieves. (Yes, thieves! Here the great system of indoctrination broke down because of the inconsistency of the local instructors.) They kept whole barracks of "nuns" there who had refused to work for the devil. (At the penalty camp for prisoners under convoy at the Pechora State Farm they held them in a penalty block up to their knees in water. In the autumn of 1941 they gave them all 58-14—economic counterrevolution—and shot them.) They sent the priest Father Viktor Shipovalnikov there on charges of conducting "religious propaganda" (he had celebrated vespers for five nurses on Easter Eve).

And often prisoners were sent to penalty compounds for refusal to become informers. The majority died there and naturally cannot speak about themselves. And the murderers from State Security are even less likely to speak of them.

There were stories of women in this context too. It is impossible to reach a sufficiently balanced judgment on these stories because some intimate element always remains hidden from us. However, here is the story of Irena Nagel as she told it herself. She worked as a typist for the Administrative Section of the Ukhta State Farm, in other words as a very comfortably established trusty. Heavy-set and imposing, she wore her hair in long braids wrapped around her head; and partly for convenience she went around in wide Oriental-type trousers and a jacket cut like a ski jacket. Whoever knows camp life will understand what an enticement this was. A security officer, Junior Lieutenant Sidorenko, expressed a desire to get more intimately acquainted. And Nagel replied: "I would rather be kissed by the lowliest thief in camp! You ought to be ashamed of yourself. I can hear your baby crying in the next room." Repulsed by her outburst, the security officer suddenly changed his expression and said: "Surely you don't really think I like you? I merely wanted *to put you to the test*. So here's the way it is: you must *collaborate* with us." She refused and was sent to a penalty camp.

Chapter 16

■

The Socially Friendly

Let my feeble pen, too, join in praise of this tribe! They have been hailed as pirates, as freebooters, as tramps, as escaped convicts. They have been lauded as noble brigands—from Robin Hood on down to operetta heroes. And we have been assured that they have sensitive hearts, that they plunder the rich and share with the poor.

And, indeed, has not all world literature glorified the thieves? It is not for us to reproach François Villon; but neither Hugo nor Balzac could avoid that path; and Pushkin, too, praised the thief principle in his Gypsies. (And what about Byron?) But never have they been so widely glorified and with such unanimity and so consistently as in Soviet literature. Who was there who was not breathless with sacred emotion in describing the thieves to us—their vivid, unreined nihilism at the beginning, and their dialectical "reforging" at the end—starting with Mayakovsky (and, in his footsteps, Shostakovich with his ballet *The Young Lady and the Hooligan)* and including Leonov, Selvinsky, Inber—and you could go on and on?

In Old Russia there existed (just as there still exists in the West) an incorrect view of thieves as incorrigible, permanent criminals (a "nucleus of criminality"). Because of this the politicals were segregated from them on prisoner transports and in prisons. In Old Russia there was just one single formula to be applied to the criminal recidivists: "Make them bow their heads beneath the iron yoke of the law!" And so it was that up to 1914 the thieves did not play the boss either in Russia as a whole or in Russian prisons.

But the shackles fell and freedom dawned. In the desertion of millions in 1917, and then in the Civil War, all human passions were largely unleashed, and those of the thieves most of all, and they no longer wished to bow their heads beneath the yoke; moreover, they were informed that they didn't have to. It was found both useful and amusing that they were enemies of private property and therefore a revolutionary force which had to be guided into the mainstream of the proletariat, yes, and this would constitute no special difficulty. Reasoning on a social basis: wasn't the *environment* to blame for everything? So let us re-educate these healthy lumpenproletarians and introduce them into the system of conscious life!

And now, when more than forty years have gone by, one can look around and begin to have doubts: Who re-educated whom? Did the Chekists re-educate the thieves, or the thieves the Chekists? The urka —the habitual thief—who adopted the Chekist faith became a *bitch,* and his fellow thieves would cut his throat. The Chekist who acquired the psychology of the thief was an *energetic* interrogator of the thirties and forties, or else a *resolute* camp chief—such men were appreciated. They got the service promotions.

And the psychology of the urki was exceedingly simple and very easy to acquire:

1. I want to live and enjoy myself; and f— the rest!
2. Whoever is the strongest is right!
3. If they aren't [beat]ing you, then don't lie down and ask for it. (In other words: As long as they're beating up someone else, don't stick up for the ones being beaten. Wait your own turn.)

Beat up your submissive enemies one at a time! Somehow this is a very familiar law. It is what Hitler did. It is what Stalin did.

Come on now, stop lying, you mercenary pens! You who have observed the Russian thieves through a steamship rail or across an interrogator's desk! You who have never encountered the thieves when you were defenseless.

The thieves—the urki—are not Robin Hoods! When they want, they steal from last-leggers! When they want they are not squeamish about—taking the last footcloths off a man freezing to death. Their great slogan is: "You today, me tomorrow!"

Here is what our laws were like for thirty years—to 1947: For robbery of the state, embezzlement of state funds, a packing case from a warehouse, for three potatoes from a collective farm—ten years! (After 1947 it was as much as twenty!) But robbery of a *free person?* Suppose they cleaned out an apartment, carting off on a truck everything the family had acquired in a lifetime. If it was not accompanied by murder, then the sentence was *up to one year,* sometimes six months.

The thieves flourished because they were encouraged.

Through its laws the Stalinist power said to the thieves clearly: Do not steal from me! Steal from private persons! You see, private property is a belch from the past. (But "personally assigned" VIP property is the hope of the future. . . .)

And the thieves . . . understood. In their intrepid stories and songs, did they go to steal where it was difficult, dangerous, where they could lose their heads? No. Greedy cowards, they pushed their way in where they were encouraged to push their way in—they stripped the clothes from solitary passers-by and stole from unguarded apartments.

How many citizens who were robbed knew that the police didn't even bother to look for the criminals, didn't even set a case in motion, so as not to spoil their record of completed cases—why should they sweat to catch a thief if he would be given only six months, and then be given three months off for good behavior? And anyway, it wasn't certain that the bandits would even be tried when caught.

Finally, sentences were bound to be reduced, and of course for habitual criminals especially. Watch out there now, witness in the courtroom! They will all be back soon, and it'll be a knife in the back of anyone who gave testimony!

Therefore, if you see someone crawling through a window, or slitting a pocket, or your neighbor's suitcase being ripped open—shut your eyes! Walk by! You didn't see anything!

That's how the thieves have trained us—the thieves and our laws!

There is one more important feature of our public life which helps thieves and bandits prosper—*fear of publicity.* Our newspapers are filled with reports on production victories which are a big bore to everyone, but you will find no reports of trials or crime in them. (After all, according to the Progressive Doctrine, criminal activity arises only from the presence of classes; we have no classes in our country, therefore there is no crime and therefore you cannot write about it in the press! We simply cannot afford to give the American newspapers evidence that we have not fallen behind the United States in criminal activity!) If there is a murder in the West, photographs of the criminal are plastered on the walls of buildings, they peer out at one from the counters of bars, the windows of streetcars, and the criminal feels himself a persecuted rat. If a brazen murder is committed here, the press is silent, there are no photographs, the murderer goes sixty miles away to another province and lives there in peace and quiet. And the

Minister of Internal Affairs will not have to answer questions in parliament as to why the criminal has not been found; after all, no one knows about the case except the inhabitants of that little town.

It was the same with criminal activity as it was with malaria. It was simply announced one day that it no longer existed in our country, and from then on it became impossible to treat it or even to diagnose it.

And there is always that sanctifying lofty theory for everything. It was by no means the least significant of our literary figures who determined that the thieves were our allies in the building of Communism. This was set forth in textbooks on Soviet corrective-labor policy (there were such textbooks, they were published!), in dissertations and scientific essays on camp management, and in the most practical way of all—in the regulations on which the high-ranking camp officials were trained. All this flowed from the One-and-Only True Teaching, which explained all the iridescent life of humanity . . . in terms of the class struggle and it alone.

And here is how it was worked out. Professional criminals can in no sense be equated with capitalist elements (i.e., engineers, students, agronomists, and "nuns"), for the latter are steadfastly hostile to the dictatorship of the proletariat, while the former are only (!) politically unstable! (A professional murderer is *only* politically unstable!) The lumpenproletarian is not a property owner, and therefore cannot ally himself with the hostile-class elements, but will much more willingly ally himself with the proletariat (you just wait!). That is why in the official terminology of Gulag they are called *socially friendly* elements. (Tell me who your friends are . . .) That is why the regulations repeated over and over again: *Trust* the recidivist criminals! That is why through the Cultural and Educational Section a consistent effort was supposed to be made to explain to the thieves the unity of their class interests with those of all the workers, to indoctrinate them in a "suspicious and hostile attitude toward the 'kulaks' and counterrevolutionaries," and the authorities were to *"place their hopes* in these attitudes"!

But when this elegant theory came down to earth in camps, here is what emerged from it: The most inveterate and hardened thieves were given unbridled power on the islands of the Archipelago, in camp districts, and in camps—power over the population of their own country, over the peasants, the petty bourgeoisie, and the intelligentsia, power they had never before had in history, never in any state in all the world, power which they couldn't even dream of out in freedom. And now they were given all other people as slaves. What bandit would ever decline such power? The *central thieves,* the top-level thieves, totally controlled the camp districts. They lived in individual "cabins" or tents with their own temporary wives. (Or arbitrarily picking over the "smooth broads" from among their subjects, they had the intellectual women 58's and the girl students to vary their menu. In Norillag, Chavdarov heard a moll offer her thief husband: "Would you like me to treat you to a sixteen-year-old collective-farm girl?" This was a peasant girl who had been sent to the North for ten years because of one kilo of grain. The girl tried to resist, but the moll soon broke her will: "I'll cut you up! Do you think . . . I'm any worse than you? I lie under him!")

People will object that it was only the *bitches* who accepted positions, while the "honest thieves" held to the thieves' law. But no matter how much I saw of one and the other, I never could see that one rabble was nobler than the other. The thieves knocked gold teeth out of Estonians' mouths with a poker. The thieves (in Kraslag, in 1941) drowned Lithuanians in the toilet for refusing to turn over a food parcel to them. The thieves used to plunder prisoners sentenced to death. Thieves would jokingly kill the first cellmate who came their way just to get a new interrogation and trial, and spend the winter in a warm place, or to get out of a hard camp into which they had fallen. So why mention such petty details as stripping the clothes or shoes from someone out in subzero temperatures? And why mention stolen rations?

No, you'll not get fruit from a stone, nor good from a thief.

The theoreticians of Gulag were indignant; the kulaks (in camp) didn't even regard the thieves as real people (thereby, so to speak, betraying their true bestial colors).

But how can you regard them as people if they tear your heart out of your body and suck on it? All their "romantic bravado" is the bravado of vampires.

Chapter 17

■

The Kids

The Archipelago had many ugly mugs and many bared fangs. No matter what side you approached it from, there wasn't one you could admire. But perhaps the most abominable of all was that maw that swallowed up *the kids*.

The kids were not at all those besprizorniki or waifs in drab tatters who scurried hither and thither thieving and warming themselves at asphalt caldrons on the streets, without whom one could not picture the urban life of the twenties. The waifs were taken from the streets—not from their families—into the colonies for juvenile delinquents (there was one attached to the People's Commissariat of Education as early as 1920), into workhouses for juveniles (which existed from 1921 to 1930 and had bars, bolts, and jailers, so that in the outworn bourgeois terminology they could have been called prisons), and also into the "Labor Communes of the OGPU" from 1924 on. They had been orphaned by the Civil War, by its famine, by social disorganization, the execution of their parents, or the death of the latter at the front, and at that time justice really did try to return these children to the mainstream of life, removing them from their street apprenticeship as thieves.

But where did the young offenders come from? They came from Article 12 of the Criminal Code of 1926, which permitted children *from the age of twelve* to be sentenced for theft, assault, mutilation, and murder (Article 58 offenses were also included under this heading), but they had to be given moderate sentences, not "the whole works" like adults. Here was the first crawl hole into the Archipelago for the future "kids"—but it was not yet a wide gate.

We are not going to omit one interesting statistic: In 1927 prisoners aged sixteen (they didn't count the younger ones) to twenty-four represented 48 percent of all prisoners.

What this amounts to is that nearly *half* the entire Archipelago in 1927 consisted of youths whom the October Revolution had caught between the ages of *six and fourteen.* Ten years after the victorious Revolution these same girls and boys turned up in prison and constituted half the prison population! This jibes poorly with the struggle against the vestiges of bourgeois consciousness which we inherited from the old society, but figures are figures. They demonstrate that the Archipelago never was short of young people.

But the question of *how* young was decided in 1935. In that year the Great Evildoer once more left his thumbprint on History's submissive clay. Among such deeds as the destruction of Leningrad and the destruction of his own Party, he did not overlook the children—the children whom he loved so well, whose Best Friend he was, and with whom he therefore had his photograph taken. Seeing no other way to bridle those insidious mischiefmakers, those washerwomen's brats, who were overrunning the country in thicker and thicker swarms and growing more and more brazen in their violations of socialist legality, he invented a gift for them: These children, from twelve years of age (by this time his beloved daughter was approaching that borderline, and he could see that age tangibly before his eyes), should be sentenced *to the whole works* in the Code. (Including capital punishment as well.)

Illiterates that we were, we scrutinized decrees very little at the time. More and more we gazed at the portraits of Stalin with a black-haired little girl in his arms. . . . Even less did the twelve-year-olds read the decrees. And the decrees kept coming out, one after another. On December 10, 1940, the sentencing of juveniles from the age of twelve for "putting various objects on railroad tracks." (This was training young diversionists.) On May 31, 1941, it was decreed that for all other varieties of crime not included in Article 12 juveniles were to be given full sentences from the age of fourteen on!

But here a small obstacle arose: the War of the Fatherland began. But the law is the law! And on July 7, 1941—four days after Stalin's panicky speech in the days when German tanks were driving toward Leningrad, Smolensk, and Kiev—one more decree of the Presidium of the Supreme Soviet was issued, and it is difficult now to say in what respect it is more interesting for us today—in its unwavering academic character, showing what important questions were being decided by the government in those flaming days, or in its actual contents. The situation was that the Prosecutor of the U.S.S.R. (Vyshinsky?) had complained to the Supreme Soviet about the Supreme Court (which means His Graciousness had heard about the matter), because the courts were applying the Decree of 1935 incorrectly and these brats were being sentenced only when they had *intentionally* committed crimes. But this was impermissible softness! And so right in the heat of war, the Presidium of the Supreme Soviet elucidated: This interpretation does not correspond to the text of the law. It introduces limitations not provided for by the law! And in agreement with the prosecutor, the Presidium issued a clarification to the Supreme Court: Children must be sentenced and the full measure of punishment applied (in other words, "the whole works"), even in cases where crimes were committed not intentionally but as a result of *carelessness.*

Now that is something! Perhaps in all world history no one has yet approached such a radical solution of the problem of children! From twelve years on for carelessness . . . up to and including execution! And that is when all the escape holes were shut off to the greedy mice! That is when, finally, all the collective-farm ears of grain were saved! And now the granaries were going to be filled to overflowing and life would flourish, and children who had been bad from birth would be set on the long path of correction.

And none of the Party prosecutors with children the same age shuddered! They found no problem in stamping the arrest warrants. And none of the Party judges shuddered either! With bright eyes they sentenced little children to three, five, eight, and ten years in general camps!

And for "shearing sheaves" these tykes got not less than eight years!

And for a pocketful of potatoes—one pocketful of potatoes in a child's trousers!—they also got eight years!

Cucumbers did not have so high a value put on them. For a dozen cucumbers Sasha Blokhin got five years.

And the hungry fourteen-year-old girl Lida, in the Chingirlau District Center of Kustanai Province, walked down the street picking up, mixed with the dust, a narrow trail of grain spilled from a truck (doomed to go to waste in any case). For this she was sentenced to only *three* years because of the alleviating circumstances that she had not taken socialist property directly from the field or from the barn. And perhaps what also inclined the judges to be less harsh was that in that same year of 1948 there had been a clarification of the Supreme Court to the effect that children need not be tried for theft which had the character of childish mischief (such as the petty theft of apples in an orchard). By analogy the court drew the conclusion here that it was possible to be just a wee bit less harsh. (But the conclusion we draw is that from 1935 to 1948 children *were* sentenced for taking apples.)

And a great many were sentenced for running away from Factory Apprenticeship Training. True, they got only six months for that. (In camp they were jokingly called *death-row prisoners.* But joke or no joke, here is a scene with some such "death-row prisoners" in a Far Eastern camp: They were assigned to dump the shit from latrines. There was a cart with two enormous wheels and an enormous barrel on it, full of stinking sludge. The "death-row prisoners" were hitched up, with many of them in the shafts and others pushing from the sides and from behind [the barrel kept swaying and splashing them]. And the crimson-cheeked *bitches* in their twill suits roared with laughter as they urged the children on with clubs. And on the prisoner transport ship from Vladivostok to Sakhalin in 1949, the *bitches used* these children at knifepoint for carnal enjoyment. So even six months was sometimes enough too.)

And it was then that the twelve-year-olds crossed the thresholds of the adult prison cells, were equated with adults as citizens possessing full rights, equated by virtue of the most savage prison terms, equated, in their whole unconscious life, by bread rations, bowls of gruel, their places on the sleeping shelves—that is when that old term of Communist re-education, "minors," somehow lost its significance, when the outlines of its meaning faded, became unclear—and Gulag itself gave birth to the ringing and impudent word "kids." And with a proud and bitter intonation these bitter citizens began to use this term to describe themselves—not yet citizens of the country but already citizens of the Archipelago.

So early and so strangely did their adulthood begin—with this step across the prison threshold!

And upon the twelve- and fourteen-year-old heads burst a life style that was too much for brave men who were experienced and mature. But the young people, by the laws of their young life, were not about to be flattened by this life style but, instead, grew into it and adapted to it. Just as new languages and new customs are learned without difficulty in childhood, so the juveniles adopted *on the run* both the language of the Archipelago—which was that of the thieves—and the philosophy of the Archipelago—and whose philosophy was that?

From this life they took for themselves all its most inhuman essence, all its poisonous rotten juice—and as readily as if it had been this liquid, and not milk, that they had sucked from their mothers' breasts in infancy.

They grew into camp life so swiftly—not in weeks even, but in days!—as if they were not in the least surprised by it, as if that life were not completely new to them, but a natural continuation of their free life of yesterday.

Even out in freedom they hadn't grown up in linens and velvets; it had not been the children of secure and powerful parents who had gone out to clip stalks of grain, filled their pockets with potatoes, been late at the factory gate, or run away from Factory Apprenticeship Training. The kids were the children of workers. Out in freedom they had understood very well that life was built upon injustice. But out there things had not been laid out stark and bare to the last extremity; some of it was dressed up in decent clothing, some of it softened by a mother's kind word. In the Archipelago the kids saw the world as it is seen by quadrupeds: Only might makes right! Only the beast of prey has the right to live! That is how we, too, in our adult years saw the Archipelago, but we were capable of counterposing to it all our experience, our thoughts, our ideals, and everything that we had read to that very day. Children accepted the Archipelago with the divine impressionability of childhood. And in a few *days* children became beasts there! And the worst kind of beasts, with no ethical concepts whatever. The kid masters the truth: If other teeth are weaker than your own, then tear the piece away from them. It belongs to you!

There were two basic methods of maintaining kids in the Archipelago: in separate children's colonies (principally the younger kids, not yet fifteen) and in mixed-category camps, most often with invalids and women (the senior kids).

Both were equally successful in developing animal viciousness. And neither rescued the kids from being educated in the spirit of the thieves' ideals.

Take Yura Yermolov. He reports that when he was only twelve years old (in 1942) he saw a great deal of fraud, thievery, and speculation going on around him, and arrived at the following judgment about life: "*The only people* who do not steal and deceive *are those who are afraid to.* As for me—I don't want to be afraid of anything! Which means that I, too, will steal and deceive and live well." And yet for a time his life somehow developed differently. He became fascinated by the shining examples whose spirit he was taught in school. However, having got a taste of the Beloved Father, at the age of fourteen he wrote a leaflet: "Down with Stalin! Hail Lenin!" They caught him on that one, beat him up, gave him 58-10, and imprisoned him with the kids and thieves. And Yura Yermolov quickly mastered the thieves' law. The spirit of his existence spiraled upward steeply—and at the age of fourteen he had executed his "negation of a negation": he had returned to the concept of thievery as the highest and the best of all existence.

And what did he see in the children's colony? "There was even more injustice than in freedom. The chiefs and the jailers lived off the state, shielded by the correctional system. Part of the kids' ration went from the kitchen into the bellies of the instructors. The kids were beaten with boots, kept in fear so that they would be silent and obedient."

The simplest reply to the overpowering injustices was to create injustices oneself! This was the easiest conclusion, and it would now become the rule of life of the kids for a long time to come (or even forever).

But here is what's interesting! In giving the cruel world battle, the kids didn't battle against one another! They didn't look on each other as enemies! They entered this struggle as *a collective,* a united group! Was this a budding socialism? The indoctrination of the instructors? Oh, come on, cut the cackle, big-mouths! This is a descent into the law of the thieves! After all, the thieves are united; after all, the thieves have their own discipline and their own ringleaders. And the juveniles were the apprentices of the thieves, they were mastering the precepts of their elders.

No one could avoid being cooked up in that mash! No boy could remain a separate individual—he would be trampled, torn apart, ostracized, if he did not immediately declare himself a thieves' apprentice. And *all of them* took that inevitable oath. . . . (Reader! Put *your own* children in their place. . . .)

Who was the enemy of the kids in the children's colonies? The jailers and the instructors. The struggle was against them!

Say they were marching a column of kids under armed guard through a city, and it seems even shameful to guard children so strictly. Far from it! They had worked out a plan. A whistle—and all who wanted to scattered in different directions! And what were the guards to do? Shoot? At whom? At children? . . . And so their prison terms came to an end. In one fell swoop 150 years ran away from the state. You don't enjoy looking silly? Then don't arrest children!

Here is one of their boastful stories about themselves, which, knowing the typical pattern of the kids' actions, I fully believe. Some excited and frightened children ran to the nurse of a children's colony and summoned her to help one of their comrades who was seriously ill.

Forgetting caution, she quickly accompanied them to their big cell for forty. And as soon as she was inside, the whole anthill went into action! Some of them barricaded the door and kept watch. Dozens of hands tore everything off her, all the clothes she had on, and toppled her over; and then some sat on her hands and on her legs; and then, everyone doing what he could and where, they raped her, kissed her, bit her. It was against orders to shoot them, and no one could rescue her until they themselves let her go, profaned and weeping.

In general, of course, interest in the female body begins early among boys, and in the kids' cells it was intensely heated up by colorful stories and boasting. And they never let a chance go by to let off steam. Here is an episode. In broad daylight in full view of everyone, four kids were sitting in the compound of Krivoshchekovo Camp No. 1, talking with a girl called Lyuba from the bookbinding shop. She retorted sharply to something they had said. The boys leaped up, grabbed her legs, and lifted them in the air. She was in a defenseless position; while she supported herself on the ground with her hands, her skirt fell over her head. The boys held her that way and caressed her with their free hands. And then they let her down—and not roughly either. Did she slap them? Did she run away from them? Not at all. She sat down just as before and continued the argument.

These were sixteen-year-old kids, and it was an adult camp, with mixed categories. (It was the same one that had the women's barracks for five hundred where all the copulation took place without curtains and which the kids used to enter importantly like men.)

In the children's colonies the kids worked for four hours and then were supposed to be in school for four. (But all that schooling was a fake.) When transferred to an adult camp, they had a ten-hour working day, except that their work norms were reduced, while their ration norms were the same as adults'. They were transferred at the age of sixteen, but their undernourishment and improper development in camp and before camp endowed them at that age with the appearance of small frail children. Their height was stunted, as were their minds and their interests.

After the children's colony their situation changed drastically. No longer did they get the children's ration which so tempted the jailers —and therefore the latter ceased to be their principal enemy. Some old men appeared in their lives on whom they could try their strength. Women appeared on whom they could try their maturity. Some real live thieves appeared, fat-faced camp storm troopers, who willingly undertook their guidance both in world outlook and in training in thievery. To learn from them was tempting—and not to learn from them was impossible.

For a *free* reader does the word "thief" perhaps sound like a reproach? In that case he has understood nothing. This word is pronounced in the underworld in the same way that the word "knight" was pronounced among the nobility—and with even greater esteem, and not loudly but softly, like a sacred word. To become a worthy thief someday . . . was the kid's dream.

On one occasion, at the Ivanovo Transit Prison, I spent the night in a cell for kids. In the next bunk to me was a thin boy just over fifteen —called Slava, I think. It appeared to me that he was going through the whole kid ritual somehow unwillingly, as if he were growing out of it or was weary of it. I thought to myself: This boy has not perished, and is more intelligent, and he will soon move away from the others. And we had a chat. The boy came from Kiev. One of his parents had died, and the other had abandoned him. Before the war, at the age of nine, Slava began to steal. He also stole "when our army came," and after the war, and, with a sad, thoughtful smile which was so old for fifteen, he explained to me that in the future, too, he intended to live only by thievery. "You know," he explained to me very reasonably, "that as a worker you can earn only bread and water. And my *childhood* was bad so I want to live well." "What did you do during the German occupation?" I asked, trying to fill in the two years he had bypassed without describing them—the two years of the occupation of Kiev. He shook his head. "Under the Germans I worked. What do you think—that I could have gone on stealing under the Germans? They shot you on the spot for that."

Here, as recounted by A. Y. Susi, are several pictures from Krivoshchekovo (Penalty) Camp No. 2 of Novosiblag. Life was lived in enormous half-dark dugouts (for five hundred each) which had been dug into the earth to a depth of five feet. The chiefs did not interfere with the life inside the compound—no slogans and no lectures. The thieves and kids held sway. Almost no one was taken out to work. Rations were correspondingly meager. On the other hand, there was a surplus of time.

One day they were bringing a breadbox from the bread-cutting room under the guard of brigade members. The kids started a fake fight in front of the box itself, started shoving one another, and tipped the box over. The brigade members hurled themselves on the bread ration to pick it up from the ground. Out of twenty rations they managed to save only fourteen. The "fighting" kids were nowhere to be seen.

The mess hall at this camp was a plank lean-to not adequate for the Siberian winter. The gruel and the bread ration had to be carried about 150 yards in the cold from the kitchen to the dugout. For the elderly invalids this was a dangerous and difficult operation. They pushed their bread ration far down inside their shirt and gripped their mess tin with freezing hands. But suddenly, with diabolical speed, two or three kids would attack from the side. They knocked one old man to the ground, six hands frisked him all over, and they made off like a whirlwind. His bread ration had been pilfered, his gruel spilled, his empty mess tin lay there on the ground, and the old man struggled to get to his knees. (And other zeks saw this—and hastily bypassed the dangerous spot, hurrying to carry their own bread rations to the dugout.) And the weaker their victim, the more merciless were the kids. They openly tore the bread ration from the hands of a very weak old man. The old man wept and implored them to give it back to him: "I am dying of starvation." "So you're going to kick the bucket soon anyway—what's the difference?" And the kids once decided to attack the invalids in the cold, empty building in front of the kitchen where there was always a mob of people. The gang would hurl their victim to the ground, sit on his hands, his legs, and his head, search his pockets, take his makhorka and his money, and then disappear.

It was enough for a careless free worker to go into the camp compound with a dog and turn his head for one second. And he could buy his dog's pelt that very same evening outside the camp compound: the dog would have been coaxed away, knifed, skinned, and cooked, all in a trice.

Their ears simply didn't admit anything that they themselves didn't need. If irritated old men started to grab them and pull them up short, the kids would hurl heavy objects at them. The kids found amusement in just about anything. They would grab the field shirt off an elderly invalid and play "Keep away"—forcing him to run back and forth just as if he were their own age. Does he become angry and leave? Then he will never see it again! They will have sold it outside the compound for a smoke! (And they will even come up to him afterward innocently: "Papasha! Give us a light! Oh, come on now, don't be angry. Why did you leave? Why didn't you stay and catch it?")

For adults, fathers and grandfathers, these boisterous games of the kids in the crowded conditions of camp could cause more anguish and be more hurtful than their robbing and their rapacious greed. It proved to be one of the most sensitive forms of humiliation for an elderly person to be made equal with a young whippersnapper—if only it were equal! But not to be turned over to the tyranny of the whippersnappers.

That is how small stubborn Fascists were trained by the joint action of Stalinist legislation, a Gulag education, and the leaven of the thieves. It was impossible to invent a better method of brutalizing children! It was quite impossible to find a quicker, stronger way of implanting all the vices of camp in tiny, immature hearts.

Even when it would have cost nothing to soften the heart of a child, the camp bosses didn't permit it. This was not the goal of *their* training. At Krivoshchekovo Camp No. 1 a boy asked to be transferred so that he could be with his father in Camp No. 2. This was not permitted. (After all, the rules required families to be broken up.) And the boy had to hide in a barrel to get from one camp to the other and lived there with his father in secret. And in their confusion they assumed he had escaped and used a stick with spikes made of nails to poke about in the latrine pits, to see whether or not he had drowned there.

Stalin's immortal laws on kids existed for twenty years—until the Decree of April 24, 1954, which relaxed them slightly: releasing those kids who had served more than one-third . . . of their *first* term! And what if there were fourteen? Twenty years, twenty harvests. And twenty different age groups had been maimed with crime and depravity.

So *who* dares cast a shadow on the memory of our Great Coryphaeus?

■

There were nimble children who managed to *catch* Article 58 very early in life. For example, Geli Pavlov got it at twelve (from 1943 to 1949 he was imprisoned in the colony in Zakovsk). For Article 58, in fact, *no minimum age* existed! That is what they said even in public lectures on jurisprudence—as, for example, in Tallinn in 1945. Dr. Usma knew a six-year-old boy imprisoned in a colony under 58. But that, evidently, is the record!

And where, if not in this chapter, are we going to mention the children orphaned by the arrest of their parents?

The children of the women of the religious commune near Khosta were fortunate. When their mothers were sent off to Solovki in 1929, the children were softheartedly left in their own homes and on their own farms. The children looked after the orchards and vegetable gardens themselves, milked their goats, assiduously studied at school, and sent their school grades to their parents on Solovki, together with assurances that they were prepared to suffer for God as their mothers had. (And, of course, the Party soon gave them this opportunity.)

Considering the instructions to "disunite" exiled children and their parents, how many of these kids must there have been even back in the twenties? And who will ever tell us of their fate? . . .

Even a superficial glance reveals one characteristic: The children, too, were destined for imprisonment; they, too, in their turn would be sent off to the promised land of the Archipelago, sometimes even at the same time as their parents. Take the eighth-grader Nina Peregud. In November, 1941, they came to arrest her father. There was a search. Suddenly Nina remembered that inside the stove lay a crumpled but not yet burned humorous rhyme. And it might have just stayed there, but out of nervousness Nina decided to tear it up at once. She reached into the firebox, and the dozing policeman grabbed her. And this horrible sacrilege, in a schoolgirl's handwriting, was revealed to the eyes of the Chekists:

> The stars in heaven are shining down
> And their light falls on the dew;
> Smolensk is already lost and gone
> And we're going to lose Moscow too.

And she expressed the desire:

> We only wish they'd bomb the school,
> We're awfully tired of studies.

Naturally these full-grown men engaged in saving their Motherland deep in the rear in Tambov, these knights with hot hearts and clean hands, had to scotch such a mortal danger. Nina was arrested. Confiscated for her interrogation were her diaries from the sixth grade and a counterrevolutionary photograph: a snapshot of the destroyed Vavarinskaya Church. "What did your father talk about?" pried the knights with the hot hearts. Nina only sobbed. They sentenced her to five years of imprisonment and three years' deprivation of civil rights (even though she couldn't lose them since she didn't yet have them).

In camp, of course, she was separated from her father. . . .

Oh, you corrupters of young souls! How prosperously you are living out your lives! You are never going to have to stand up somewhere, blushing and tongue-tied, and confess what slops you poured over souls!

But Zoya Leshcheva managed to outdo her whole family. And here is how. Her father, her mother, her grandfather, her grandmother, and her elder adolescent brothers had all been scattered to distant camps because of their faith in God. But Zoya was a mere ten years old. They took her to an orphanage in Ivanovo Province. And there she declared she would never remove the cross from around her neck, the cross which her mother had hung there when she said farewell. And she tied the knot of the cord tighter so they would not be able to remove it when she was asleep. The struggle went on and on for a long time. Zoya became enraged: "You can strangle me and then take it off a corpse!" Then she was sent to an orphanage *for retarded children*— because she would not submit to their training. And in that orphanage were the dregs, a category of kids worse than anything described in this chapter. The struggle for the cross went on and on. Zoya stood her ground. Even here she refused to learn to steal or to curse. "A mother as sacred as mine must never have a daughter who is a criminal. I would rather be a political, like my whole family."

And she became a political! And the more her instructors and the radio praised Stalin, the more clearly she saw in him the culprit responsible for all their misfortunes. And, refusing to give in to the criminals, she now began to win them over to her views! In the court-yard stood one of those mass-produced plaster statues of Stalin. And mocking and indecent graffiti began to appear on it. (Kids love sport! The important thing is to point them in the right direction.) The administration kept repainting the statue, kept watch over it, and reported the situation to the MGB. And the graffiti kept on appearing, and the kids kept on laughing. Finally one morning they found that the statue's head had been knocked off and turned upside down, and inside it were feces.

This was a terrorist act! The MGB came. And began, in accordance with all their rules, their interrogations and threats: "Turn over the gang of terrorists to us, otherwise *we are going to shoot the lot of you* for terrorism!" (And there would have been nothing remarkable if they had: so what, 150 children shot! If He Himself had known about it, he would himself have given the order.)

It's not known whether the kids would have stood up to them or given in, but Zoya Leshcheva declared: "I did it all myself! What else is the head of that papa good for?"

And she was tried. And she was sentenced *to the supreme measure,* no joke. But because of the intolerable humanitarianism of the 1950 law on the restoration of capital punishment the execution of a fourteen-year-old was forbidden. And therefore they gave her a "tenner" (it's surprising it wasn't twenty-five). Up to the age of eighteen she was in ordinary camps, and from the age of eighteen on she was in Special

Camps. For her directness and her language she got a second camp sentence and, it seems, a third one as well.

Zoya's parents had already been freed and her brothers too, but Zoya languished on in camp.

Long live our tolerance of religion!

Long live our children, the masters of Communism!

And let any country speak up that can say it has loved its children as we have ours!

Chapter 18

■

The Muses in Gulag

This chapter recounts the attempts of the Gulag's Cultural and Educational Section (the KVCL) to re-educate zeks, which included organizing such groups as propagandists, artists, sculptors, poets, and actors.

Chapter 19

■

The Zeks as a Nation

AN ETHNOGRAPHICAL ESSAY BY FAN FANYCH

This chapter is a grimly ironic tour de force, *a mock-serious anthropological treatise which describes the zeks as though they were a separate race.*

Chapter 20

■

The Dogs' Service

The title of this chapter was not intended as an intentionally scathing insult, but it is our duty to uphold the camp tradition. If you think about it, they themselves chose this lot: their service is the same as that of guard dogs, and their service is connected with dogs. And there even exists a special statute on service with dogs, and there are whole officers' committees which monitor the *work* of an individual dog, fostering *a good viciousness* in the dog. And if the maintenance of one pup for a year costs the people 11,000 pre-Khrushchev rubles (police dogs are fed better than prisoners), then the maintenance of each officer must cost even more.

And then throughout this book we have also had the difficulty of knowing what to call them in general. "The administration," "the chiefs," are too generalized and relate to freedom as well, to the whole life of the whole country, and they are shopworn terms anyway. "The bosses"—likewise. "The camp managers"? But this is a circumlocution that only demonstrates our impotence. Should they be named straightforwardly in accordance with camp tradition? That would seem crude, profane. It would be fully in the spirit of the language to call them *lagershchiki*—"camp keepers." . . . And it expresses an exact and unique sense: those who manage and govern the camps.

And that is what this chapter is about: the "camp keepers" (and the "prison keepers" with them). We could begin with the generals— and it would be a marvelous thing to do—but we don't have any material. It was quite impossible for us worms and slaves to learn about them and to see them close up. And when we did see them, we were dazzled by the glitter of gold braid and couldn't make anything out.

So we really know nothing at all about the chiefs of Gulag who followed one another in turn—those tsars of the Archipelago.

In this chapter we are going to cover those from colonel down. Our limitation is this: when you are confined in prison or in camp, the personality of the prison keepers interests you only to the extent that it helps you evade their threats and exploit their weaknesses. As far as anything else is concerned, you couldn't care less. They are unworthy of your attention.

And then later, too late, you suddenly realize that you didn't observe them closely enough.

Without even discussing the question of talent, can a person become a jailer in prison or camp if he is capable of the very least kind of useful activity? Let us ask: On the whole, can a camp keeper be a good human being? What system of moral selection does life arrange for them? The first selection takes place on assignment to the MVD armies, MVD schools, or MVD courses. Every man with the slightest speck of spiritual training, with a minimally circumspect conscience, or capacity to distinguish good from evil, is instinctively going to back out and use every available means to avoid joining this dark legion. But let us concede that he did not succeed in backing out. A second selection comes during training and the first service assignment, when the bosses themselves take a close look and eliminate all those who manifest laxity (kindness) instead of strong will and firmness (cruelty and mercilessness). And then a third selection takes place over a period of many years: All those who had not visualized where and into what they were getting themselves now come to understand and are horrified. To be constantly a weapon of violence, a constant participant in evil! Not everyone can bring himself to this, and certainly not right off. You see, you are trampling on others' lives. And inside yourself something tightens and bursts. You can't go on this way any longer! And although it is belated, men can still begin to fight their way out, report themselves ill, get disability certificates, accept lower pay, take off their shoulder boards—anything just to get out, get out, get out!

Does that mean the rest of them have got used to it? Yes. The rest of them have got used to it, and their life already seems normal to them. And useful too, of course. And even honorable. And some didn't have to get used to it; they had been that way from the start.

Thanks to this process of selection one can conclude that the percentage of the merciless and cruel among the camp keepers is much higher than in a random sample of the population. And the longer, the more constantly, and the more notably a person serves in the *Organs,* the more likely it becomes that he is a scoundrel.

Do a similarity of paths in life and a similarity of situations give rise to a similarity in characters? As a general thing it doesn't. For people with strong minds and spirits of their own it does not. They have their own solutions, their own special traits, and they can be very surprising. But among the camp keepers, who have passed through a severe negative-selection process—both in morality and mentality—the similarity is astonishing, and we can, in all likelihood, describe without difficulty their basic *universal* characteristics.

Arrogance. The camp chief lives on a separate island, flimsily connected with the remote external power, and on this island he is without qualification the first. No limits are set to his power, and it admits to no mistakes; every person complaining is always proven wrong (repressed). He has the best house on the island. The best means of transportation. The camp keepers immediately below him in rank are also raised extremely high. And since their whole preceding life has not given birth to any spark of critical capacity inside them, it is impossible for them to see themselves as other than a special race—of born rulers. Out of the fact that no one is capable of resisting them, they draw the conclusion that they rule very wisely, that this is their talent ("organizational"). Every day each ordinary event permits them visibly to observe their superiority: people rise before them, stand at attention, bow; at their summons people do not just approach but run up to them; on their orders people do not simply leave but run out.

Stupidity always follows on the heels of smugness. Deified alive, each knows everything inside out, doesn't need to read or learn, and no one can tell him anything worth pausing over. If Kudlaty, the chief of one of the Ust-Vym work parties, decided that the 100 percent fulfillment of state work norms was not 100 percent at all, and that instead what had to be fulfilled was his own daily norm (taken out of his head) and that otherwise he would put everyone on a penalty ration—there was no way to get him to change his mind. So, having fulfilled 100 percent, they all got penalty rations. In Kudlaty's office there were whole piles of volumes of Lenin. He summoned V. G. Vlasov and unctuously informed him: "Lenin writes what attitude one must take toward parasites." (He understood parasites to mean prisoners who had fulfilled the work norm by only 100 percent, and he understood by the term "proletariat" . . . himself. These two things fitted into their heads simultaneously: Here is my estate, and I am a proletarian.)

But the old serf-owning gentry were incomparably better educated; many of them had studied in St. Petersburg, and some of them even in Göttingen. From them, after all, came the Aksakovs, the Radishchevs, the Turgenevs. But no one ever emerged from our MVD men, and no one ever will.

Autocracy. Autotyranny. In this respect the camp keepers were fully the equals of the very worst of the serf owners of the eighteenth and nineteenth centuries. Innumerable are the examples of senseless orders, the sole purpose of which was to demonstrate their power. And the farther into Siberia and the North they were, the truer this was.

A sense of possessing a patrimonial estate was typical of all camp chiefs. They perceived their camp not as a part of some state system but as a patrimonial estate entrusted them indivisibly for as long as they occupied their positions. Hence came all the tyranny over lives, over personalities, and hence also came the bragging among themselves. The chief of one of the Kengir camps said: "I've got a professor working in the baths!" But the chief of another camp, Captain Stadnikov, put him down with "And in my camp I've got an academician barracks orderly who carries out latrine barrels."

Greed and *money-grubbing.* Among the camp keepers this was the most *widespread* trait of all. Not every one was stupid, and not every one was a petty tyrant—but every last one was engaged in attempting to enrich himself from the free labor of the zeks and from state property, whether he was the chief in that camp or one of his aides. Neither I nor any one of my friends could recollect any disinterested camp keeper, nor have any of the zeks who have been corresponding with me ever named one.

In their greed to grab as much as possible, none of their multitudinous legitimate monetary advantages and privileges could satisfy them. Neither high pay (with double and triple bonuses for work "in the Arctic," "in remote areas," "for dangerous work"). Nor prize money (provided management executives of camp by Article 79 of the Corrective Labor Code of 1933—that same code that did not hinder them from establishing a twelve-hour workday without any Sundays for the prisoners). Nor the exceptionally advantageous calculation of their seniority (in the North, where half the Archipelago was located, one year of work counted as two, and the total required for "military personnel" to earn a pension was twenty years; thus an MVD officer on completing MVD school at age twenty-two could retire on a full pension and go to live at Sochi at thirty-two!).

It was not enough for the camp bosses that both they and their families were clothed and shod by the camp craftsmen. It was not enough for them that they had their own furniture manufactured there, as well as all other kinds of household supplies. It was not enough for them that their pigs were fattened by the camp kitchen. It was too little! They were distinguished from the old serf owners because their power was not for a lifetime and not hereditary. And because of this difference the serf owners did not have to steal from themselves, but the camp keepers had their heads occupied with one thing—how to steal something from their own enterprise.

And what happened when they laid their hands on "American gifts" (collected by people in the United States for the Soviet people)? According to T. Sgovio's description, in 1943 in Ust-Nera the chief of the camp Colonel Nagorny, the head of the political section Goloulin, the chief of the Indigir administrative section Bykov, and the chief of the geological section Rakovsky, together with their wives, personally opened the boxes containing the gifts, picking what they wanted and fighting over the spoils. What they did not take for themselves they distributed as premiums to the camp employees. As late as 1948 the chiefs' orderlies were selling on the black market what remained of the gifts from the United States.

Lasciviousness. This was not true of each and every one of them, of course, and it was closely tied to individual physiology. But the situation of camp chief and the absoluteness of his rights allowed harem inclinations full sway. The chief of the Burepolom Camp, Grinberg, had each comely young woman brought to him immediately on arrival. (And what other choice did she have except death?) In Kochmes the camp chief Podlesny enjoyed nighttime roundups in the women's barracks (of the same sort we have already seen in Khovrino). He himself personally pulled the blankets off the women.

Malice, cruelty. There was no curb, either practical or moral, to restrain these traits. Unlimited power in the hands of limited people always leads to cruelty. (And we cite here all this similarity in vices to those of the serf owners not merely for eloquent argument. This similarity, alas, demonstrates that the nature of our compatriots has not changed in the slightest in two hundred years; give as much power as that and there will be all the same vices!)

They are still going about among us today. They may well turn out to be next to us in a train (though not in anything less than a first-class compartment). Or in a plane. An oaken cruelty is etched into their faces, and they always have a gloomy, dissatisfied expression. It would seem as if everything was going well in their lives, but there is that expression of dissatisfaction. Perhaps they have the feeling that they are missing out on something better? Or perhaps God has marked them out infallibly for all their evildoings?

In 1962 I traveled through Siberia on a train for the first time as a free man. And it just had to happen! In my compartment there was a young MVD man. . . . I pretended to be a sympathetic idiot, and he told me how they went through probationary work in contemporary camps, and how impudent, feelingless, and hopeless the prisoners were. On his face there had not yet set in that constant, permanent cruelty, but he triumphantly showed me a photo of the third graduating class at Tavda, in which there were not only boys—but also veteran camp keepers finishing up their education (in training dogs, in criminal investigation, in camp management and in Marxism-Leninism) more for the sake of their pensions than for the sake of service. And even though I had been around, nonetheless I exclaimed! Their blackness of heart stands out on their faces! How adroitly they pick them out from all humanity!

But I feel that my tale is becoming monotonous: Does it seem that I am repeating myself? Or is it that we have already read about this here, there, and elsewhere?

I hear objections! I hear objections! Yes, there were such individual facts. . . . But for the most part under Beria . . . But why don't you give some of the bright examples? Just describe some of the good ones for us! Show us our dearly beloved fathers. . . .

But no! Let those who saw them show them. I didn't see them. I have already deduced the generalized judgment that a camp keeper *could not be a decent person*—either he had to change direction or they got rid of him.

■

The camp custodial staff was considered the junior command staff of the MVD. These were the Gulag noncoms. And that was the kind of assignment they had—to worry the prey and not let go. They were on the same Gulag ladder, only lower. Year by year they coarsened in the service, and you couldn't observe in them the least cloudlet of pity toward the soaked, freezing, hungry, tired, and dying prisoners. And they could vent their malice and display cruelty—they encountered no barriers. The jailers willingly copied their officers in their conduct, and in character traits too—but they didn't have that gold on their shoulder boards and their greatcoats were dirtyish, and they went everywhere on foot, and they were not allowed prisoner-servants, and they dug in their own gardens, and looked after their own farm animals. Well, of course, they did manage *to haul off* a zek to their places for half a day now and then—to chop wood, to wash floors—this they could do, but not on a lavish scale. At work you could make the zek do a small task for you —solder, cook, hammer, or sharpen something. Anything larger than a stool you'd not always manage to carry out. This limitation on thievery deeply affronted the jailers, especially their wives, and because of this there was much bitterness against the chiefs, because of this life seemed very unjust, and within the jailer's breast there stirred not so much sensitive heart strings as a sense of unfulfillment, an emptiness echoing a human groan. And sometimes the lower-ranking jailers were capable of talking sympathetically with the zeks. Not so often, but not all that rarely either. In any case, among both prison and camp jailers it was possible to find human beings. Every prisoner encountered more than one in his career. In an officer it was virtually impossible.

This, properly speaking, was the universal law of the inverse ratio between social position and humaneness.

■

Self-guarding constitutes a special area in the history of the camp guard. Back, indeed, in the first postrevolutionary years it was proclaimed that *self-watch* was a duty of Soviet prisoners.

We will not affirm that this was a special, diabolical plan for the moral disintegration of the people. As always in the half-century of our most recent modern history, a lofty, bright theory and creeping moral vileness somehow got naturally interwoven, and were easily transformed into one another. But from the stories of the old zeks it has become known that the prisoner trusty guards were cruel to their own brothers, strove to curry favor and to hold on to their dogs' duties, and sometimes settled old accounts with a bullet on the spot.

And this has also been noted in our literature on jurisprudence: "In many cases those who were deprived of freedom carried out their duties of guarding the colonies and maintaining order *better* than the staff jailers."

And so tell me—what bad is there that one cannot teach a nation? Or people? Or all humanity?

Chapter 21

■

Campside

Like a piece of rotten meat which not only stinks right on its own surface but also surrounds itself with a stinking molecular cloud of stink, so, too, each island of the Archipelago created and supported a zone of stink around itself. This zone, more extensive than the Archipelago itself, was the intermediate transmission zone between the small zone of each individual island and the Big Zone—the Big Camp Compound—comprising the entire country.

Everything of the most infectious nature in the Archipelago—in human relations, morals, views, and language—in compliance with the universal law of osmosis in plant and animal tissue, seeped first into this transmission zone and then dispersed through the entire country. It was right here, in the transmission zone, that those elements of camp ideology and culture worthy of entering into the nationwide culture underwent trial and selection. And when camp expressions ring in the corridors of the new building of the Moscow State University, or when an independent woman in the capital delivers a verdict wholly from out of camp on the essence of life—don't be surprised: it got there via the transmission zone, via campside.

Thus it is that the Archipelago takes its vengeance on the Soviet Union for its creation.

Thus it is that no cruelty whatsoever passes by without impact.

Thus it is that we always pay dearly for chasing after what is cheap.

■

To give a list of these places, these hick towns, these settlements, would be almost the same as recapitulating the geography of the entire Archipelago. Sometimes entire districts were infected and belonged to the world of campside. And when a camp was injected into the body of a big city, even Moscow itself, campside also existed. . . .

Who lived in campside? (1) The basic indigenous local inhabitants (there might be none). (2) The VOKhR—the Militarized Camp Guards. (3) The camp officers and their families. (4) The jailers and their families—and jailers, as distinct from camp guards, always lived with their families, even when they were listed as on military service. (5) Former zeks (released from the local camp or one nearby). (6) Various restricted persons—"half-repressed" people, people with "unclean" passports. (7) The works administration. These were highly placed people and constituted in all only a few people in a big settlement. (8) Then, too, there were the *free employees* proper, all the tramps and riffraff—all kinds of strays and good-for-nothings and seekers after easy money. After all, in these remote death traps you could work three times as poorly as in the metropolis and get four times the wages—with bonuses for the Arctic, for remoteness, and for hardship, and you could also steal the work of the prisoners. And, in addition, many flocked there under recruitment programs and on contract, receiving moving and traveling expenses as well. For those able to pan the gold out of the work sheets campside was a real Klondike. People swarmed there with forged diplomas; adventurers, rascals, and money-grabbers poured in. But there was also quite a different category among the free employees: the elderly, who had already lived in campside for whole decades and had become so assimilated to its atmosphere that they no longer needed some other, sweeter world. If their camp shut down, or if the administration stopped paying them what they demanded—they left. But invariably they would move to some other such zone near a camp. They knew no other way to live.

Chapter 22

■

We Are Building

After everything that has been said about the camps, the question simply bursts out: That's enough! But was the prisoners' labor profitable to the state? And if it was not profitable—then was it worthwhile undertaking the whole Archipelago?

In the camps themselves both points of view on this were to be found among the zeks, and we used to love to argue about it.

Of course, if one believed the leaders, there was nothing to argue about. On the subject of the use of the prisoners' labor, Comrade Molotov, once the second-ranking man in the state, declared at the Sixth Congress of the Soviets of the U.S.S.R.: "We did this earlier. We are doing it now. And we are going to go on doing it in the future. It is profitable to society. It is useful to the criminals."

Not profitable to the state, note that! But to society itself. And useful to the criminals. And we will go on doing it in the future! So what is there to argue about?

Yes, the entire system of the Stalin decades, when first the construction projects were planned, and only afterward the recruitment of criminals to man them took place, confirms that the government evidently had no doubt of the economic profitability of the camps. Economics went before justice.

But it is quite evident that the question posed needs to be made more precise and to be split into parts:

- Did the camps justify themselves in a political and social sense?
- Did they justify themselves economically?
- Did they pay for themselves (despite the apparent similarity of the second and third questions, there is a difference)?

It is not difficult to answer the first question: For Stalin's purposes the camps were a wonderful place into which he could herd millions as a form of intimidation. And so it appears that they justified themselves politically. The camps were also profitable in lucre to an enormous social stratum—the countless number of camp officers; they gave them "military service" safely in the rear, special rations, pay, uniforms, apartments, and a position in society. Likewise they sheltered throngs of jailers and hard-head guards who dozed atop camp towers (while thirteen-year-old boys were driven into trade schools). And all these parasites upheld the Archipelago with all their strength—as a nest of serf exploitation. They feared a universal amnesty like the plague.

But we have already understood that by no means only those with different ideas, by no means only those who had got off the trodden path marked out by Stalin, were in the camps. The recruitment into camps obviously and clearly exceeded political needs, exceeded the needs of terror. It was proportionate (although perhaps in Stalin's head alone) to economic plans. Yes, and had not the camps (and exile) arisen out of the crisis unemployment of the twenties? From 1930 on, it was not that the digging of canals was invented for dozing camps, but that camps were urgently scraped together for the envisioned canals. It was not the number of genuine "criminals" (or even "doubtful persons") which determined the intensity of the courts' activities—but the requisitions of the economic establishment. At the beginning of the Belomor Canal there was an immediate shortage of Solovetsky Islands zeks, and it became clear that three years was too short and too unprofitable a sentence for the 58's, that they had to serve out two Five-Year plans taken together.

The reason why the camps proved economically profitable had been foreseen as far back as Thomas More, the great-grandfather of socialism, in his *Utopia*. The labor of the zeks was needed for degrading and particularly heavy work, which no one, under socialism, would wish to perform. For work in remote and primitive localities where it would not be possible to construct housing, schools, hospitals, and stores for many years to come. For work with pick and spade—in the flowering of the twentieth century. For the erection of the great construction projects of socialism, when the economic means for them did not yet exist.

On the great Belomor Canal even an automobile was a rarity. Everything was created, as they say in camp, with "fart power."

On the even larger Moscow-Volga Canal (seven times bigger in scale of work than the Belomor Canal and comparable to the Panama

Canal and the Suez Canal), 80 miles of canal were dug to a depth of over sixteen feet and a top width of 280 feet. And almost all of it with pick, shovel, and wheelbarrow. The future bottom of the Rybinsk Sea was covered with forest expanses. All of them were cut down by hand, and nary an electric saw was seen there, and the branches and brushwood were burned by total invalids.

Who, except prisoners, would have worked at logging ten hours a day, in addition to marching four miles through the woods in predawn darkness and the same distance back at night, in a temperature of minus 20, and knowing in a year no other rest days than May 1 and November 7? (Volgolag, 1937.)

And who other than the Archipelago natives would have grubbed out stumps in winter? Or hauled on their backs the boxes of mined ore in the open goldfields of the Kolyma? Or have dragged cut timber a half-mile from the Koin River (a tributary of the Vym) through deep snow on Finnish timber-sledge runners, harnessed up in pairs in a horse collar (the collar bows upholstered with tatters of rotten clothing to make them softer and the horse collar worn over one shoulder)?

True, the authorized journalist Y. Zhukov assures us the Komsomols built the city of Komsomolsk-on-the-Amur (in 1932) thus: They cut down trees without axes, having no smithies, got no bread, and died from scurvy. And he is delighted: Oh, how heroically we built! And would it not be more to the point to be indignant? Who was it, hating their own people, who sent them to build in such conditions?

And how was it then that the camps were economically unprofitable?

Read, read, in the "The Dead Road" by Pobozhi, that description of how they disembarked from and unloaded the lighters on the Taz River, that Arctic Iliad of the Stalinist epoch: how in the savage tundra, where human foot had never trod, the antlike prisoners, guarded by an antlike convoy, dragged thousands of shipped-in logs on their backs, and built wharves, and laid down rails, and rolled into that tundra locomotives and freight cars which were fated never to leave there under their own power. They slept five hours a day on the bare ground, surrounded by signboards saying, "CAMP COMPOUND."

And he describes further how the prisoners laid a telephone line through the tundra: They lived in lean-tos made of branches and moss, the mosquitoes devoured their unprotected bodies, their clothing never dried out from the swamp mud, still less their footgear. The route had been surveyed hit or miss and poorly laid out (and was doomed to be redrawn); and there was no timber nearby for poles, and they had to go off to either side on two- and three-day expeditions (!) in order to drag in poles from out there.

There was unfortunately no other Pobozhi to tell how before the war another railroad was built—from Kotlas to Vorkuta, where beneath each tie two heads were left.

Now who would have done that without prisoners? And how on earth could the camps not be profitable?

The question of the camps' *paying for themselves* was, however, a different question. The state's saliva had been flowing over this for a long time. They so wanted to have their little camps—and free too! From 1929 all the corrective-labor institutions of the country were included in the economic plan.

But no matter how they huffed and puffed and broke all their nails on the crags, no matter how they corrected the plan fulfillment sheets twenty times over, and wore them down to holes in the paper, the Archipelago did not pay its own way, and it never will! The income from it would never equal the expenses, and our young workers' and peasants' state (subsequently the elderly state of all the people) is forced to haul this filthy bloody bag along on its back.

And here's why. The first and principal cause was the lack of conscientiousness of the prisoners, the negligence of those stupid slaves. Not only couldn't you expect any socialist self-sacrifice of them, but they didn't even manifest simple capitalist diligence. All they were on the lookout for was ways to spoil their footgear—and not go out to work; how to wreck a crane, to buckle a wheel, to break a spade, to sink a pail—anything for a pretext to sit down and smoke. All that the camp inmates made for their own dear state was openly and blatantly botched: you could break the bricks they made with your bare hands; the paint would peel off the panels; the plaster would fall off; posts would fall down; tables rock; legs fall out; handles come off. Carelessness and mistakes were everywhere. And it could happen that you had to tear off a roof already nailed on, redig the ditch they had filled in, demolish with crowbar and drill a wall they had already built. In the fifties they brought a new Swedish turbine to Steplag. It came in a frame made of logs like a hut. It was winter, and it was cold, and so the cursed zeks crawled into this frame between the beams and the turbine and started a bonfire to get warm. The silver soldering on the blades melted—and they threw the turbine out. It cost 3,700,000 rubles. Now that's being self-supporting for you!

And in the presence of the zeks—and this was a second reason—the free employees didn't care either, as though they were working not for themselves but for some stranger or other, and they stole a lot, they stole a great, great deal. (They were building an apartment building, and the free employees stole several bathtubs. But the tubs had been supplied to match the number of apartments. So how could they hand over the apartment building as completed? They could not confess to the construction superintendent, of course—he was triumphantly showing the official acceptance committee around the first stair landing, yes, and he did not omit to take them into every bathroom too and show them each tub. And then he took the committee to the second-floor landing, and the third, not hurrying there either, and kept going into all the bathrooms—and meanwhile the adroit and experienced zeks, under the leadership of an experienced foreman plumber, broke bathtubs out of the apartments on the first landing, hauled them upstairs on tiptoe to the fourth floor and hurriedly installed and puttied them in before the committee's arrival. This ought to be shown in a film comedy, but they wouldn't allow it: there is nothing funny in our life; everything funny takes place in the West!)

The third cause was the zeks' lack of independence, their inability to live without jailers, without a camp administration, without guards, without a camp perimeter complete with watchtowers, without a Planning and Production Section, a Records and Classification Section, a Security Operations Section, a Cultural and Educational Section, and without higher camp administrations right up to Gulag itself; without censorship, and without penalty isolators, without Strict Regimen Brigades, without trusties, without stockroom clerks and warehouses; their inability to move around without convoy and dogs. And so the state had to maintain at least one custodian for each working native (and every custodian had a family!). Well and good that it was so, but what were all these custodians to live on?

And there were some bright engineers who pointed out a fourth reason as well: that, so they claimed, the necessity of setting up a perimeter fence at every step, of strengthening the convoy, of allotting a supplementary convoy, interfered with their, the engineers', technical maneuverability, as, for example, during the disembarkation on the River Taz; and because of this, so they claimed, everything was done late and cost more. But this was already an *objective* reason, this was a pretext! Summon them to the Party bureau, give them a good scolding, and the cause will disappear. Let them break their heads; they'll find a solution.

And then, beyond all these reasons, there were the natural and fully condonable miscalculations of the Leadership itself. As Comrade

Railroad from Salekhard to Igarka

Lenin said: Only the person who does nothing makes no mistakes.

For example, no matter how earth-moving work was planned—rarely did it take place in the summer, but always for some reason in the autumn and winter, in mud and in freezing weather.

Or at the Zarosshy Spring in the Shturmovoi goldfields (the Kolyma) in March, 1938, they sent out five hundred people to drive prospecting shafts to a depth of twenty-five to thirty feet in the permafrost. They completed them. (Half the zeks kicked the bucket.) It was time to start blasting, but they changed their minds: the metal content was low. They abandoned it. In May the prospecting shafts thawed, all the work was lost. And two years later, again in March, in the Kolyma frosts, they had another brainstorm: to drive prospecting shafts! In the very same place! Urgently! Don't spare lives!

Well, that's what superfluous expenditures are. . . .

Or, on the Sukhona River near the settlement of Opoki—the prisoners hauled earth and built a dam. And the spring freshets carried it away immediately. And that was that—gone.

Or, for example, the Talaga logging operation of the Archangel administration was given a plan to produce furniture, but the authorities forgot to assign supplies of lumber with which to make the furniture. But a plan is a plan, and has to be fulfilled! Talaga had to send out special brigades to fish driftwood out of the river—logs which had fallen behind the timber rafting. There was not enough. Then, in hit-and-run raids, they began to break up whole rafts and carry them off. But, after all, those rafts belonged to someone else in the plan, and now they wouldn't have enough. And also it was quite impossible for Talaga to write up work sheets to pay those bold young fellows who had grabbed the timber; after all, it was thievery. So that's what self-support is. . . .

Well, of course, these small mistakes are inevitable in any work. No Leader is immune to them.

What about the railroad from Salekhard to Igarka? Hundreds of kilometers of dikes had been laid across the swamps. By the time Stalin died, only 300 additional kilometers were needed to join both ends. That, too, was abandoned. But then one quails to say *whose* mistake that was. It was, after all, His Own. . . .

Due to all these causes not only does the Archipelago not pay its own way, but the nation has to pay dearly for the additional satisfaction of having it.

PART IV

The Soul and Barbed Wire

■

"Behold, I shew you a mystery; we shall not all sleep, but we shall all be changed."

I Corinthians, 15:51

Chapter 1

■

The Ascent

And the years go by. . . .

Not in swift staccato, as they joke in camp—"winter-summer, winter-summer"—but a long-drawn-out autumn, an endless winter, an unwilling spring, and only a summer that is short.

Even one mere year, whew, how long it lasts! Even in one year how much time is left for you to think! For 365 days you stomp out to line-up in a drizzling, slushy rain, and in a piercing blizzard, and in a biting and still subzero cold. For 365 days you work away at hateful, alien work with your mind unoccupied. For 365 evenings you squinch up, wet, chilled, in the end-of-work line-up, waiting for the convoy to assemble from the distant watchtowers. And then there is the march out. And the march back. And bending down over 730 bowls of gruel, over 730 portions of grits. Yes, and waking up and going to sleep on your multiple bunk. And neither radio nor books to distract you. There are none, and thank God.

And that is only one year. And there are ten. There are twenty-five. . . .

And then, too, when you are lying in the hospital with dystrophy —that, too, is a good time—*to think*.

Think! Draw some conclusions from misfortune.

And all that endless time, after all, the prisoners' brains and souls are not inactive?! In the mass and from a distance they seem like swarming lice, but they are the crown of creation, right? After all, once upon a time a weak little spark of God was breathed into them too— is it not true? So what has become of it now?

For centuries it was considered that a criminal was given a *sen-*

tence for precisely this purpose, to think about his crime for the whole period of his sentence, be conscience-stricken, repent, and gradually reform.

But the Gulag Archipelago knows no pangs of conscience! Out of one hundred natives—five are thieves, and their transgressions are no reproach in their own eyes, but a mark of valor. They dream of carrying out such feats in the future even more brazenly and cleverly. They have nothing to repent. Another five . . . *stole* on a big scale, but not from people; in our times, the only place where one can steal on a big scale is from the state, which itself squanders the people's money without pity or sense—so what was there for such types to repent of? Maybe that they had not stolen more and divvied up—and thus remained free? And, so far as another 85 percent of the natives were concerned—they had never committed any crimes whatever. What were they supposed to repent of? That they had thought what they thought?

No, not only do you not repent, but your clean conscience, like a clear mountain lake, shines in your eyes. (And your eyes, purified by suffering, infallibly perceive the least haze in other eyes; for example, they infallibly pick out stool pigeons. And the Cheka-GB is not aware of this capacity of ours to see with the eyes of truth—it is our "secret weapon" against that institution.)

It was in this nearly unanimous consciousness of our innocence that the main distinction arose between us and the hard-labor prisoners of Dostoyevsky. There they were conscious of being doomed renegades, whereas we were confidently aware that they could haul in any free person at all in just the same way they had hauled us in; that barbed wire was only a nominal dividing line between us. In earlier times there had been among the majority . . . the unconditional consciousness of personal guilt, and among us . . . the consciousness of disaster on a mammoth scale.

Just not to perish from the disaster! It had to be survived.

Wasn't this the root cause of the astounding rarity of camp suicides? Yes, rarity, although every ex-prisoner could in all probability recall a case of suicide. But he could recall even more escapes. There were certainly more escapes than suicides! (Admirers of socialist realism can praise me: I am pursuing an optimistic line.) And there were far more self-inflicted injuries, too, than there were suicides! But this, too, is an act indicating love of life—a straightforward calculation of sacrificing a portion to save the whole. I even imagine that, statistically speaking, there were fewer suicides per thousand of the population in camp than in freedom. I have no way of verifying this, of course.

It is a very spectacular idea to imagine all the innocently outraged millions beginning to commit suicide en masse, causing double vexation to the government—both by demonstrating their innocence and by depriving the government of free manpower. And maybe the government would have had to soften up and begin to take pity on its subjects? —well, hardly! Stalin wouldn't have been stopped by that. He would have merely picked up another twenty million people from freedom.

But it did not happen! People died by the hundreds of thousands and millions, driven, it would seem, to the extremity of extremities— but for some reason there were no suicides! Condemned to a misshapen existence, to waste away from starvation, to exhaustion from labor— they did not put an end to themselves!

And thinking the whole thing over, I found that proof to be the stronger. A suicide is always a bankrupt, always a human being in a blind alley, a human being who has gambled his life and lost and is without the will to continue the struggle. If these millions of helpless and pitiful vermin still did not put an end to themselves—this meant some kind of invincible feeling was alive inside them. Some very powerful idea.

This was their feeling of universal innocence. It was the sense of an ordeal of the entire people—like the Tatar yoke.

■

But what if one has nothing to repent of—what then, what then does the prisoner think about all the time? "Poverty and prison . . . give wisdom." They do. But—where is it to be directed?

Here is how it was with many others, not just with me. Our initial, first prison sky consisted of black swirling storm clouds and black pillars of volcanic eruptions—this was the heaven of Pompeii, the heaven of the Day of Judgment, because it was not just anyone who had been arrested, but I—the center of this world.

Our last prison sky was infinitely high, infinitely clear, even paler than sky-blue.

We all (except religious believers) began from one point: we tried to tear our hair from our head, but our hair had been clipped close! . . . How could we? How could we not have seen those who informed against us?! How could we not have seen our enemies? (And how we hated them! How could we avenge ourselves on them?) And what recklessness! What blindness! How many errors! How can they be corrected? They must be corrected all the more swiftly! We must write. . . . We must speak out. . . . We must communicate. . . .

But—there is nothing that we can do. And nothing is going to save us!

Then there begins the period of transit prisons. Interspersed with our thoughts about our future camp, we now love to recall our past: How well we used to live! (Even if we lived badly.) But how many unused opportunities there were! When will we now make up for it? If I only manage to survive—oh, how differently, how wisely, I am going to live! The day of our future *release?* It shines like a rising sun!

And the conclusion is: Survive to reach it! Survive! At any price!

This is simply a turn of phrase, a sort of habit of speech: "at any price."

But then the words swell up with their full meaning, and an awesome vow takes shape: to survive *at any price.*

And whoever takes that vow, whoever does not blink before its crimson burst—allows his own misfortune to overshadow both the entire common misfortune and the whole world.

This is the great fork of camp life. From this point the roads go to the right and to the left. One of them will rise and the other will descend. If you go to the right—you lose your life, and if you go to the left—you lose your conscience.

One's own order to oneself, *"Survive!,"* is the natural splash of a living person. Who does not wish to survive? Who does not have the right to survive? Straining all the strength of our body! An order to all our cells: Survive! A powerful charge is introduced into the chest cavity, and the heart is surrounded by an electrical cloud so as not to stop beating. They lead thirty emaciated but wiry zeks three miles across the Arctic ice to a bathhouse. The bath is not worth even a warm word. Six men at a time wash themselves in five shifts, and the door opens straight into the subzero temperature, and four shifts are obliged to stand there before or after bathing—because they cannot be left without convoy. And not only does none of them get pneumonia. They don't even catch cold. (And for ten years one old man had his bath just like that, serving out his term from age fifty to sixty. But then he was released, he was at home. Warm and cared for, he burned up in one month's time. That order—"Survive!"—was not there. . . .)

But simply "to survive" does not yet mean "at any price." "At any price" means: at the price of someone else.

Let us admit the truth: At that great fork in the camp road, at that great divider of souls, it was not the majority of the prisoners that turned to the right. Alas, not the majority. But fortunately neither was it just a few. There are many of them—human beings—who made this choice. But they did not shout about themselves. You had to look closely to see them. Dozens of times this same choice had arisen before them too, but they always knew, and knew their own stand.

Take Arnold Susi, who was sent to camp at the age of about fifty. He had never been a believer, but he had always been fundamentally decent, he had never led any other kind of life—and he was not about to begin any other. He was a "Westerner." And what that meant was that he was doubly unprepared, and kept putting his foot into it all the time, and getting into serious difficulties. He worked at general work. And he was imprisoned in a penalty camp—and he still managed to survive; he survived as exactly the same kind of person he had been when he came to camp. I knew him at the very beginning, and I knew him . . . afterward, and I can testify personally. True, there were three seriously mitigating circumstances which accompanied him through-out his camp life: He was classified as an invalid. For several years he received parcels. And thanks to his musical abilities, he got some addi-tional nourishment out of amateur theatricals. But these three circum-stances only explain why he survived. If they had not existed, he would have died. But he would not have changed. (And perhaps those who died did die because they did not change?)

And Tarashkevich, a perfectly ordinary, straightforward person, recalls: "There were many prisoners prepared to grovel for a bread ration or a puff of makhorka smoke. I was dying, but I kept my soul pure: I always called a spade a spade."

It has been known for many centuries that prison causes the profound rebirth of a human being. The examples are innumerable—such as that of Silvio Pellico: Through serving eight years he was transformed from a furious Carbonaro to a meek Roman Catholic. In our country they always mention Dostoyevsky in this respect. These transformations always proceed in the direction of deepening the soul. Ibsen wrote: "From lack of oxygen even the conscience will wither."

By no means! It is not by any means so simple! In fact, it is the opposite! Take General Gorbatov: He had fought from his very youth, advanced through the ranks of the army, and had no time at all in which to think about things. But he was imprisoned, and how good it was—various events awakened within his recollection, such as his hav-ing suspected an innocent man of espionage; or his having ordered by mistake the execution of a quite innocent Pole. (Well, when else would he have remembered this? After rehabilitation he did not remember such things very much?) Enough has been written about prisoners' changes of heart to raise it to the level of penological theory. For example, in the prerevolutionary *Prison Herald* Luchenetsky wrote: "Darkness renders a person more sensitive to light; involuntary inactivity in imprisonment arouses in him a thirst for life, movement, work; the quiet compels profound pondering over his own 'I,' over surrounding conditions, over his own past and present, and forces him to think about his future."

Our teachers, who had never served time themselves, felt for prisoners only the natural sympathy of the outsider; Dostoyevsky, however, who served time himself, was a proponent of punishment! And this is something worth thinking about.

The proverb says: "Freedom spoils, and lack of freedom teaches."

But Pellico and Luchenetsky wrote about *prison*. But Dostoyevsky demanded punishment—in prison. But *what kind of* lack of freedom is it that educates?

Camp?

That is something to think about.

Of course, in comparison with prison our camps are poisonous and harmful.

Of course, they were not concerned with our souls when they pumped up the Archipelago. But nonetheless: is it really hopeless to stand fast in camp?

And more than that: was it really impossible for one's soul to rise in camp?

Here is E.K., who was born around 1940, one of those boys who, under Khrushchev, gathered to read poems on Mayakovsky Square, but were hauled off instead in Black Marias. From camp, from a Potma camp, he writes to his girl: "Here all the trivia and fuss have decreased. . . . I have experienced a turning point. . . . Here you harken to that voice deep inside you, which amid the surfeit and vanity used to be stifled by the roar from outside."

At the Samarka Camp in 1946 a group of intellectuals had reached the very brink of death: They were worn down by hunger, cold, and work beyond their powers. And they were even deprived of sleep. They had nowhere to lie down. Dugout barracks had not yet been built. Did they go and steal? Or squeal? Or whimper about their ruined lives? No! Foreseeing the approach of death in days rather than weeks, here is how they spent their last sleepless leisure, sitting up against the wall: Timofeyev-Ressovsky gathered them into a "seminar," and they hastened to share with one another what one of them knew and the others did not —they delivered their last lectures to each other. Father Savely—spoke of "unshameful death," a priest academician—about patristics, one of the Uniate fathers—about something in the area of dogmatics and canonical writings, an electrical engineer—on the principles of the energetics of the future, and a Leningrad economist—on how the effort to create principles of Soviet economics had failed for lack of new ideas. Timofeyev-Ressovsky himself talked about the principles of microphysics. From one session to the next, participants were missing—they were already in the morgue.

That is the sort of person who can be interested in all this while already growing numb with approaching death—now that is an intellectual!

Pardon me, you . . . love life? You, you! You who exclaim and sing over and over and dance it too: "I love you, life! Oh, I love you, life!" Do you? Well, go on, love it! Camp life—love that too! It, too, is life!

> There where there is no struggle with fate,
> There you will resurrect your soul. . . .

You haven't understood a thing. When you get there, you'll collapse.

Along our chosen road are twists and turns and twists and turns. Uphill? Or up into the heavens? Let's go, let's stumble and stagger.

The day of liberation! What can it give us after so many years? We will change unrecognizably and so will our near and dear ones—and places which once were dear to us will seem stranger than strange.

And the thought of freedom after a time even becomes a forced thought. Farfetched. Strange.

The day of "liberation"! As if there were any liberty in this country! Or as if it were possible to liberate anyone who has not first become liberated in his own soul.

The stones roll down from under our feet. Downward, into the past! They are the ashes of the past!

And we ascend!

■

It is a good thing *to think* in prison, but it is not bad in camp either. Because, and this is the main thing, there are no *meetings.* For ten years you are free from all kinds of meetings! Is that not mountain air? While they openly claim your labor and your body, to the point of exhaustion and even death, the camp keepers do not encroach at all on your thoughts. They do not try to screw down your brains and to fasten them in place. And this results in a sensation of freedom of much greater magnitude than the freedom of one's feet to run along on the level.

No one tries to persuade you *to apply* for Party membership. No one comes around to squeeze membership dues out of you in *voluntary* societies. There is no trade union—the same kind of protector of your interests as an official lawyer before a tribunal. And there are no "production meetings." You cannot be elected to any position. You cannot be appointed some kind of delegate. And the really important thing is ... that they cannot compel you to be a propagandist. Nor—to listen to propaganda. Nor—will they ever drag you off to the electoral precinct to vote freely and secretly for a single candidate. No one requires any "socialist undertakings" of you. Nor—self-criticism of your mistakes. Nor—articles in the wall newspaper. Nor—an interview with a provincial correspondent.

A free head—now is that not an advantage of life in the Archipelago?

And there is one more freedom: No one can deprive you of your family and property—you have already been deprived of them. What does not exist—not even God can take away. And this is a basic freedom.

It is good to think in imprisonment. And the most insignificant cause gives you a push in the direction of extended and important thoughts. Once in a long, long while, once in three years maybe, they brought a movie to camp. The film turned out to be—the cheapest kind of "sports" comedy. It was a bore. But from the screen they kept drumming into the audience the moral of the film:

The result is what counts, and the result is not in your favor.

On the screen they kept laughing. In the hall the audience kept laughing too. But blinking as you came out into the sunlit camp yard, you kept thinking about this phrase. And during the evening you kept thinking about it on your bunk. And Monday morning out in line-up. And you could keep thinking about it as long as you wanted. And where else could you have concentrated on it like that? And slow clarity descended into your brain.

This was no joke. This was an infectious thought. It has long since been inculcated in our Fatherland—and they keep on inculcating it over and over. The concept that only the material result counts has become so much a part of us that when, for example, some Tukhachevsky, Yagoda, or Zinoviev was proclaimed ... a traitor who had sidled up to the enemy, people only exclaimed in a chorus of astonishment: *"What more could he want?"*

Now that is a high moral plane for you! Now that is a real unit of measure for you! "What more could he want?" Since he had a belly full of chow, and twenty suits, and two country homes, and an automobile, and an airplane, and fame—what more could he want?!! Millions of our compatriots find it unthinkable to imagine that a human being (and I am not speaking here of this particular trio) might have been motivated by something other than material gain!

To such an extent has everyone been indoctrinated with and absorbed the slogan: "The result is what counts."

Whence did this come to us?

If we look back at our history, maybe about three hundred years —could anything of the kind have taken place in the Russia of Old Believers?

All this came to us from Peter I, from the glory of our banners and the so-called "honor of our Fatherland." We were crushing our neighbors; we were expanding. And in our Fatherland it became well established that: The result is what counts.

And then from our Demidovs, Kabans and Tsybukins. They clambered up, without looking behind them to see whose ears they were smashing with their jackboots. And ever more firmly it became established among a once pious and openhearted people: The result is what counts.

And then—from all kinds of socialists, and most of all from the most modern, infallible, and intolerant Teaching, which consists of this one thing only: The result is what counts! It is important to forge a fighting Party! And to seize power! And to hold on to power! And to remove all enemies! And to conquer in pig iron and steel! And to launch rockets!

And though for this industry and for these rockets it was necessary to sacrifice the way of life, and the integrity of the family, and the spiritual health of the people, and the very soul of our fields and forests and rivers—to hell with them! The result is what counts!!!

But that is a lie! Here we have been breaking our backs for years at All-Union hard labor. Here in slow annual spirals we have been climbing up to an understanding of life—and from this height it can all be seen so clearly: It is not the result that counts! It is not the result —but *the spirit!* Not *what*—but *how.* Not what has been attained—but at what price.

And so it is with us the prisoners—if it is the result which counts, then it is also true that one must survive at any price. And what that means is: One must become a stool pigeon, betray one's comrades. And thereby get oneself set up comfortably. And perhaps even get time off sentence. In the light of the Infallible Teaching there is, evidently, nothing reprehensible in this. After all, if one does that, then the result will be in our favor, and the result is what counts.

No one is going to argue. It is pleasant to win. But not at the price of losing one's human countenance.

If it is the result which counts—you must strain every nerve and sinew to avoid *general work*. You must bend down, be servile, act meanly—yet hang on to your position as a trusty. And by this means . . . survive.

If it is the essence that counts, then the time has come to reconcile yourself to *general work*. To tatters. To torn skin on the hands. To a piece of bread which is smaller and worse. And perhaps . . . to death. But while you're alive, you drag your way along proudly with an aching back. And that is when—when you have ceased to be afraid of threats and are not chasing after rewards—you become the most dangerous character in the owl-like view of the bosses. Because . . . what hold do they have on you?

And as soon as you have renounced that aim of "surviving at any price," and gone where the calm and simple people go—then imprisonment begins to transform your former character in an astonishing way. To transform it in a direction most unexpected to you.

And it would seem that in this situation feelings of malice, the disturbance of being oppressed, aimless hate, irritability, and nervousness ought to multiply. But you yourself do not notice how, with the impalpable flow of time, slavery nurtures in you the shoots of contradictory feelings.

Once upon a time you were sharply intolerant. You were constantly in a rush. And you were constantly short of time. And now you have time with interest. You are surfeited with it, with its months and its years, behind you and ahead of you—and a beneficial calming fluid pours through your blood vessels—patience.

You are ascending. . . .

Formerly you never forgave anyone. You judged people without mercy. And you praised people with equal lack of moderation. And now an understanding mildness has become the basis of your uncategorical judgments. You have come to realize your own weakness

—and you can therefore understand the weakness of others. And be astonished at another's strength. And wish to possess it yourself.

The stones rustle beneath our feet. We are ascending. . . .

With the years, armor-plated restraint covers your heart and all your skin. You do not hasten to question and you do not hasten to answer. Your tongue has lost its flexible capacity for easy oscillation. Your eyes do not flash with gladness over good tidings nor do they darken with grief.

For you still have to verify whether that's how it is going to be. And you also have to work out—what is gladness and what is grief.

And now the rule of your life is this: Do not rejoice when you have found, do not weep when you have lost.

Your soul, which formerly was dry, now ripens from suffering. And even if you haven't come to love your neighbors in the Christian sense, you are at least learning to love those close to you.

Those close to you in spirit who surround you in slavery. And how many of us come to realize: It is particularly in slavery that for the first time we have learned to recognize genuine friendship!

And also those close to you in blood, who surrounded you in your former life, who loved you—while you played the tyrant over them . . .

Here is a rewarding and inexhaustible direction for your thoughts: Reconsider all your previous life. Remember everything you did that was bad and shameful and take thought—can't you possibly correct it now?

Yes, you have been imprisoned for nothing. You have nothing to repent of before the state and its laws.

But . . . before your own conscience? But . . . in relation to other individuals?

. . . Following an operation, I am lying in the surgical ward of a camp hospital. I cannot move. I am hot and feverish, but nonetheless my thoughts do not dissolve into delirium—and I am grateful to Dr. Boris Nikolayevich Kornfeld, who is sitting beside my cot and talking to me all evening. The light has been turned out—so it will not hurt my eyes. He and I—and there is no one else in the ward.

Fervently he tells me the long story of his conversion from Judaism to Christianity. This conversion was accomplished by an educated, cultivated person, one of his cellmates, some good-natured old fellow like Platon Karatayev. I am astonished at the conviction of the new convert, at the ardor of his words.

We know each other very slightly, and he was not the one responsible for my treatment, but there was simply no one here with whom he could share his feelings. He was a gentle and well-mannered person.

It is already late. All the hospital is asleep. Kornfeld is ending up his story thus:

"And on the whole, do you know, I have become convinced that there is no punishment that comes to us in this life on earth which is undeserved. Superficially it can have nothing to do with what we are guilty of in actual fact, but if you go over your life with a fine-tooth comb and ponder it deeply, you will always be able to hunt down that transgression of yours for which you have now received this blow."

I cannot see his face. Through the window come only the scattered reflections of the lights of the perimeter outside. And the door from the corridor gleams in a yellow electrical glow. But there is such mystical knowledge in his voice that I shudder.

These were the last words of Boris Kornfeld. Noiselessly he went out into the nighttime corridor and into one of the nearby wards and there lay down to sleep. Everyone slept. And there was no one with whom he could speak even one word. And I went off to sleep myself.

And I was wakened in the morning by running about and tramping in the corridor; the orderlies were carrying Kornfeld's body to the operating room. He had been dealt eight blows on the skull with a plasterer's mallet while he still slept. (In our camp it was the custom to kill immediately after rising time, when the barracks were all unlocked and open and when no one yet had got up, when no one was stirring.) And he died on the operating table, without regaining consciousness.

And so it happened that Kornfeld's prophetic words were his last words on earth. And, directed to me, they lay upon me as an inheritance. You cannot brush off that kind of inheritance by shrugging your shoulders.

But by that time I myself had matured to similar thoughts.

I would have been inclined to endow his words with the significance of a universal law of life. However, one can get all tangled up that way. One would have to admit that on that basis those who had been punished even more cruelly than with prison—those shot, burned at the stake—were some sort of super-evildoers. (And yet . . . the innocent are those who get punished most zealously of all.) And what would one then have to say about our so evident torturers: Why does not fate punish *them?* Why do they prosper?

(And the only solution to this would be that the meaning of earthly existence lies not, as we have grown used to thinking, in prospering, but

. . . in the development of the soul. From *that* point of view our torturers have been punished most horribly of all: they are turning into swine, they are departing downward from humanity. From that point of view punishment is inflicted on those whose development . . . *holds out hope.*)

But there was something in Kornfeld's last words that touched a sensitive chord, and that I accept quite completely *for myself.* And many will accept the same for themselves.

In the seventh year of my imprisonment I had gone over and re-examined my life quite enough and had come to understand why everything had happened to me: both prison and, as an additional piece of ballast, my malignant tumor. And I would not have murmured even if all that punishment had been considered inadequate.

Punishment? But . . . whose?

Well, just think about that—*whose?*

I lay there a long time in that recovery room from which Kornfeld had gone forth to his death, and all alone during sleepless nights I pondered with astonishment my own life and the turns it had taken. In accordance with my established camp custom I set down my thoughts in rhymed verses—so as to remember them. And the most accurate thing is to cite them here—just as they came from the pillow of a hospital patient, when the hard-labor camp was still shuddering outside the windows in the wake of a revolt.

> When was it that I completely
> Scattered the good seeds, one and all?
> For after all I spent my boyhood
> In the bright singing of Thy temples.
>
> Bookish subtleties sparkled brightly,
> Piercing my arrogant brain,
> The secrets of the world were . . . in my grasp,
> Life's destiny . . . as pliable as wax.
>
> Blood seethed—and every swirl
> Gleamed iridescently before me,
> Without a rumble the building of my faith
> Quietly crumbled within my heart.
>
> But passing here between being and nothingness,
> Stumbling and clutching at the edge,
> I look behind me with a grateful tremor
> Upon the life that I have lived.

Not with good judgment nor with desire
Are its twists and turns illumined.
But with the even glow of the Higher Meaning
Which became apparent to me only later on.

And now with measuring cup returned to me,
Scooping up the living water,
God of the Universe! I believe again!
Though I renounced You, You were with me!

Looking back, I saw that for my whole conscious life I had not understood either myself or my strivings. What had seemed for so long to be beneficial now turned out in actuality to be fatal, and I had been striving to go in the opposite direction to that which was truly necessary to me. But just as the waves of the sea knock the inexperienced swimmer off his feet and keep tossing him back onto the shore, so also was I painfully tossed back on dry land by the blows of misfortune. And it was only because of this that I was able to travel the path which I had always really wanted to travel.

It was granted me to carry away from my prison years on my bent back, which nearly broke beneath its load, this essential experience: *how* a human being becomes evil and *how* good. In the intoxication of youthful successes I had felt myself to be infallible, and I was therefore cruel. In the surfeit of power I was a murderer, and an oppressor. In my most evil moments I was convinced that I was doing good, and I was well supplied with systematic arguments. And it was only when I lay there on rotting prison straw that I sensed within myself the first stirrings of good. Gradually it was disclosed to me that the line separating good and evil passes not through states, nor between classes, nor between political parties either—but right through every human heart —and through all human hearts. This line shifts. Inside us, it oscillates with the years. And even within hearts overwhelmed by evil, one small bridgehead of good is retained. And even in the best of all hearts, there remains . . . an unuprooted small corner of evil.

Since then I have come to understand the truth of all the religions of the world: They struggle with the *evil inside a human being* (inside every human being). It is impossible to expel evil from the world in its entirety, but it is possible to constrict it within each person.

And since that time I have come to understand the falsehood of all the revolutions in history: They destroy only *those carriers* of evil contemporary with them (and also fail, out of haste, to discriminate the carriers of good as well). And they then take to themselves as their heritage the actual evil itself, magnified still more.

The Nuremberg Trials have to be regarded as one of the special achievements of the twentieth century: they killed the very idea of evil, though they killed very few of the people who had been infected with it. (Of course, Stalin deserves no credit here. He would have preferred to explain less and shoot more.) And if by the twenty-first century humanity has not yet blown itself up and has not suffocated itself—perhaps it is this direction that will triumph?

Yes, and if it does not triumph—then all humanity's history will have turned out to be an empty exercise in marking time, without the tiniest mite of meaning! Whither and to what end will we otherwise be moving? To beat the enemy over the head with a club—even cavemen knew that.

"Know thyself!" There is nothing that so aids and assists the awakening of omniscience within us as insistent thoughts about one's own transgressions, errors, mistakes. After the difficult cycles of such ponderings over many years, whenever I mentioned the heartlessness of our highest-ranking bureaucrats, the cruelty of our executioners, I remember myself in my captain's shoulder boards and the forward march of my battery through East Prussia, enshrouded in fire, and I say: "So were *we* any better?"

When people express vexation, in my presence, over the West's tendency to crumble, its political shortsightedness, its divisiveness, its confusion—I recall too: "Were we, before passing through the Archipelago, more steadfast? Firmer in our thoughts?"

And that is why I turn back to the years of my imprisonment and say, sometimes to the astonishment of those about me: *"Bless you, prison!"*

Lev Tolstoi was right when he *dreamed* of being put in prison. At a certain moment that giant began to dry up. He actually needed prison as a drought needs a shower of rain!

All the writers who wrote about prison but who did not themselves serve time there considered it their duty to express sympathy for prisoners and to curse prison. I . . . have served enough time there. I nourished my soul there, and I say without hesitation:

"Bless you, prison, for having been in my life!"

(And from beyond the grave come replies: It is very well for you to say that—when you came out of it alive!)

Chapter 2

■

Or Corruption?

But I have been brought up short: You are *not talking about the subject* at all! You have got off the track again—onto prison! And what you are supposed to be talking about is *camp*.

But I was also, I thought, talking about camp. Well, all right, I'll shut up. I shall give some space to contrary opinions. Many camp inmates will object to what I have said and will say that they did not observe any "ascent" of the soul, that this is nonsense, and that corruption took place at every step.

More insistent and more significant than others (because he had already written about all this) was Shalamov's objection:

In the camp situation human beings never remain human beings—the camps were created to this end.

All human emotions—love, friendship, envy, love of one's fellows, mercy, thirst for fame, honesty—fell away from us along with the meat of our muscles. . . . We had no pride, no vanity, and even jealousy and passion seemed to be Martian concepts. . . . The only thing left was anger—the most enduring of human emotions.

We came to understand that truth and falsehood were kin sisters.

There is only one distinction here to which Shalamov agrees: Ascent, growth in profundity, the development of human beings, is possible in *prison*. But

. . . camp—is wholly and consistently a negative school of life. There is nothing either necessary or useful that anyone derives from it. The prisoner learns flattery, falsehood, and petty and large-scale meanness. . . . When he returns home, he sees not only that he has not grown during his time in camp, but that his interests have become meager and crude.

Y. Ginzburg also agrees with this distinction: "Prison ennobled people, while camp corrupted them."

And how can one object to that?

In prison, both in solitary confinement and outside solitary too, a human being confronts his grief face to face. This grief is a mountain, but he has to find space inside himself for it, to familiarize himself with it, to digest it, and it him. This is the highest form of moral effort, which has always ennobled every human being. A duel with years and with walls constitutes moral work and a path upward (if you can climb it). If you share those years with a comrade, it is never in a situation in which you are called on to die in order to save his life, nor is it necessary for him to die in order for you to survive. You have the possibility of entering not into conflict but into mutual support and enrichment.

But in camp, it would appear, you do not have that path. Bread is not issued in equal pieces, but thrown onto a pile—go grab! Knock down your neighbors, and tear it out of their hands! The quantity of bread issued is such that one or two people have to die for each who survives. The bread is hung high up on a pine tree—go fell it. The bread is deposited in a coal mine—go down and mine it. Can you think about your own grief, about the past and the future, about humanity and God? Your mind is absorbed in vain calculations which for the present moment cut you off from the heavens—and tomorrow are worth nothing. You *hate* labor—it is your principal enemy. You hate your companions—rivals in life and death. You are reduced to a frazzle by intense *envy* and alarm lest somewhere behind your back others are right now dividing up that bread which could be yours, that somewhere on the other side of the wall a tiny potato is being ladled out of the pot which could have ended up in your own bowl.

Camp life was organized in such a way that envy pecked at your soul from all sides, even the best-defended soul.

And in addition you are constantly gripped by *fear:* of slipping off even that pitifully low level to which you are clinging, of losing your work which is still not the hardest, of coming a cropper on a prisoner transport, of ending up in a Strict Regimen Camp. And on top of that, you got beaten if you were weaker than all the rest, or else you yourself beat up those weaker than you. And wasn't this corruption? *Soul mange* is what A. Rubailo, an old camp veteran, called this swift decay under external pressure.

Amid these vicious feelings and tense petty calculations, when and on what foundation could you ascend?

So isn't it the right time not to object, and not to rise to the defense of some sort of alleged camp "ascent," but to describe hundreds, thousands of cases of genuine soul corruption? To cite examples of how no one could resist the camp philosophy of Yashka, the Dzhezkazgan work assigner: "The more you spit on people, the more they'll esteem you." To tell how newly arrived front-line soldiers (in Kraslag in 1942) had no sooner scented the thieves' atmosphere than they themselves undertook *to play the thief—to plunder* the Lithuanians and to fatten up off their foodstuffs and possessions: You greenhorns can go die! Or how certain Vlasov men began *to pass for thieves* out of the conviction that that was the only way to survive in camp. Or about that assistant professor of literature who became a thief Ringleader.

And how much corruption was introduced by that democratic and progressive system of "trusty watchmen"—which in our zek terminology became converted to *self-guarding*—introduced back in 1918? After all, this was one of the main streams of camp corruption: the enlistment of prisoners in the trusty guards!

And . . . he grows proud. And . . . he tightens his grip on his gun stock. And . . . he shoots. And . . . he is even more severe than the free guards. (How is one to understand this: Was it really a purblind faith in social initiative? Or was it just an icy, contemptuous calculation based on the lowest human feelings?)

After all, it was not just a matter of "self-guarding" either. There were also "self-supervision," and "self-oppression"—right up to the situation in the thirties when all of them, all the way up to the camp chief, were zeks. Including the transport chief. The production chief. Yes, and even *security chiefs* were zeks too. One could not have carried "self-supervision" any further than that: The zeks were conducting interrogations of themselves. They were recruiting stool pigeons to denounce themselves.

Yes, yes. But I am not going to examine those countless cases of corruption here. They are well known to everyone. They have already been described, and they will be described again. It is quite enough to admit they took place. This is the general trend, this is as it should be.

Why repeat about each and every house that in subzero weather it loses its warmth? It is much more surprising to note that there are houses which retain their warmth even in subzero weather.

And how is it that genuine religious believers survived in camp (as we mentioned more than once)? In the course of this book we have already mentioned their self-confident procession through the Archipelago—a sort of silent religious procession with invisible candles. How some among them were mowed down by machine guns and those next in line continued their march. A steadfastness unheard of in the twentieth century! And it was not in the least for show, and there weren't any declamations. Take some Aunt Dusya Chmil, a round-faced, calm, and quite illiterate old woman. The convoy guards called out to her: "Chmil! What is your article?"

And she gently, good-naturedly replied: "Why are you asking, my boy? It's all written down there. I can't remember them all." (She had a bouquet of sections under Article 58.)

"Your term!"

Auntie Dusya sighed. She wasn't giving such contradictory answers in order to annoy the convoy. In her own simplehearted way she pondered this question: Her term? Did they really think it was given to human beings to know their terms?

"What term! . . . Till God forgives my sins—till then I'll be serving time."

"You are a silly, you! A silly!" The convoy guards laughed. "Fifteen years you've got, and you'll serve them all, and maybe some more besides."

But after two and a half years of her term had passed, even though she had sent no petitions—all of a sudden a piece of paper came: release!

How could one not envy those people? Were circumstances more favorable for them? By no means! It is a well-known fact that the "nuns" were kept only with prostitutes and thieves at penalty camps. And yet who was there among the religious believers whose soul was corrupted? They died—most certainly, but . . . they were not corrupted.

And how can one explain that certain unstable people found faith right there in camp, that they were strengthened by it, and that they survived uncorrupted?

And many more, scattered about and unnoticed, came to their allotted turning point and made no mistake in their choice. Those who managed to see that things were not only bad for them, but even worse, even harder, for their neighbors.

And all those who, under the threat of a penalty zone and a new term of imprisonment, refused to become stoolies?

How, in general, can one explain Grigory Ivanovich Grigoryev, a soil scientist? A scientist who volunteered for the People's Volunteer

Corps in 1941—and the rest of the story is a familiar one. Taken prisoner near Vyazma, he spent his whole captivity in a German camp. And the subsequent story is also familiar. When he returned, he was arrested by us and given a tenner. I came to know him in winter, engaged in general work in Ekibastuz. His forthrightness gleamed from his big quiet eyes, some sort of unwavering forthrightness. This man was never able to bow in spirit. And he didn't bow in camp either, even though he worked only two of his ten years in his own field of specialization, and didn't receive food parcels from home for nearly the whole term. He was subjected on all sides to the camp philosophy, to the camp corruption of soul, but he was incapable of adopting it. In the Kemerovo camps (Antibess) the security chief kept trying to recruit him as a stoolie. Grigoryev replied to him quite honestly and candidly: "I find it quite *repulsive* to talk to you. You will find many willing without me." "You bastard, you'll crawl on all fours." "I would be better off hanging myself on the first branch." And so he was sent off to a penalty situation. He stood it for half a year. And he made *mistakes* which were even more unforgivable: When he was sent on an agricultural work party, he refused (as a soil scientist) to accept the post of brigadier offered him. He hoed and scythed with enthusiasm. And even more stupidly: in Ekibastuz at the stone quarry he refused to be a work checker—only because he would have had to pad the work sheets for the sloggers, for which, later on, when they caught up with it, the eternally drunk free foreman would have to pay the penalty. (But would he?) And so he went to break rocks! His honesty was so monstrously unnatural that when he went out to process potatoes with the vegetable storeroom brigade, he did not steal any, though everyone else did. When he was in a good post, in the privileged repair-shop brigade at the pumping-station equipment, he left simply because he refused to wash the socks of the free bachelor construction supervisor, Treivish. (His fellow brigade members tried to persuade him: Come on now, isn't it all the same, the kind of work you do? But no, it turned out it was not at all the same to him!) How many times did he select the worst and hardest lot, just so as not to have to offend against conscience—and he didn't, not in the least, and I am a witness. And even more: because of the astounding influence on his body of his bright and spotless human spirit (though no one today believes in any such influence, no one understands it) the organism of Grigory Ivanovich, who was no longer young (close to fifty), grew stronger in camp; his earlier rheumatism of the joints disappeared completely, and he became particularly healthy after the typhus from which he recovered: in winter he went out in cotton sacks, making holes in them for his head and his arms—and he did not catch cold!

So wouldn't it be more correct to say that no camp can corrupt those who have a stable nucleus, who do not accept that pitiful ideology which holds that "human beings are created for happiness," an ideology which is done in by the first blow of the work assigner's cudgel?

Those people became corrupted in camp who before camp had not been enriched by any morality at all or by any spiritual upbringing. (This is not at all a theoretical matter—since during our glorious half-century millions of them grew up.)

Those people became corrupted in camp who had already been corrupted out in freedom or who were ready for it. Because people are corrupted in freedom too, sometimes even more effectively than in camp.

If a person went swiftly bad in camp, what it might mean was that he had not just gone bad, but that that inner foulness which had not previously been needed had disclosed itself.

M. A. Voichenko has his opinion: "In camp, existence did not determine consciousness, but just the opposite: consciousness and steadfast faith in the human essence decided whether you became an animal or remained a human being."

Yes, camp corruption was a mass phenomenon. But not only because the camps were awful, but because in addition we Soviet people stepped upon the soil of the Archipelago spiritually disarmed—long since prepared to be corrupted, already tinged by it out in freedom, and we strained our ears to hear from the old camp veterans "how to live in camp."

But we ought to have known how to live (and how to die) without any camp.

Yes, the camps were calculated and intended to corrupt. But this didn't mean that they succeeded in crushing *everyone.*

Just as in nature the process of oxidation never occurs without an accompanying reduction (one substance oxidizes while at the same time another reduces), so in camp, too (and everywhere in life), there is no corruption without ascent. They exist alongside one another.

In the next part I hope still to show how in other camps, in the Special Camps, a different *environment* was created after a certain time: the process of corruption was greatly hampered and the process of ascent became attractive even to the camp careerists.

Chapter 3

■

Our Muzzled Freedom

But even when all the main things about the Gulag Archipelago are written, read, and understood, will there be anyone even then who grasps what our *freedom* was like? What sort of a country it was that for whole decades dragged that Archipelago about inside itself?

It was my fate to carry inside me a tumor the size of a large man's fist. This tumor swelled and distorted my stomach, hindered my eating and sleeping, and I was always conscious of it (though it did not constitute even one-half of one percent of my body, whereas within the country as a whole the Archipelago constituted 8 percent). But the horrifying thing was not that this tumor pressed upon and displaced adjacent organs. What was most terrifying about it was that it exuded poisons and infected the whole body.

And in this same way our whole country was infected by the poisons of the Archipelago. And whether it will ever be able to get rid of them someday, only God knows.

Can we, *dare* we, describe the full loathsomeness of the state in which we lived (not so remote from that of today)? And if we do not show that loathsomeness in its entirety, then we at once have a lie. For this reason I consider that *literature did not exist* in our country in the thirties, forties, and fifties. Because without the *full* truth it is not literature. And today they show this loathsomeness according to the fashion of the moment—by inference, an inserted phrase, an after-thought, or hint—and the result is again a lie.

This is not the task of our book, but let us try to enumerate briefly those traits of *free* life which were determined by the closeness of the Archipelago or which were in the same style.

1. *Constant Fear.* As the reader has already seen, the roster of the waves of recruitment into the Archipelago is not exhausted with 1935, or 1937, or 1949. The recruitment went on *all the time.* Just as there is no minute when people are not dying or being born, so there was no minute when people were not being arrested. Sometimes this came close to a person, sometimes it was further off; sometimes a person deceived himself into thinking that nothing threatened him, and sometimes he himself became an executioner, and thus the threat to him diminished. But any adult inhabitant of this country, from a collective farmer up to a member of the Politburo, always knew that it would take only one careless word or gesture and he would fly off irrevocably into the abyss.

Just as in the Archipelago beneath every trusty lay the chasm (and death) of general work, so beneath every inhabitant lay the chasm (and death) of the Archipelago. In appearance the country was much bigger than its Archipelago, but all of it and all its inhabitants hung phantom-like above the latter's gaping maw.

Fear was not always the fear of arrest. There were intermediate threats: purges, inspections, the completion of security questionnaires —routine or extraordinary ones—dismissal from work, deprivation of residence permit, expulsion or exile. The security questionnaires were so detailed and so inquisitive that more than half the inhabitants of the country had a bad conscience and were constantly and permanently tormented by the approach of the period when they had to be filled out. Once people had invented a false life story for these questionnaires, they had to try not to get tangled up in it.

The aggregate fear led to a correct consciousness of one's own insignificance and of the lack of any kind of *rights.*

Nadezhda Mandelstam speaks truly when she remarks that our life is so permeated with prison that simple meaningful words like "they took," or "they put inside," or "he is inside," or "they let out," are understood by everyone in our country in only one sense, even without a context.

Peace of mind is something our citizens have never known.

2. *Servitude.* If it had been easy to change your place of residence, to leave a place that had become dangerous for you and thus shake off fear and refresh yourself, people would have behaved more boldly, and they might have taken some risks. But for long decades we were shackled by that same system under which no worker could quit work of his own accord. And the passport regulations also fastened everyone to

particular places. And the housing, which could not be sold, nor exchanged, nor rented. And because of this it was an insane piece of daring *to protest* in the place where you lived or worked.

3. *Secrecy and Mistrust.* These feelings replaced our former openhearted cordiality and hospitality (which had still not been destroyed in the twenties). These feelings were the natural defense of any family and every person, particularly because no one could ever quit work or leave, and every little detail was kept in sight and within earshot for years. The secretiveness of the Soviet person is by no means superfluous, but is absolutely necessary, even though to a foreigner it may at times seem superhuman. The former Tsarist officer K.U. survived and was never arrested only because when he got married he did not tell his *wife* about his past. His brother, N.U., was arrested—and the wife of the arrested man, taking advantage of the fact that they lived in different cities at the time of his arrest, hid his arrest from her own *father and mother*—so they would not blurt it out. She preferred telling them and everyone else that her husband had abandoned her, and then playing that role a long time! Now these were the secrets of one family which I was told thirty years later. And what urban family did not have such secrets?

4. *Universal Ignorance.* Hiding things from each other, and not trusting each other, we ourselves helped implement that *absolute secrecy,* absolute misinformation, among us which was *the cause of causes* of everything that took place—including both the millions of arrests and the mass approval of them also. Informing one another of nothing, neither shouting nor groaning, and learning nothing from one another, we were completely in the hands of the newspapers and the official orators.

5. *Squealing* was developed to a mind-boggling extent. Hundreds of thousands of Security officers in their official offices, in the innocent rooms of official buildings, and in prearranged apartments, sparing neither paper nor their unoccupied time, tirelessly recruited and summoned stool pigeons to give reports, and this in such enormous numbers as they could never have found necessary for collecting information. One of the purposes of such extensive recruitment was, evidently, to make each subject feel the breath of the stool pigeons on his own skin. So that in every group of people, in every office, in every apartment,

either there would be an informer or else the people there would be afraid there was.

I will give my own superficial speculative estimate: Out of every four to five city dwellers there would most certainly be one who at least once in his life had received a proposal to become an informer. And it might even have been more widespread than that. Quite recently I carried out my own spot check, both among groups of ex-prisoners and among groups of those who have always been free. I asked which out of the group they had tried to recruit and when and how. And it turned out that out of several people at a table *all* had received such proposals at one time or another!

Nadezhda Mandelstam correctly concludes: Beyond the purpose of weakening ties between people, there was another purpose as well. Any person who had let himself be recruited would, out of fear of public exposure, be very much interested in the continuing stability of the regime.

6. *Betrayal as a Form of Existence.* Given this constant fear over a period of many years—for oneself and one's family—a human being became a vassal of fear, subjected to it. And it turned out that the least dangerous form of existence was constant betrayal.

The mildest and at the same time most widespread form of betrayal was not to do anything bad directly, but just not to notice the doomed person next to one, not to help him, to turn away one's face, to shrink back. They had arrested a neighbor, your comrade at work, or even your close friend. You kept silence. You acted as if you had not noticed. (For you could not afford to lose your current job!) And then it was announced at work, at the general meeting, that the person who had disappeared the day before was . . . an inveterate enemy of the people. And you, who had bent your back beside him for twenty years at the same desk, now by your noble silence (or even by your condemning speech!), had to show how hostile you were to his crimes. (You had to make this sacrifice for the sake of your own dear family, for your own dear ones! What right had you not to think *about them?*) But the person arrested had left behind him a wife, a mother, children, and perhaps they at least ought to be helped? No, no, that would be dangerous: after all, these were the wife of an *enemy* and the mother of an enemy, and they were the children of an enemy (and your own children had a long education ahead of them)!

And one who concealed an enemy was also an enemy! And one

who abetted an enemy was also an enemy! And one who continued his friendship with an enemy was also an enemy. And the telephone of the accursed family fell silent. And in the hustle of a big city people felt as if they were in a desert.

And that was precisely what Stalin needed! And he laughed in his mustaches, the shoeshine boy!

In evaluating 1937 for the Archipelago, we refused it the title of the crowning glory. But here, in talking about *freedom,* we have to grant it this corroded crown of betrayal; one has to admit that this was the particular year that broke the soul of our *freedom* and opened it wide to corruption on a mass scale.

Yet even this was not yet the end of our society! (As we see today, the end never did come—the living thread of Russia survived, hung on until better times came in 1956, and it is now less than ever likely to die.) The resistance was not overt. It did not beautify the epoch of the universal fall, but with its invisible warm veins its heart kept on beating, beating, beating, beating.

Every act of resistance to the government required heroism quite out of proportion to the magnitude of the act. It was safer to keep dynamite during the rule of Alexander II than it was to shelter the orphan of an enemy of the people under Stalin. Nonetheless, how many such children were taken in and saved . . . Let the children themselves tell their stories. And secret assistance to families . . . did occur. And there was someone who took the place of an arrested person's wife who had been in a hopeless line for three days, so that she could go in to get warm and get some sleep. And there was also someone who went off with pounding heart to warn someone else that an ambush was waiting for him at his apartment and that he must not return there. And there was someone who gave a fugitive shelter, even though he himself did not sleep that night.

Nowadays it is quite convenient to declare that *arrest* was a lottery (Ehrenburg). Yes, it was a lottery all right, but some of the numbers were "fixed." They threw out a general dragnet and arrested in accordance with assigned quota figures, yes, but every person who *objected publicly* they grabbed that very minute! And it turned into a *selection on the basis of soul,* not a lottery! Those who were bold fell beneath the ax, were sent off to the Archipelago—and the picture of the monotonously obedient *freedom* remained unruffled. All those who were purer and better could not stay in that society; and without them it kept getting more and more trashy. You would not notice these quiet departures at all. But they were, in fact, the dying of the soul of the people.

7. *Corruption.* In a situation of fear and betrayal over many years people survive unharmed only in a superficial, bodily sense. And inside . . . they become corrupt.

So many millions of people agreed to become stool pigeons. And, after all, if some forty to fifty million people served long sentences in the Archipelago during the course of the thirty-five years up to 1953, including those who died—and this is a modest estimate, being only three or four times the population of Gulag at any one time, and, after all, during the war the death rate there was running *one percent per day* —then we can assume that at least every third or at least every fifth case was the consequence of somebody's denunciation and that somebody was willing to provide evidence as a witness! All of them, all those murderers with ink, are still among us today. And most often they are prospering. And we still rejoice that they are "our ordinary Soviet people."

8. *The Lie as a Form of Existence.* Whether giving in to fear, or influenced by material self-interest or envy, people can't nonetheless become stupid so swiftly. Their souls may be thoroughly muddied, but they still have a sufficiently clear mind. They cannot believe that all the genius of the world has suddenly concentrated itself in one head with a flattened, low-hanging forehead. They simply cannot believe the stupid and silly images of themselves which they hear over the radio, see in films, and read in the newspapers. Nothing forces them to speak the truth in reply, but no one allows them to keep silent! They have to *talk!* And what else but a lie? They have to applaud madly, and no one requires honesty of them.

The permanent lie becomes the only safe form of existence, in the same way as betrayal. Every wag of the tongue can be overheard by someone, every facial expression observed by someone. Therefore every word, if it does not have to be a direct lie, is nonetheless obliged not to contradict the general, common lie. There exists a collection of ready-made phrases, of labels, a selection of ready-made lies. And not one single speech nor one single essay or article nor one single book— be it scientific, journalistic, critical, or "literary," so-called—can exist without the use of these primary clichés. In the most scientific of texts it is required that someone's false authority or false priority be upheld somewhere, and that someone be cursed for telling the truth; without this lie even an academic work cannot see the light of day. And what can be said about those shrill meetings and trashy lunch-break gatherings where you are compelled to vote against your own opinion, to pretend to be glad over what distresses you?

In prison Tenno recalled with shame how two weeks before his own arrest he had lectured the sailors on "The Stalinist Constitution —The Most Democratic in the World." And of course not one word of it was sincere.

There is no man who has typed even one page . . . without lying. There is no man who has spoken from a rostum . . . without lying. There is no man who has spoken into a microphone . . . without lying.

But if only it had all ended there! After all, it went further than that: every conversation with the management, every conversation in the Personnel Section, every conversation of any kind with any other Soviet person called for lies. And if your idiot interlocutor said to you face to face that the Colorado beetles had been dropped on us by the Americans—it was necessary to agree! (And a shake of the head instead of a nod might well cost you resettlement in the Archipelago. Remember the arrest of Chulpenyov, in Part I, Chapter 7.)

But that was not all: Your children were growing up! And if the children were still little, then you had to decide what was the best way to bring them up; whether to start them off on lies instead of the truth (so that it would be *easier* for them to live) and then to lie forevermore in front of them too; or to tell them the truth, with the risk that they might make a slip, that they might let it out, which meant that you had to instill into them from the start that the truth was murderous, that beyond the threshold of the house you had to lie, only lie, just like papa and mama.

The choice was really such that you would rather not have any children.

9. *Cruelty.* And where among all the preceding qualities was there any place left for kindheartedness? How could one possibly preserve one's kindness while pushing away the hands of those who were drowning? Once you have been steeped in blood, you can only become more cruel. And, anyway, cruelty ("class cruelty") was praised and instilled, and you would soon lose track, probably, of just where between bad and good that trait lay. And when you add that kindness was ridiculed, that pity was ridiculed, that mercy was ridiculed—you'd never be able to chain all those who were drunk on blood!

10. *Slave Psychology.*

In various parts of our country we find a certain piece of sculpture: a plaster guard with a police dog which is straining forward in order to sink its teeth into someone. In Tashkent there is one right in front of the NKVD school, and in Ryazan it is like a symbol of the city, the one and only monument to be seen if you approach from the direction of Mikhailov.

And we do not even shudder in revulsion. We have become accustomed to these figures setting dogs onto people as if they were the most natural things in the world.

Setting the dogs onto us.

PART V

Katorga

■

Chapter 1

■

The Doomed

Revolution is often rash in its generosity. It is in such a hurry to disown so much. Take the word *katorga,* for instance. Now, *katorga* is a good word, a word with some weight in it. (It means "hard labor" or "penal servitude," and it was in use during the time of the Tsars.) *Katorga* descends from the judicial bench like the blade of a guillotine, stops short of beheading the prisoner but breaks his spine, shatters all hope there and then in the courtroom.

Stalin was very fond of old words. And twenty-six years after the February Revolution had abolished *"katorga,"* Stalin reintroduced it.

Little attempt was made to conceal the purpose of these *katorga* camps: the *katorzhane* were to be done to death. These were, undisguisedly, murder camps: but in the Gulag tradition murder was protracted, so that the doomed would suffer longer and put a little work in before they died.

They were housed in "tents," seven meters by twenty, of the kind common in the north. Surrounded with boards and sprinkled with sawdust, the tent became a sort of flimsy hut. It was meant to hold eighty people, if they were on bunk beds, or one hundred on sleeping platforms. But *katorzhane* were put into them two hundred at a time.

Yet there was no reduction of average living space—just a rational utilization of accommodation. The *katorzhane* were put on a twelve-hour working day with two shifts, and no rest days, so that there were always one hundred at work and one hundred in the hut.

At work they were cordoned off by guards with dogs, beaten whenever anybody felt like it, urged on to greater efforts by Tommy guns. On their way back to the living area their ranks might be raked with Tommy-gun fire for no good reason, and the soldiers would not have to answer for the casualties. Even at a distance a column of exhausted *katorzhane* was easily identified—no ordinary prisoners dragged themselves so hopelessly, so painfully along.

Their twelve working hours were measured out in full to the last tedious minute.

Those quarrying stone for roadmaking in the polar blizzards of Norilsk were allowed ten minutes for a warm-up once in the course of a twelve-hour shift. And then their twelve-hour rest was wasted in the silliest way imaginable. Part of these twelve hours went into moving them from one camp area to another, parading them, searching them. Once in the living area, they were immediately taken into a "tent" which was never ventilated—a windowless hut—and locked in. In winter a foul sour stench hung so heavy in the damp air that no one unused to it could endure it for two minutes. The living area was even less accessible to the *katorzhane* than the camp work area. They were never allowed to go to the latrine, nor to the mess hut, nor to the Medical Section. All their needs were served by the latrine bucket and the feeding hatch. Such was Stalin's *katorga* as it took shape in 1943–1944: a combination of all that was worst in the camps with all that was worst in the prisons.

Their twelve hours of *rest* also included inspections, morning and evening—a full and formal roll call. Then again, food was distributed twice in the course of these twelve hours: mess tins were passed through the feeding hatch, and through the feeding hatch they were collected again.

According to the camp records, which were not meant to preserve for history the fact that political prisoners were also starved to death, they were entitled to supplementary "miner's rations" and "bonus dishes," which were miserable enough even before three lots of thieves got at them. This was another lengthy procedure conducted through the feeding hatch—names were called out one by one, and dishes exchanged for coupons. And when at last you were about to collapse onto the sleeping platform and fall asleep, the hatch would drop again, once again names were called, and they would start reissuing the same coupons for use the next day.

So that out of twelve leisure hours in the cell, barely four remained for undisturbed sleep.

Then again, *katorzhane* were of course paid no money, nor had they any right to receive parcels or letters.

The *katorzhane* responded nicely to this treatment and quickly died.

But I can already hear angry cries from my compatriots and contemporaries. Stop! *Who are* these people of whom you dare to speak? Yes! They were there to be destroyed—and rightly so! Why, these were traitors, Polizei, burgomasters! They got what they asked for! Surely you are not sorry for them? And the women there were German bedstraw, I hear *women's* voices crying.

First, a few words about our women. Did not the whole of world literature (before Stalin) rapturously proclaim that love could not be contained by national boundaries? By the will of generals and diplomats? But once again we have adopted Stalin's yardstick: except as decreed by the Supreme Soviet, thou shalt not mate! Your body is, first and foremost, the property of the Fatherland.

Before we go any further, how old were these women when they closed with the enemy in bed instead of in battle? Certainly under thirty, and often no more than twenty-five. Which means that from their first childhood impression onward they had been educated *after* the October Revolution, been brought up in Soviet schools and on Soviet ideology! So that our anger was for the work of our own hands? Some of these girls had taken to heart what we had tirelessly dinned into them for fifteen years on end—that there is no such thing as one's own country, that the Fatherland is a reactionary fiction. Others had grown a little bored with our puritanical Lenten fare of meetings, conferences, and demonstrations, of films without kisses and dancing at arm's length. And yet others were simply hungry—yes, hungry in the most primitive sense: they had nothing to put in their bellies.

But who is really to blame for all this? Who? I ask you. Those women? Or—fellow countrymen, contemporaries—we ourselves, all of us? What was it in *us* that made the occupying troops much more attractive to our women? Was this not one of the innumerable penalties which we are continually paying, and will be paying for a long time yet, for the path we so hastily chose and have so stumblingly followed, with never a look back at our losses, never a cautious look ahead?

"All right, then, but the men at least were in for good reasons? They were traitors to their country, and to their class."

Since we have begun, let us go on.

What about the schoolteachers? Those whom our army in its panicky recoil abandoned with their schools, and pupils, for a year. For two years, or even for three. The quartermasters had been stupid, the generals no good—so what must the teachers do now? Teach their children or not teach them? And what were the kids to do—not kids of fifteen, who could earn a wage, or join the partisans, but the little kids? Learn their lessons, or live like sheep for two or three years to atone for the Supreme Commander's mistakes? If daddy doesn't give you a cap you let your ears freeze—is that it?

For some reason no such question ever arose either in Denmark or in Norway or in Belgium or in France. In those countries it was not felt that a people placed under German rule by its own foolish government or by force of overwhelming circumstances must thereupon stop living altogether. In those countries schools went on working, as did railways and local government.

Somebody's brains (theirs, of course!) are 180 degrees out of true. Because in our country teachers received anonymous letters from the partisans: "Don't dare teach! You will be made to pay for it!" Working on the railways also became collaboration with the enemy. As for participation in local administration—that was treason, unprecedented in its enormity.

Everybody knows that a child who once drops out of school may never return to it. Just because the greatest strategic genius of all times and all nations had made a blooper, was the grass to wither till he righted it or could it keep growing? Should children be taught in the meantime, or shouldn't they?

Of course, a price would have to be paid. Pictures of the big mustache would have to be taken out of school, and pictures of the little mustache perhaps brought in. The children would gather round the tree at Christmas instead of New Year's, and at this ceremony (as also on some imperial anniversary substituted for that of the October Revolution) the headmaster would have to deliver a speech in praise of the splendid new life, however bad things really were. But similar speeches had been made in the past—and life had been just as bad then.

Or rather, you had to be more of a hypocrite before, had to tell the children many more lies—because the lies had had time to mature, and to permeate the syllabus in versions painstakingly elaborated by experts on teaching technique and by school inspectors. In every lesson, whether it was pertinent or not, whether you were studying the anatomy of worms or the use of conjunctions in complex sentences, you were required to take a kick at God (even if you yourself believed in Him); you could not omit singing the praises of our boundless freedom (even if you had lain awake expecting a knock in the night); whether you were reading Turgenev to the class or tracing the course of the

Dnieper with your ruler, you had to anathematize the poverty-stricken past and hymn our present plenty (though long before the war you and the children had watched whole villages dying of hunger, and in the towns a child's ration had been 300 grams).

None of this was considered a sin against the truth, against the soul of the child, or against the Holy Ghost.

Whereas now, under the temporary and still unsettled occupation regime, far fewer lies had to be told—but they stood the old ones on their heads, that was the trouble! So it was that the voice of the Fatherland, and the pencil of the underground Party Committee, forbade you to teach children their native language, geography, arithmetic, and science. Twenty years of *katorga* for work of that sort!

Fellow countrymen, nod your heads in agreement! There they go, guards with dogs alongside, marching to the barracks with their night pails. Stone them—they taught your children.

But my fellow countrymen (particularly former members of specially privileged government departments, retired on pension at forty-five) advance on me with raised fists: Who is it that I am defending? Those who served the Germans as burgomasters? As village headmen? As Polizei? As interpreters? All kinds of filth and scum?

Well, let us go a little deeper. We have done far too much damage by looking at people as entries in a table. Whether we like it or not, the future will force us to reflect on the reasons for their behavior.

When they started playing and singing "Let Noble Rage"—what spine did not tingle? Our natural patriotism, long banned, howled down, under fire, anathematized, was suddenly permitted, encouraged, praised as *sacred*—what Russian heart did not leap up, swell with grateful longing for unity. How could we, with our natural magnanimity, help forgiving in spite of everything the native butchers as the foreign butchers drew near? Later, the need to drown half-conscious misgivings about our impulsive generosity made us all the more unanimous and violent in cursing the traitors—people plainly worse than ourselves, people incapable of forgiveness.

Russia has stood for eleven centuries, known many foes, waged many wars. But—have there been many traitors in Russia? Did traitors ever leave the country in *crowds?* I think not. I do not think that even their foes ever accused the Russians of being traitors, turncoats, renegades, though they lived under a regime inimical to ordinary working people.

Then came the most righteous war in our history, to a country with a supremely just social order—and tens and hundreds of thousands of our people stood revealed as traitors.

Where did they all come from? And why?

Perhaps the unextinguished embers of the Civil War had flared up again? Perhaps these were Whites who had not escaped extermination? No! I have mentioned before that many White émigrés took sides with the Soviet Union and against Hitler. They had freedom of choice—and that is what they chose.

These tens and hundreds of thousands—Polizei and executioners, headmen and interpreters—were all ordinary Soviet citizens. And there were many young people among them, who had grown up since the Revolution.

What made them do it? . . . What sort of people were they?

For the most part, people who had fallen, themselves and their families, under the caterpillar tracks of the twenties and thirties. People who had lost parents, relatives, loved ones in the turbid streams of our sewage system. Or who themselves had time and again sunk and struggled to the surface in camps and places of banishment. People who knew well enough what it was to stand with feet numb and frostbitten in the queue at the parcels window. People who in those cruel decades had found themselves severed, brutally cut off from the most precious thing on earth, the land itself—though it had been promised to them, incidentally, by the great Decree of 1917, and though they had been called upon to shed their blood for it in the Civil War. (Quite another matter are the country residences bought and bequeathed by Soviet officers, the fenced-in manorial domains outside Moscow: that's ours, so it's all right.) Then some people had been seized for snipping ears of wheat or rye. And some deprived of the right to live where they wished. Or the right to follow a long-practiced and well-loved trade (no one now remembers how fanatically we persecuted craftsmen).

All such people are spoken of nowadays (especially by professional agitators and the proletarian vigilantes of *Oktyabr*) with a contemptuous compression of the lips: "people with a grudge against the Soviet state," "formerly repressed persons," "sons of the former kulak class," "people secretly harboring black resentment of the Soviet power."

One says it—and another nods his head. As though it explained anything. As though the people's state had the right to offend its citizens. As though this were the essential defect, the root of the evil: "people with a grudge," "secretly resentful" . . .

And no one cries out: How can you! Damn your insolence! Do you or do you not hold that being determines consciousness? Or only when it suits you? And when it doesn't suit you does it cease to be true?

Then again, some of us are very good at saying—and a shadow flits over our faces—"Well, yes, certain errors were committed." Always the same disingenuously innocent, impersonal form: "were committed" —only nobody knows by whom. You might almost think that it was by ordinary workers, by men who shift heavy loads, by collective farmers. Nobody has the courage to say: "The Party committed them! Our irremovable and irresponsible leaders committed them!" Yet by whom, except those who had power, could such errors be "committed"? Lump all the blame on Stalin? Have you no sense of humor? If Stalin committed all these errors—where were you at the time, you ruling millions?

In any case, even these mistakes have faded in our eyes to a dim, shapeless blur, and they are no longer regarded as the result of stupidity, fanaticism, and malice; they are all subsumed in the only mistake acknowledged—that Communists jailed Communists. If 15 to 17 million peasants were ruined, sent off for destruction, scattered about the country without the right to remember their parents or mention them by name—that was apparently no mistake. And all the tributary streams of the sewage system surveyed at the beginning of this book were also, it seems, no mistake. That they were utterly unprepared for war with Hitler, emptily vainglorious, that they retreated shamefully, changing their slogans as they ran, that only Ivan fighting for Holy Russia halted the Germans on the Volga—all this turns out to be not a silly blunder, but possibly Stalin's greatest achievement.

In the space of two months we abandoned very nearly one-third of our population to the enemy—including all those incompletely destroyed families; including camps with several thousand inmates, who scattered as soon as their guards ran for it; including prisons in the Ukraine and the Baltic States, where smoke still hung in the air after the mass shooting of political prisoners.

As long as we were strong, we smothered these unfortunates, hounded them, denied them work, drove them from their homes, hurried them into their graves. When our weakness was revealed, we immediately demanded that they should forget all the harm done them, forget the parents and children who had died of hunger in the tundra, forget the executions, forget how we ruined them, forget our ingratitude to them, forget interrogation and torture at the hands of the NKVD, forget the starvation camps—and immediately join the partisans, go underground to defend the Homeland, with no thought for their lives. (There was no need for *us* to change! And no one held out the hope that when we came back we should treat them any differently, no longer hounding, harassing, jailing, and shooting them.)

Given this state of affairs, should we be surprised that too many people welcomed the arrival of the Germans? Or surprised that there were so few who did?

And the believers? For twenty years on end, religious belief was persecuted and churches closed down. The Germans came—and churches began to open their doors. (Our masters lacked the nerve to shut them again immediately after the German withdrawal.) In Rostov-on-the-Don, for instance, the ceremonial opening of the churches was an occasion for mass rejoicing and great crowds gathered. Were they nonetheless supposed to curse the Germans for this?

In Rostov again, in the first days of the war, Aleksandr Petrovich M——, an engineer, was arrested and died in a cell under interrogation. For several anxious months his wife expected to be arrested herself. Only when the Germans came could she go to bed with a quiet mind. "Now at least I can get some sleep!" Should she instead have prayed for the return of her tormentors?

In May, 1943, while the Germans were in Vinnitsa, men digging in an orchard on Podlesnaya Street (which the city soviet had surrounded with a high fence early in 1939 and declared a "restricted area under the People's Commissariat of Defense") found themselves uncovering graves which had previously escaped notice because they were overgrown with luxuriant grass. They found thirty-nine mass graves, 3.5 meters deep, 3 meters wide, 4 meters long. In each grave they found first a layer of outer garments belonging to the deceased, then bodies laid alternately head first or feet first. The hands of all of them were tied with rope, and they had all been shot by small-bore pistols in the back of the head. They had evidently been executed in prison and carted out for burial by night. Documents which had not decayed made it possible to identify people who had been sentenced to "20 years without the right to correspond" in 1938. In one picture of the excavation site: inhabitants of Vinnitsa have come to view the bodies or identify their relatives. There was more to come. In June they began digging near the Orthodox cemetery, outside the Pirogov Hospital, and discovered another forty-two graves. Next the Gorky Park of Culture and Rest—where, under the swings and carrousel, the "funhouse," the games area, and the dance floor, fourteen more mass graves were found. Altogether, 9,439 corpses in ninety-five graves. This was in Vinnitsa alone, and the discoveries were accidental. How many lie successfully hidden in other towns? After viewing these corpses, were the population supposed to rush off and join the partisans?

Relatives identifying the corpses of those executed at Vinnitsa

Perhaps in fairness we should at least admit that if you and I suffer when we and all we hold dear are trodden underfoot, those *we* tread on feel no less pain. Perhaps in fairness we should at last admit that those whom we seek to destroy have a right to hate us. Or have they no such right? Are they supposed to die gratefully?

We attribute deep-seated if not indeed congenital malice to these Polizei, these burgomasters—but we ourselves planted their malice in them, they were "waste products" of our making. How does Krylenko's dictum go? "In our eyes every crime is the product of a particular social system!" In this case—of your system, comrades! Don't forget your own doctrine!

Let us not forget either that among those of our fellow countrymen who took up the sword against us or attacked us in words, some were completely disinterested. No property had been taken from them (they had had none to begin with), they had never been imprisoned in the camps (nor yet had any of their kin), but they had long ago been sickened by our whole system: its contempt for the fate of the individual; the persecution of people for their beliefs; that cynical song "There's no land where men can breathe so freely"; the kowtowing of the devout to the Leader; the nervous twitching of pencils as everyone hurries to sign up for the state loan; the obligatory applause rising to an ovation! Cannot we realize that these perfectly normal people could not breathe our fetid air? (Father Fyodor Florya's accusers asked him how he had dared talk about Stalin's foul deeds when the Rumanians were on the spot. "How could I say anything different about you?" he answered. "I only told them what I knew. I only told them what had happened." What we ask is something different: lie, go against your conscience, perish—just so long as it helps us! But this, unless I'm mistaken, is hardly materialism.)

Mankind is almost incapable of dispassionate, unemotional thinking. In something which he has recognized as evil man can seldom force himself to see also what is good. Not everything in our lives was foul, not every word in the papers was false, but the minority, downtrodden, bullied, beset by stool pigeons, saw life in our country as an abomination from top to bottom, saw every page in the newspapers as one long lie.

We have been talking about the towns, but we should not forget the countryside. Liberals nowadays commonly reproach the village with its political obtuseness and conservatism. But before the war the village to a man, or overwhelmingly, was sober, much more sober than the town: it took no part at all in the deification of Daddy Stalin (and needless to say had no time for world revolution either). The village was, quite simply, sane and remembered clearly how it had been promised land, then robbed of it; how it had lived, eaten, and dressed before and after collectivization; how calves, ewes, and even hens had been taken away from the peasant's yard; how churches had been desecrated and defiled. Even in 1941 the radio's nasal bray was not yet heard in peasant huts, and not every village had even one person able to read the newspapers, so that to the Russian countryside all those Chang Tso-lins, MacDonalds, and Hitlers were indistinguishably strange and meaningless lay figures.

In a village in Ryazan Province on July 3, 1941, peasants gathered near the smithy were listening to Stalin's speech relayed by a loudspeaker. The man of iron, hitherto unmoved by the tears of Russian peasants, was now a bewildered old gaffer almost in tears himself, and as soon as he blurted out his humbugging "Brothers and Sisters," one of the peasants answered the black paper mouthpiece. "This is what you want, you bastard," and he made in the direction of the loudspeaker a rude gesture much favored by Russians: one hand grips the opposite elbow, and the forearm rises and falls in a pumping motion.

The peasants all roared with laughter.

If we questioned eyewitnesses in every village, we should learn of ten thousand such incidents, some still more pungent.

Such was the mood of the Russian village at the beginning of the war—the mood, then, of the reservists drinking the last half-liter and dancing in the dust with their kinsmen while they waited at some wayside halt for a train. On top of all this came a defeat without precedent in Russian memories, as vast rural areas stretching to the outskirts of both capitals and to the Volga, as many millions of peasants, slipped from under kolkhoz rule, and—why go on lying and prettifying history?—it turned out that the republics only wanted independence, the village only wanted freedom from the kolkhoz! The workers freedom from feudal decrees! But now, since further postponement is impossible, should I not also talk about those who even before 1941 had only one dream—to take up arms and blaze away at those Red commissars, Chekists, and collectivizers? Remember Lenin's words: "An oppressed class which did not aspire to possess arms and learn how to handle them would deserve only to be treated as slaves." There is, then, reason to be proud if the Soviet-German war showed that we are not such slaves as all those studies by liberal historians contemptuously make us out to be. There was nothing slavish about those who reached for their sabers to cut off Daddy Stalin's head (nor about those on the other side, who straightened their backs for the first time when they put on Red Army greatcoats—in a strange brief interval of freedom which no student of society could have foreseen).

These people, who had experienced on their own hides twenty-four years of Communist happiness, knew by 1941 what as yet no one else in the world knew: that nowhere on the planet, nowhere in history, was there a regime more vicious, more bloodthirsty, and at the same time more cunning and ingenious than the Bolshevik, the self-styled Soviet regime. That no other regime on earth could compare with it either in the number of those it had done to death, in hardiness, in the range of its ambitions, in its thoroughgoing and unmitigated totalitarianism— no, not even the regime of its pupil Hitler, which at that time blinded Western eyes to all else. Came the time when weapons were put in the hands of these people, should they have curbed their passions, allowed Bolshevism to outlive itself, steeled themselves to cruel oppression again—and only then begun the struggle with it (a struggle which has still hardly started anywhere in the world)? No, the natural thing was to copy the methods of Bolshevism itself: it had eaten into the body of a Russia sapped by the First World War, and it must be defeated at a similar moment in the Second.

The Germans were met with bread and salt in the villages on the Don. The pre-1941 population of the Soviet Union naturally imagined that the coming of a foreign army meant the overthrow of the Communist regime—otherwise it could have no meaning for us at all. People expected a political program which would liberate them from Bolshevism.

From where we were, separated from them by the wilderness of Soviet propaganda, by the dense mass of Hitler's army—how could we readily believe that the Western allies had entered this war not for the sake of freedom in general, but for their own Western European freedom, only against Nazism, intending to take full advantage of the Soviet armies and leave it at that? Was it not more natural for us to believe that our allies were true to the very *principle* of freedom and that they would not abandon us to a worse tyranny? ... True, these were the same allies for whom Russians had died in the First World War, and who then, too, had abandoned our army in the moment of collapse, hastening back to their comforts. But this was a lesson too cruel for the heart to learn.

Even in 1943 tens of thousands of refugees from the Soviet provinces trailed along behind the retreating German army—anything was better than remaining under Communism.

I will go so far as to say that our folk would have been worth nothing at all, a nation of abject slaves, if it had gone through that war without brandishing a rifle at Stalin's government even from afar, if it had missed its chance to shake its fist and fling a ripe oath at the *Father of the Peoples.* The Germans had their generals' plot—but what did we have? Our generals were (and remain to this day) nonentities, corrupted by Party ideology and greed, and have not preserved in their own persons the spirit of the nation, as happens in other countries. So that those who raised their hands and struck were almost to a man from the lowest levels of society—the number of former gentry émigrés, former members of the wealthier strata, and intellectuals taking part was microscopically small.

Chekhov complained that we had no "legal definition of *katorga,* or of its purpose."

But that was in the enlightened nineteenth century! In the middle of the twentieth, the cave man's century, we didn't even feel the need to understand and define. Old Man Stalin had decided that it would be so—and that was all the definition necessary.

We just nodded our heads in understanding.

Chapter 2

■

The First Whiff of Revolution

This chapter, which touches on several themes, including Russian-Ukrainian relations, in the end focuses on the increasing spirit of resistance ("the vague mutinous anticipations which had grown in us") among the prisoners.

Chapter 3

■

Chains, Chains . . .

Our eager hopes, our leaping expectations, were soon crushed. The wind of change was blowing only in drafty corridors—in the transit prisons. Here, behind the tall fences of the Special Camps, its breath did not reach us. And although there were only political prisoners in these camps, no mutinous leaflets hung on posts.

They say that at Minlag the blacksmiths refused to forge bars for hut windows. All glory to those as yet nameless heroes!

The Special Camps began with that uncomplaining, indeed eager submission to which prisoners had been trained by three generations of Corrective Labor Camps.

Prisoners brought in from the Polar North had no cause to be grateful for the Kazakh sunshine. At Novorudnoye station they jumped down from the red boxcars onto ground no less red. This was the famous Dzhezkazgan copper, and the lungs of those who mined it never held out more than four months. There and then the warders joyfully demonstrated their new weapon on the first prisoners to step out of line: handcuffs, which had not been used in the Corrective Labor Camps, gleaming nickel handcuffs, which went into mass production in the Soviet Union to mark the thirtieth anniversary of the October Revolution. These handcuffs were remarkable in that they could be clamped on very tight. Serrated metal plates were let into them, so that when a camp guard banged a man's handcuffed wrists against his knee, more of the teeth would slip into the lock, causing the prisoner greater pain. In this way the handcuffs became an instrument of torture instead of a mere device to inhibit activity: they crushed the wrists, causing constant acute pain, and prisoners were kept like that for hours, always with their hands behind their backs, palms outward. The warders also perfected the practice of trapping four fingers in the handcuffs, which caused acute pain in the finger joints.

In Berlag the handcuffs were used religiously: for every trifle, even for failure to take off your cap to a warder, they put on the handcuffs (hands behind the back) and stood you by the guardhouse. The hands became swollen and numb, and grown men wept: "I won't do it again, sir! Please take the cuffs off!"

It was easy enough for someone to scribble the order: "Establish Special Camps! Submit draft regulations by such and such a date!" But somewhere hard-working penologists (and psychologists, and connoisseurs of camp life) had to think out the details: How could screws already galling be made yet tighter? How could burdens already back-breaking be made yet heavier? How could the lives of Gulag's denizens, already far from easy, be made harder yet? Transferred from Corrective Labor Camps to Special Camps, these animals must be aware at once of their strictness and harshness—but obviously someone must first devise a detailed program!

Then again, they quite blatantly borrowed from the Nazis a practice which had proved valuable to them—the substitution of a number for the prisoner's name, his "I," his human individuality, so that the difference between one man and another was a digit more or less in an otherwise identical row of figures. This measure, too, could be a great hardship, provided it was implemented consistently and fully. This they tried to do. Every new recruit, when he "played the piano" in the Special Section (i.e., had his fingerprints taken, as was the practice in ordinary prisons, but not in Corrective Labor Camps), had to hang around his neck a board suspended from a rope. And in this guise he had his picture taken by the Special Section's photographer. (All those photographs are still preserved somewhere! One of these days we shall see them!)

In work rolls, too, it was the rule to write numbers before names. Why before and not instead of names? They were afraid to give up names altogether! However you look at it, a name is a reliable handle, a man is pegged to his name forever, whereas a number is blown away at a puff. If only the numbers were branded or picked out on the man himself, that would be something! But they never got around to it. Though they might easily have done so; they came close enough.

All this was in 1949 (the year one thousand nine hundred and forty-nine), the thirty-second year after the October Revolution, four years after the war, with its harsh imperatives, had ended, three years after the conclusion of the Nuremberg Trials, where mankind at large had learned about the horrors of the Nazi camps and said with a sigh of relief: "It can never happen again."

If you remember all this, it may not surprise you to hear that making him wear numbers was not the most hurtful and effective way of damaging a prisoner's self-respect.

But there were people for whom the numbers were indeed the most diabolical of the camp's devices. Among them were the devout women members of certain religious sects. These women refused to wear numbers—the mark of Satan! Nor would they give signed receipts (to Satan, of course) in return for regulation dress. The camp authorities showed laudable firmness! They gave orders that the women should be *stripped to their shifts,* and have their shoes taken from them, thus enlisting winter's help in forcing these senseless fanatics to accept regulation dress and sew on their numbers. But even with the temperature below freezing, the women walked about the camp in their shifts and barefoot, refusing to surrender their souls to Satan!

Faced with this spirit (the spirit of reaction, needless to say; enlightened people like ourselves would never protest so strongly about such a thing!), the administration capitulated and gave their clothing back to the sectarians, who put it on without numbers!

Chapter 4

■

Why Did We Stand For It?

In the conventional Cadet (let alone socialist) interpretation, the whole of Russian history is a succession of tyrannies. The Tatar tyranny. The tyranny of the Moscow princes. Five centuries of indigenous tyranny on the Oriental model, and of a social order firmly and frankly rooted in slavery. (Forget about the Assemblies of the Land, the village commune, the free Cossacks, the free peasantry of the North.) Whether it is Ivan the Terrible, Alexis the Gentle, heavy-handed Peter, velvety Catherine, or even Alexander II—until the Great February Revolution, all the Tsars right up to the Crimean War knew one thing only—how to *crush.*

Only . . . only . . . Crushed, yes, but the word needs qualification. Not crushed in our modern technical sense. After the war with Napoleon, when our army came back from Europe, the first breath of freedom passed over Russian society. Faint as it was, the Tsar had to reckon with it. The common soldiers, for instance, who took part in the Decembrist rising—was a single one of them strung up? Was a single one shot? And in our day would a single one of them have been left alive? Neither Pushkin nor Lermontov could be simply put inside for a *tenner*—roundabout ways of dealing with them had to be found. Whereas all of us have felt on our own hides the workings of a mechanized judicial system.

Seven attempts were made on the life of Alexander II himself. What did he do about it? Ruin and banish half Petersburg, as happened after Kirov's murder? You know very well that such a thing could never enter his head. Did he apply the methods of prophylactic mass terror? Total terror, as in 1918? Take *hostages?* The concept didn't exist.

Imprison *dubious persons?* It simply wasn't possible. . . . Execute thousands? They executed . . . five. Fewer than three hundred were convicted by the courts in this period. (If just *one* such attempt had been made on Stalin, how many million lives would it have cost us?)

The Bolshevik Olminsky writes that in 1891 he was the only *political prisoner* in the whole Kresty Prison. Transferred to Moscow, he was the only one in the Taganka. It was only in the Butyrki, awaiting deportation, that a small party of them was assembled. And a quarter of a century later the February Revolution revealed the presence of seven political prisoners in the Odessa castle prison and another three such prisoners in Mogilev.

The Tsars persecuted revolutionaries just sufficiently to broaden their circle of acquaintance in prisons, toughen them, and ring their heads with haloes. We now have an accurate yardstick to establish the scale of these phenomena—and we can safely say that the Tsarist government did not persecute revolutionaries but tenderly nurtured them, for its own destruction. The uncertainty, half-heartedness, and feebleness of the Tsarist government are obvious to all who have experienced an infallible judicial system.

Let us examine, for instance, some generally known biographical facts about Lenin. In spring, 1887, his brother was executed for an attempt on the life of Alexander III. And what happened to him? In the autumn of that very year Vladimir Ulyanov was admitted to the Imperial University at Kazan, and what is more, to the Law Faculty! Surprising, isn't it?

True, Vladimir Ulyanov was expelled from the university in the same academic year. But this was for organizing a student demonstration against the government. The younger brother of a would-be regicide inciting students to insubordination? What would he have got for that in our day? He would certainly have been shot! Whereas he was merely expelled. Such cruelty! Yes, but he was also banished. . . . To Sakhalin? No, to the family estate of Kokushkino, where he intended to spend the summer anyway. He wanted to work—so they gave him an opportunity. . . . To fell trees in the frozen north? No, to practice law in Samara, where he was simultaneously active in illegal political circles. After this he was allowed to take his examinations at St. Petersburg University as an external student. (With his curriculum vitae? What was the Special Section thinking of?)

Then a few years later this same young revolutionary was arrested for founding in the capital a "League of Struggle for the Liberation of the Working Class"—no less! He had repeatedly made "seditious" speeches to workers, had written political leaflets. Was he tortured, starved? No, they created for him conditions conducive to intellectual work. In the Petersburg investigation prison, where he was held for a year, and where he was allowed to receive the dozens of books he needed, he wrote the greater part of *The Development of Capitalism in Russia,* and, moreover, forwarded—legally, through the Prosecutor's Office—his *Economic Essays* to the Marxist journal *Novoye Slovo.* While in prison, he followed a prescribed diet, could have dinners sent in at his own expense, buy milk, buy mineral water from a chemist's shop, and receive parcels from home three times a week. (Trotsky, too, was able to put the first draft of his theory of permanent revolution on paper in the Peter and Paul Fortress.)

But then, of course, he was condemned by a three-man tribunal and shot? No, he wasn't even jailed, only banished. To Yakutya, then, for life? No, to a land of plenty, Minusinsk, and for three years. He was taken there in handcuffs? In a prison train? Not at all! He traveled like a free man, went around Petersburg for three days without interference, then did the same in Moscow—he had to leave instructions for clandestine correspondence, establish connections, hold a conference of revolutionaries still at large. He was even allowed to go into exile at his own expense—that is, to travel with free passengers. Lenin never sampled a single convict train or a single transit prison on his way out to Siberia or, of course, on the return journey. Then, in Krasnoyarsk, two more months' work in the library saw *The Development of Capitalism* finished, and this book, written by a political exile, appeared in print without obstruction from the censorship. (Measure that by our yardstick!) But what would he live on in that remote village, where he would obviously find no work? He asked for an allowance from the state, and they paid him more than he needed. It would have been impossible to create better conditions than Lenin enjoyed in his one and only period of banishment. A healthy diet, at extremely low prices, plenty of meat (a sheep every week), milk, vegetables; he could hunt to his heart's content (when he was dissatisfied with his dog, friends seriously considered sending him one from Petersburg; when mosquitoes bit him while he was out hunting, he ordered kid gloves); he was cured of his gastric disorders and the other illnesses of his youth, and rapidly put on weight. He had no obligations, no work to do, no duties, nor did his womenfolk exert themselves; for two and a half rubles a month, a fifteen-year-old peasant girl did all the rough work about the house. Lenin had no need to write for money, turned down offers of paid work from Petersburg, and wrote only things which could bring him literary fame.

He served his term of banishment (he could have "escaped" without difficulty, but was too circumspect for that). Was his sentence automatically extended? Converted to deportation for life? How could it be—that would have been illegal. He was given permission to reside in Pskov, on condition that he did not visit the capital. He did visit Riga and Smolensk. He was not under surveillance. Then he and his friend (Martov) took a basket of forbidden literature to the capital, traveling via Tsarskoye Selo, where there were particularly strict controls (they had been too clever by half). He was picked up in Petersburg. True, he no longer had the basket, but he did have a letter to Plekhanov in invisible ink with the whole plan for launching *Iskra*. The police, though, could not put themselves to all that trouble; he was under arrest and in a cell for three weeks, the letter was in their hands—and it remained undeciphered.

What was the result of this unauthorized absence from Pskov? Twenty years' hard labor, as it would have been in our time? No, just those three weeks under arrest! After which he was freed completely, to travel around Russia setting up distribution centers for *Iskra,* then abroad, to arrange publication ("the police see no objection" to granting him a passport for foreign travel!).

But this was the least of it! As an émigré he would send home to Russia an article on Marx for the *Granat Encyclopedia!* And it would be printed. Nor was it the only one!

Finally, he carried on subversive activity from a little town in Austrian Poland, near the Russian frontier, but no one sent undercover thugs to abduct him and bring him back alive. Though it would have been the easiest thing in the world.

Tsardom was always weak and irresolute in pursuit of its enemies.

The most important special feature of persecution (if you can call it that) in Tsarist times was perhaps just this: that the revolutionary's relatives never suffered in the least. Any member of the Ulyanov family (though nearly all of them were arrested at one time or another) could readily obtain permission to go abroad at any moment. When Lenin was on the "wanted" list for his exhortations to armed uprising, his sister Anna legally and regularly transferred money to his account with the Crédit Lyonnais in Paris.

Such were the circumstances in which Tolstoi came to believe that only moral self-improvement was necessary, not political freedom.

Of course, no one is in need of freedom if he already has it. We can agree with him that political freedom is not what matters in the end. The goal of human evolution is not freedom for the sake of freedom.

Nor is it the building of an ideal polity. What matter, of course, are the moral foundations of society. But that is in the long run: what about the beginning? What about the first step? Yasnaya Polyana in those days was an open club for thinkers. But if it had been blockaded as Akhmatova's apartment was when every visitor was asked for his passport, if Tolstoi had been pressed as hard as we all were in Stalin's time, when three men feared to come together under one roof, even he would have demanded political freedom.

Russian public opinion by the beginning of the century constituted a marvelous force, was creating a climate of freedom. The defeat of Tsarism came not when Kolchak was routed, not when the February Revolution was raging, but much earlier! It was overthrown without hope of restoration once Russian literature adopted the convention that anyone who depicted a gendarme or policeman with any hint of sympathy was a lickspittle and a reactionary thug; when you didn't have to shake a policeman's hand, cultivate his acquaintance, nod to him in the street, but merely brush sleeves with him in passing to consider yourself disgraced.

Whereas we have butchers who—because they are now redundant and because their qualifications are right—are in charge of literature and culture. They order us to extol *them* as legendary heroes. And to do so is for some reason called . . . patriotism.

The reason why we put up with it all in the camps is that there was no public opinion *outside*.

What conceivable ways has the prisoner of resisting the regime to which he is subjected? Obviously, they are:

1. Protest.
2. Hunger strike.
3. Escape.
4. Mutiny.

So, then, it is *obvious to anybody,* as the Great Deceased liked to say (and if it isn't, we'll ram it into him), that if the first two have some force (and if the jailers fear them), it is *only* because of public opinion! Without that behind us we can protest and fast as much as we like and they will laugh in our faces!

It is a very dramatic way of obtaining your demands—standing before the prison authorities and tearing open your shirt, as Dzerzhinsky did. But only where public opinion exists. Without it—you'll be gagged with the tatters and pay for a government-issue shirt into the bargain!

Let me remind you of a celebrated event which took place in the Kara hard-labor prison at the end of the last century. Political prisoners were informed that in future they would be liable to corporal punishment. Nadezhda Sigida was due to be thrashed first (she had slapped the commandant's face . . . to force him to resign!). She took poison and died rather than submit to the birch. Three other women then poisoned themselves—and also died! In the men's barracks fourteen prisoners volunteered to commit suicide, though not all of them succeeded. As a result, corporal punishment was abolished outright and forever! The prisoners had counted on frightening the prison authorities. For news of the tragedy at Kara would reach Russia, and the whole world.

You see from what a lofty plane prison behavior has declined. And how low we have fallen. And how by the same token our jailers have risen in the world! No, these are not the bumpkins of Kara! Even if we had plucked up our courage and risen above ourselves—four women and fourteen men—we should all have been shot before we got at any poison. If you did manage to poison yourself, you would only make the task of the authorities easier. And the rest would be treated to a dose of the birch for not denouncing you. And needless to say, no word of the occurrence would ever leak through the boundary wires.

This is the point, this is where their power lies: no news could leak out. If some muffled rumor did, with no confirmation from newspapers, with informers busily nosing it out, it would not get far enough to matter: there would be no outburst of public indignation. So what is there to fear? So why should they lend an ear to our *protests?* If you want to poison yourselves—get on with it.

Escape, then? History has preserved for us accounts of some major escapes from Tsarist prisons. All of them, let us note, were engineered and directed *from outside*—by other revolutionaries, Party comrades of the escapers, with incidental help from many sympathizers. Many people were involved in the escape itself, in concealing the escapers afterward, and in slipping them across the frontier. Perhaps it was all a jolly game, and a legal one? Fluttering your handkerchief from a window, letting a runaway share your bedroom, helping him with his disguise? These were not indictable offenses. When Pyotr Lavrov ran away from his place of banishment, the governor of Vologda [Khominsky] gave his civil-law wife permission to leave and catch up with her man. . . . Even for forging passports you could just be rusticated to your own farm, as we saw. People *were not afraid.*

I have at present no access to information about security at the principal locations of the Tsarist *katorga;* but if escape from them was ever as desperately difficult as it was from their Soviet counterparts, with one chance in 100,000 of success, I have never heard it. There was obviously no reason for prisoners to take great risks: they were not threatened with premature death from exhaustion by hard labor, nor with extensions of sentence which they had done nothing to deserve: the second half of their term they served not in prison but in places of banishment, and they usually put off escapes till then.

Laziness would seem to be the only reason for not escaping from Tsarist places of banishment. Escape in our time has always been an enterprise for giants among men, but for doomed giants. Such daring, such ingenuity, such will power never went into prerevolutionary escape attempts—yet they were very often successful, and ours hardly ever.

The reason for our failure was that success depends in the later stages of the attempt on the attitude of the population. And our population was *afraid* to help escapers, or even *betrayed* them, for mercenary or ideological reasons.

"So much for public opinion! . . ."

As for prison mutinies, involving as many as three, five, or eight thousand men—the history of our three revolutions knew nothing of them.

Yet we did.

But the same curse was upon them, and very great efforts, very great sacrifices, produced the most trivial results.

Because society was not ready. Because without a response from public opinion, a mutiny even in a huge camp has no scope for development.

So that when we are asked: "Why did you put up with it?" it is time to answer: "But we didn't!" Read on and you will see that we didn't put up with it at all.

In the Special Camps we raised the banner of the *politicals*—and politicals we became.

Chapter 5

■

Poetry Under a Tombstone, Truth Under a Stone

At the beginning of my camp career I was very anxious to avoid general duties, but did not know how. When I arrived at Ekibastuz in the sixth year of my imprisonment I had changed completely, and set out at once to cleanse my mind of the camp prejudices, intrigues, and schemes, which leave it no time for deeper matters. So that instead of resigning myself to the grueling existence of a general laborer until I was lucky enough to become a trusty, as educated people usually have to, I resolved to acquire a skill, there and then, in *katorga*. When we joined Boronyuk's team (Oleg Ivanov and I), a suitable trade (that of bricklayer) came our way. Later my fortunes took a different turn and I was for some time a smelter.

I was anxious and unsure of myself to begin with. Could I keep it up? We were unhandy cerebral creatures, and the same amount of work was harder for us than for our teammates. But the day when I deliberately let myself sink to the bottom and felt it firm under my feet —the hard, rocky bottom which is the same for all—was the beginning of the most important years in my life, the years which put the finishing touches to my character. From then onward there seem to have been no upheavals in my life, and I have been faithful to the views and habits acquired at that time.

I needed an unmuddled mind because I had been trying to write a poem for two years past. This was very rewarding, in that it helped me not to notice what was being done with my body. Sometimes in a sullen work party with Tommy-gunners barking about me, lines and images crowded in so urgently that I felt myself borne through the air, overleaping the column in my hurry to reach the work site and find a corner to write. At such moments I was both free and happy.

But how could I *write* in a Special Camp?

Memory was the only hidey-hole in which you could keep what you had written and carry it through all the searches and journeys under escort. In the early days I had little confidence in the powers of memory and decided therefore to write in verse. It was of course an abuse of the genre. I discovered later that prose, too, can be quite satisfactorily tamped down into the deep hidden layers of what we carry in our head. No longer burdened with frivolous and superfluous knowledge, a prisoner's memory is astonishingly capacious, and can expand indefinitely. We have too little faith in memory!

I started breaking matches into little pieces and arranging them on my cigarette case in two rows (of ten each, one representing units and the other tens). As I recited the verses to myself, I displaced one bit of broken match from the units row for every line. When I had shifted ten units I displaced one of the "tens." Every fiftieth and every hundredth line I memorized with special care, to help me keep count. Once a month I recited all that I had written. If the wrong line came out in place of one of the hundreds or fifties, I went over it all again and again until I caught the slippery fugitives.

In the Kuibyshev Transit Prison I saw Catholics (Lithuanians) busy making themselves rosaries for prison use. They made them by soaking bread, kneading beads from it, coloring them (black ones with burnt rubber, white ones with tooth powder, red ones with red germicide), stringing them while still moist on several strands of thread twisted together and thoroughly soaped, and letting them dry on the window ledge. I joined them and said that I, too, wanted to say my prayers with a rosary but that in my particular religion I needed one hundred beads in a ring (later, when I realized that twenty would suffice, and indeed be more convenient, I made them myself from cork), that every tenth bead must be cubic, not spherical, and that the fiftieth and the hundredth beads must be distinguishable at a touch. The Lithuanians were amazed by my religious zeal, but with true brotherly love helped me to put together a rosary such as I had described, making the hundredth bead in the form of a dark red heart. I never afterward parted with this marvelous present of theirs; I fingered and counted my beads inside my wide mittens—at work line-up, on the march to and from work, at all waiting times; I could do it standing up, and freezing cold was no hindrance. I carried it safely through the search points, in the padding of my mittens, where it could not be felt. The warders found it on various occasions, but supposed that it was for praying and let me keep it. Until the end of my sentence (by which time I had accumulated 12,000 lines) and after that in my place of banishment, this necklace helped me to write and remember.

I realized that I was not the only one, that I was party to a great secret, a secret maturing in other lonely breasts like mine on the scattered islands of the Archipelago, to reveal itself in years to come, perhaps when we were dead, and to merge into the Russian literature of the future.

How many of us were there? Many more, I think, than have come to the surface in the intervening years. Not all of them were to survive. Some buried manuscripts in bottles, without telling anyone where. Some put their work in careless or, on the contrary, in excessively cautious hands for safekeeping. Some could not write their work down in time.

Even on the isle of Ekibastuz, could we really get to know each other? encourage each other? support each other? Like wolves, we hid from everyone, and that meant from each other, too. Yet even so I was to discover a few others in Ekibastuz.

Meeting the religious poet Anatoly Vasilyevich Silin was a surprise which I owed to the Baptists. Day in and day out he was meek and gentle with everyone, but reserved. Only when we began talking to each other freely, and strolling about the camp for hours at a stretch on our Sundays off, while he recited his very long religious poems to me (like me, he had written them right there in the camp), I was startled not for the first time or the last to realize what far from ordinary souls are concealed within deceptively ordinary exteriors.

A homeless child, brought up an atheist in a children's home, he had come across some religious books in a German prisoner-of-war camp, and had been carried away by them. From then on he was not only a believer, but a philosopher and theologian! "From then on" he had also been in prison or in camps without a break, and so had spent his whole theological career in isolation, rediscovering for himself things already discovered by others, perhaps going astray, since he had never had either books or advisers. Now he was working as a manual laborer and ditchdigger, struggling to fulfill an impossible norm, returning from work with bent knees and trembling hands—but night and day the poems, which he composed from end to end without writing a word down, in iambic tetrameters with an irregular rhyme scheme, went round and round in his head. He must have known some twenty thousand lines by that time. He, too, had a utilitarian attitude to them: they were a way of remembering and of transmitting thoughts.

His sensitive response to the riches of nature lent warmth and beauty to his view of the world. Bending over one of the rare blades of grass which grew illegally in our barren camp, he exclaimed:

"How beautiful are the grasses of the earth! But even these the Creator has given to man for a carpet under his feet. How much more beautiful, then, must we be than they!"

"But what about 'Love not this world and the things that are of this world'?" (A saying which the sectarians often repeated.)

He smiled apologetically. He could disarm anyone with that smile.

"Why, even earthly, carnal love is a manifestation of a lofty aspiration to Union!"

His theodicy, that is to say his justification of the existence of evil in the world, he formulated like this:

> Does God, who is Perfect Love, allow
> This imperfection in our lives?
> The soul must suffer first, to know
> The perfect bliss of paradise. . . .
> Harsh is the law, but to obey
> Is for weak men the only way
> To win eternal peace.

Christ's sufferings in the flesh he daringly explained not only by the need to atone for human sins, but also by God's desire to *feel* earthly suffering to the full.

"God always *knew* these sufferings, but never before had he *felt* them," Silin boldly asserted. Even of the Antichrist, who had

> Corrupted man's Free Will—perverted
> His yearning toward the One True Light

Silin found something fresh and humane to say:

> The bliss that God had given him
> That angel haughtily rejected:
> He nothing knew of human pain;
> He loved not with the love of men—
> By grief alone is love perfected.

Thinking so freely himself, Silin found a warm place in his generous heart for all shades of Christian belief.

This is the crux:
That though Christ's teaching is its theme
Genius must ever speak with its own voice.

The atheist's impatient refusal to believe that spirit could beget matter only made Silin smile.

"Why don't they ask themselves how crude matter could beget spirit? That way round, it would surely be a miracle. Yes, a still greater miracle!"

My brain was full of my own verses, and these fragments are all that I have succeeded in preserving of the poems I heard from Silin— fearing perhaps that he himself would preserve nothing. A doomed and exhausted slave, with four number patches on his clothes, this poet had more in his heart to say to living human beings than the whole tribe of hacks firmly established in journals, in publishing houses, in radio —and of no use to anyone except themselves.

Silin ate from the same pot as the Baptists, shared his bread and warm victuals with them. Of course, he needed appreciative listeners, people with whom he could join in reading and interpreting the Gospel, and in concealing the little book itself. But Orthodox Christians he either did not seek out (suspecting that they would reject him as a heretic), or did not find. The Baptists, however, seemed to respect Silin, listened to him; they even considered him one of their own: but they, too, disliked all that was heretical in him, and hoped in time to bend him to their ways. Silin was subdued when he talked to me in their presence, and blossomed out when they were not there—it was difficult for him to force himself into their mold, though their faith was firm, pure, and ardent, helping them to endure *katorga* without wavering, and without spiritual collapse. They were all honest, free from anger, hard-working, quick to help others, devoted to Christ.

That is why they are being rooted out with such determination. In the years 1948–1950 several hundred of them were sentenced to twenty-five years' imprisonment and dispatched to Special Camps *for no other reason* than that they belonged to Baptist communes (a commune is of course an *organization*).

■

The camp is different from the Great Outside. Outside, everyone uninhibitedly tries to express and emphasize his personality in his outward behavior. In prison, on the contrary, all are depersonalized—identical haircuts, identical fuzz on their cheeks, identical caps, identical padded jackets. The face presents an image of the soul distorted by wind and sun and dirt and heavy toil. Discerning the light of the soul beneath this depersonalized and degraded exterior is an acquired skill.

But the sparks of the spirit cannot be kept from spreading, breaking through to each other. Like recognizes and is gathered to like in a manner none can explain.

Chapter 6

■

The Committed Escaper

When Georgi Pavlovich Tenno talks nowadays about past escapes
—his own, those of comrades, and those of which he knows only by
repute—his words of praise for the most uncompromising and persis-
tent heroes—Ivan Vorobyov, Mikhail Khaidarov, Grigory Kudla,
Hafiz Hafizov—are these:

"There was a *committed* escaper!"

A committed escaper! One who never for a minute doubts that a
man cannot live behind bars—not even as the most comfortable of
trusties, in the accounts office, in the Culture and Education Section,
or in charge of the bread ration. One who once he lands in prison spends
every waking hour thinking about escape and dreams of escape at night.
One who has vowed never to resign himself, and subordinates every
action to his need to escape. One for whom a day in prison can never
be just another day; there are only days of preparation for escape, days
on the run, and days in the punishment cells after recapture and a
beating.

A committed escaper! This means one who knows what he is
undertaking. One who has seen the bullet-riddled bodies of other escapers on display along the central tract. He has also seen those brought
back alive—like the man who was taken from hut to hut, black and blue
and coughing blood, and made to shout: "Prisoners! Look what happened to me! It can happen to you, too!" He knows that a runaway's
body is usually too heavy to be delivered to the camp. And that therefore the head alone is brought back in a duffel bag, sometimes (this is
more reliable proof, according to the rulebook) together with the right

arm, chopped off at the elbow, so that the Special Section can check the fingerprints and write the man off.

A committed escaper! It is for his benefit that window bars are set in cement, that the camp area is encircled with dozens of strands of barbed wire, towers, fences, reinforced barriers, that ambushes and booby traps are set, that red meat is fed to gray dogs.

The committed escaper is also one who refuses to be undermined by the reproaches of the average prisoner: You escapers make it worse for the rest! Discipline will be stiffer! Ten inspections a day! Thinner gruel! He ignores the whispered suggestions of other prisoners—not only those who urge resignation ("Life's not so bad even in a camp, especially if you get parcels"), but those who want him to join in protests or hunger strikes, because all that is not struggle but self-deception. Of all possible means of struggle, he has eyes only for one, believes only in one, devotes himself only to one—escape!

He cannot do otherwise! That is how he is made. A bird cannot renounce seasonal migration, and a committed escaper cannot help running away.

In the intervals between unsuccessful attempts, peaceful prisoners would ask Tenno: "Why can't you just sit still? Why do you keep running? What do you expect to find on the Outside—especially now?" Tenno was amazed. "What d'you mean—what do I expect to find? Freedom, of course! A whole day in the taiga without chains—that's what I call freedom!"

That was Tenno for you. In each new camp (he was transferred frequently) he was depressed and miserable until his next escape plan matured. Once he had a plan, Tenno was radiant, and a smile of triumph never left his lips.

■

There is no room in this book for his complicated life story. But the urge to escape had been with him from birth. As a small boy he had run away from boarding school in Bryansk to "America"—down the Desna in a rowboat. He had climbed the iron gates of the Pyatigorsk orphanage in his underwear in midwinter, and run away to his grandmother. He was a very unusual amalgam of sailor and circus performer. He had gone through a school for seamen, served before the mast on an icebreaker, as boatswain on a trawler, as navigation officer in the merchant navy. He had graduated from the army's Institute of Foreign Languages, spent the war with the Northern Fleet, sailed to Iceland and England as liaison officer with British convoys. But he had also, from

I went right through Ekibastuz with the number Shch #232 until my last few months, when I was ordered to change it to Shch #262. I smuggled patches with this number on them out of Ekibastuz, and I have kept them to this day.

The author in 1953, right after his release from the special camp

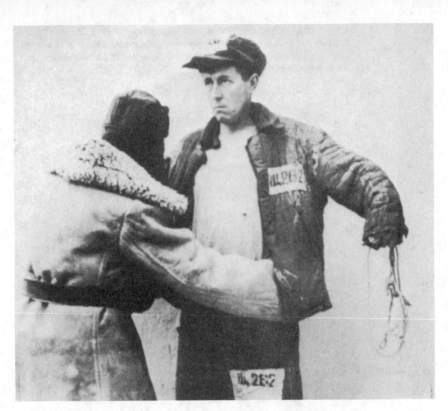

Body search

Georgi Pavlovich Tenno

his childhood on, practiced acrobatics; he had appeared in circuses, had trained gymnasts on the beam, performed as a memory man (memorizing masses of words and figures) and as a mind reader. The circus, and living in seaports, had led to some slight contact with the criminal world: he had picked up something of their language, their adventurousness, their quick-wittedness, their daredeviltry. Later on, serving time with thieves in numerous Disciplinary Barracks, he had absorbed more and more from them. This, too, would come in handy for the committed escaper.

A man is the product of his whole experience—that is how we come to be what we are.

In 1948 he was suddenly demobilized. This was a signal from the other world (he knew languages, had sailed on an English vessel, and was, moreover, an Estonian, though it is true a Petersburg Estonian), but if we are to live we must hope against hope. On Christmas Eve that year in Riga, where Christmas still feels like Christmas, like a holiday, he was arrested and taken to a cellar on Amatu Street, next door to the conservatory. As he entered his first cell he couldn't resist the temptation to tell the apathetically silent warder, "My wife and I had tickets for *The Count of Monte Cristo* and should be watching it right now. He fought for freedom, and I shall never accept defeat."

But it was too early yet to start fighting. We are always at the mercy of our assumption that a *mistake* has been made. Prison? For what? It's impossible! *They'll soon get it sorted out.* Indeed, before his transfer to Moscow they deliberately reassured him (this is done as a safety measure when prisoners are in transit). Colonel Morshchinin, chief of counterespionage, even came to the station to see him off and shook hands with him. "Have a good journey!" There were four of them, Tenno and his special escort, and they traveled in a separate first-class compartment. The rhythm of the carriage wheels was soothing. We can fill their rattle with any meaning, any prophecy we please. It filled Tenno with hope that they would "get it sorted out." And so he had no serious intention of running away. He was only sizing up the best way to do it. (Later on he would often remember that night and cluck with annoyance. Never again would it be so easy to run, never again would freedom be so near!)

The luxury of a special escort came to an end at the station in Moscow. The admission routine, sleepless nights, solitary confinement, more solitary confinement. A naïve request to be called for interrogation soon. The warder yawned. "Don't be in such a hurry; you'll get more than you want shortly."

At last, the interrogator. "Right, tell me about your criminal activities." "I'm absolutely innocent!" "Only Pope Pius is absolutely innocent."

In his cell he was tête-à-tête with a stool pigeon. Trying to *box him in*. Come on, tell me what really happened. A few interrogations and it was all quite clear: they'd never straighten it out, never let him go. So he must escape!

The world fame of the Lefortovo Prison did not daunt Tenno. Perhaps he was like a soldier new to the front who has experienced nothing and therefore fears nothing? It was the interrogator, Anatoly Levshin, who inspired Tenno's escape plan. By turning mean and arousing his hatred.

People and peoples have different criteria. So many millions had endured beatings within those walls, without even calling it torture. But for Tenno the realization that he could be beaten with impunity was intolerable. It was an outrage, and he would sooner die than suffer it. So when Levshin, after verbal threats, first advanced on him and raised his fist, Tenno jumped up and answered with trembling fury: "Look, my life's worth nothing anyway! But I can gouge one or both of your eyes out right now! That much I can do!"

The interrogator retreated. One rotten prisoner's life in exchange for a good eye was not much of a bargain. Next he tried to wear Tenno down in the punishment cells, to sap his strength. Then he put on a show, pretending that a woman screaming with pain in the next office was Tenno's wife, and that if he did not confess she would undergo still worse tortures.

Again he had misjudged his man! If a blow from a fist was hard for Tenno to bear, the idea that his wife was being interrogated was no less so. It became increasingly obvious to the prisoner that the interrogator must be killed. This and his escape were combined in a single plan. Major Levshin, too, wore naval uniform, was tall and fair-haired. As far as the sentry on the interrogation block was concerned, Tenno could very easily pass for Levshin. True, Levshin's face was round and sleek whereas Tenno had grown thin.

In the meantime they had removed the useless stoolie from the cell. His bed was left there and Tenno examined it. A metal crosspiece was rusted through at the point where it was fixed to one of the legs. It was about 70 centimeters long. How could he break it off?

First he must . . . perfect his skill in counting seconds precisely. Then calculate for each warder the interval between two peeps through the spy hole.

During one such interval he tried his strength, and the metal bar cracked off at the rusted end. Breaking the other, solid end was harder. He would have to stand on it with both feet—but then it would crash onto the floor. So in the interval between two visits he must make time to put a pillow on the cement floor, stand on the bed frame, break it, replace the pillow, and hide the bar for the time being; say, in his bed. And all the time he must be counting seconds.

It broke. The trick was done!

But the problem was only half solved: if they came in and found it, he would be rotting in the punishment cells. Twenty days of that and he would lose the strength he needed to escape, or even to defend himself against the interrogator. Yes, that was it: he would tear the mattress with his fingernails. Extract a little of the flock. Wrap flock around the ends of the bar, and put it back where it had been. Counting the seconds! Right—it was there!

But this was still good only for a short time. Once every ten days you went to the bathhouse, and while you were away your cell was searched. They might discover the breakage. So he must act quickly. How was he to take the bar from the cell to the interrogation room? When they let you out of the cellblock there was no search. They only slapped your clothing when you came back from interrogation, and then only your sides and chest, where there were pockets. They were looking for a blade, to prevent suicide.

Under his naval jacket Tenno wore the traditional sailor's striped jersey—it warms body and soul alike. "The sailor leaves his troubles ashore." He asked a warder for a needle (they will give you one at certain fixed times), as if to sew on buttons made of bread. He undid his jacket, undid his trousers, pulled out the edge of his jersey, and turned it up and stitched it so that it formed a little pocket (for the lower end of the rod). He had previously snapped off a bit of tape from his underpants. Now he pretended to be sewing a button on his jacket and stitched this tape to the inside of his jersey at chest level, so that it formed a loop to hold the rod steady.

Next he put the jersey on back to front, and began practicing day after day. The rod was set in position down his back and under his jersey: it was pushed through the loop at the top until it rested in the pocket down below. The upper end of the rod came up to his neck, under his tunic collar. His training routine went like this: In the short time between two inspections he would have to fling his hand to the back of his neck, seize the end of the rod while bending his trunk backward, then with a reverse movement straighten like a released bowstring while simultaneously drawing the rod—and strike the investigating officer a smart blow on the head. Then he would put everything back in place.

He also provided himself with two wads of flock, from the same mattress. These he could insert between his gums and his cheeks to make his face fuller.

He must also, of course, be clean-shaven on the day—and they scraped you with blunt razors only once a week. So that the day must be chosen carefully.

How was he to put some color into his cheeks? He would rub just a little blood on them. *That fellow's* blood.

Mustn't forget anything important, and must pack it all into four or five minutes. When he's lying there, knocked out, I must:

1. Slip off my jacket and put on his newer one with shoulder tabs.
2. Remove his shoelaces and lace my own floppy shoes up—that will take time.
3. Tuck his razor blade into a specially prepared place in the heel of my shoe (if they catch me and sling me into the nearest cell I can cut my veins).
4. Examine all his documents and take what I need.
5. Memorize the license plate number and find the ignition key.
6. Shove my own dossier into his bulky briefcase and take it with me.
7. Remove his watch.
8. Redden my cheeks with blood.
9. Drag his body behind the desk or screen, so that anybody coming in will think he's left and not raise a hue and cry.
10. Roll the flock into little balls and put them in my cheeks.
11. Put on his coat and cap.
12. Disconnect the wires to the light switch. If anybody comes soon afterward, finds it dark, and tries the switch, he will be sure to think that the bulb has burned out and that's why the interrogator has gone to another office. Even if they screw another bulb in, they won't immediately realize what has happened.

That makes twelve things to be done, and the escape itself will be number thirteen. . . .

Of course, the odds were against him. For the moment, he gave himself a 3 to 5 percent chance of success: the outer guardroom was completely unknown and he had no real hope of getting past it. But he couldn't die there like a slave!

So Tenno turned up, freshly shaven, for one nocturnal interrogation, with the iron bar behind his back. The interrogator questioned, abused, threatened, and all the time Tenno looked at him in surprise: couldn't he sense that his hours were numbered!

Tenno's heart thumped. A day of rejoicing was at hand. Or perhaps his last day.

Things turned out quite differently. Around midnight another interrogator hurried into the room and began whispering in Levshin's ear. This had never happened before. Levshin made hurried preparations to leave, pressed the button for the warder to come and remove the prisoner.

That was that. Tenno went back to his cell and replaced the iron bar.

Then came a daytime interrogation. And it took a strange turn: the interrogator refrained from yelling, and weakened his resolve by predicting that he would get five to seven years, so that there was no need to be downhearted. Somehow Tenno no longer felt angry enough to split his head open. Tenno's wrath was not the sort that lasted.

The mood of high excitement had passed. It seemed to him now that the odds were too great, that it was too much of a gamble.

The escaper's moods are perhaps even more capricious than those of the artist.

All his lengthy preparations had gone for nothing. . . .

But the escaper must be ready for this, too. He had brandished his bar in the air a hundred times, killed a hundred interrogators. A dozen times he had lived through every minute of his escape in detail—in the office, past the square window, along to the guardroom, beyond the guardroom. He had worn himself out with an escape which he would after all not be making.

Soon afterward they changed his interrogator and transferred him to the Lubyanka. There Tenno did not actively prepare to escape (his heart was not in it now that his interrogation seemed to have taken a more hopeful turn).

Vague hopes of clemency and reasonableness clouded Tenno's resolution. Only in the Butyrki Prison was he relieved of this burden: his sentence, read out from a piece of paper with a Special Board stamp, was confinement in camps for twenty-five years. He signed his name and felt relieved, found himself smiling, felt his legs carrying him easily to the cell for twenty-five-year prisoners. That sentence released him from humiliation, from the temptation to compromise, from humble submission, from truckling, from promises of five to seven years bestowed like alms on a beggar. Twenty-five is it, you bastards? Right; if that's all we can expect from you—we escape!!

Other people in the cell might talk about what they liked, but Tenno wanted stories about escape attempts and those who took part in them.

From the Kuibyshev Transit Prison they were taking prisoners to the station in open trucks—making up a long train of red prison cars. In the transit prison Tenno obtained from a local sneak thief who "respected escapers" two local addresses to which he might go for initial support. He shared these addresses with two other would-be runaways and they concerted a plan; all three would try to sit in the back row and when the truck slowed down at a turning they would jump, all three of them at once—right, left, and rear—past the guards, knocking them over if necessary. The guards would open fire, but they would not hit all three. They might not shoot at all—there would be people in the streets. Would they give chase? No, they couldn't abandon the other prisoners in the truck. So they would just shout and fire into the air. If the runaways were stopped, it would be by ordinary people, our Soviet people, passers-by. To frighten them off, the runaways must pretend to be holding knives! (They had no knives.)

The three of them maneuvered at the search point and hung back so that they would get onto the last truck and not leave before dusk. The last truck arrived, but . . . it was not a shallow three-tonner, like its predecessors, but a Studebaker with high sides. When he sat down even Tenno found that the top of his head was below the rim. The Studebaker moved quickly. Here was the turn! Tenno looked around at his comrades in arms. There was terror on their faces. No, they wouldn't jump. No, they were not committed escapers. ("But can you be sure of yourself?" he wondered.)

In the dark, with lanterns to light their way, to a confused accompaniment of barking, yelling, cursing, clanking, they were installed in cattle cars. Here Tenno let himself down—he was too slow to inspect the outside of the car (and your committed escaper must see everything while the seeing is good; he is not allowed to miss anything at all!).

At stops the guards anxiously sounded the cars with mallets. They sounded every single plank. They were afraid of something, then—but of what? Afraid that a plank might be sawn through. So that was the thing to do!

A small piece broken off a hacksaw and sharpened was produced (by the thieves). They decided to cut through a solid plank under the bottom bed shelf. Then when the train slowed down, to lower themselves through the gap, drop onto the line, and lie still until the cars had passed over them. True, the experts said that at the end of a cattle train carrying prisoners there was usually a *drag*—a metal scraper, with teeth which passed close to the ties, caught the body of anyone trying to escape, and dragged him over the ties to his death.

All night long they took turns slipping under the bed shelf and sawing away at a plank in the wall, gripping the blade, which was only a few centimeters long, with a piece of rag. It was hard going. Nonetheless, the first breach was made. The plank began to give a little. Loosening it, they saw in what was now the morning light white, unplaned boards outside their car. Why white? The reason was that an additional footboard for guards had been built onto their car. Right there, by the breach they had made, stood a sentry. It was impossible to go on sawing till the board came away.

Prison escapes, like all forms of human activity, have their own history, and their own theory. It's as well to know about them before you try your own hand.

The history is that of previous escapes. You can learn the history from others who once escaped and were recaptured. Their experience has been dearly bought—with blood, with suffering, almost at the cost of their lives. But to inquire in detail, step by step, about the attempts of one escaper, then a third, then a fifth, is no laughing matter; it can be very dangerous. It is not much less dangerous than asking whether anyone knows whom you should see about joining an underground organization.

Still, as he moved from prison to prison, camp to camp, Tenno eagerly interrogated escapers. He carried out escapes himself, he was caught, he had other escapers for cellmates in the camp jails—and that was his chance to question them.

As for the theory of escape—it is very simple. You do it any way you can. If you get away—that shows you know your theory. If you're caught—you haven't yet mastered it. The elementary principles are as follows. You can escape from a work site or you can escape from the living area. It is easier from work sites: there are many of them, the security measures are less rigid, and the escaper has tools to hand. You can run away alone—it is more difficult, but no one will betray you. Or you can run away in a group, which is easier, but then everything depends on whether you are a well-matched team. Theory further prescribes that you should know the geography as well as if you had an illuminated map in front of you. But you will never catch sight of a map in the camp. A further precept: you must know the people through whose region your escape route lies. Then there is the following general advice as to method: you must constantly prepare to escape according to plan, but be ready at any minute to do it quite differently, to seize a *chance*.

Tenno's first camp was Novorudnoye, near Dzhezkazgan. Now you're in the very place where they have doomed you to die. This is the place of all places from which you must escape! All around there is desert—salt flats and dunes, or firmer ground held together by tufted grass or prickly camel weed. In some parts of the plain Kazakhs roam with their herds, in others there is not a soul. There are no rivers, and you are very unlikely to come upon a well. The best time for flight is April or May, while melting snow still lingers here and there in puddles. But the camp guards are very well aware of this. At this time of year the search of prisoners going out to work becomes stricter, and they are not allowed to take with them a single bite or a single rag more than is necessary.

During his time there Tenno had got to know a lot of people in the camp and he now quickly assembled a team of four: Misha Khaidarov (he had been with the marines in North Korea, had crossed the 38th parallel to avoid a court-martial; not wishing to spoil the good relations firmly established in Korea, the Americans had handed him back and he had got a *quarter*); Jazdik, a Polish driver from the Anders army (he vividly summarized his life story with the help of his unmatching boots—"one from Hitler, one from Stalin"); and, lastly, Sergei, a railwayman from Kuibyshev.

Then a lorry arrived with real posts and rolls of barbed wire for a boundary fence—just as the dinner break was beginning. Tenno's team, loving forced labor as they did, especially when it was to make their prison more secure, volunteered to unload the lorry in the rest period. They scrambled onto the back. But since it was, after all, dinnertime, they took their time while they thought things over. The driver had moved away from his vehicle. The prisoners were lying all over the place, basking in the sun.

Should they run for it or not? They had nothing ready—no knife, no equipment, no food, no plan. But Tenno knew from his little map that if they were driving they must make a dash for Dzhezdy and then to Ulutau. The lads were eager to try it: this was their *chance!* Their lucky chance!

From where they were to the sentry at the "gate," the way was downhill. Just beyond the gate the road rounded a hill. If they drove out fast they'd soon be safe from marksmen. And the sentries could not leave their posts!

They finished unloading before the break was over. Jazdik was to drive. He jumped off, and puttered about the lorry while the other three lazily lay down in the rear, out of sight—hoping that some of the sentries hadn't seen where they had got to. Jazdik brought the driver over. We haven't kept you waiting—so let's have a smoke. They lit up. Right, wind her up! The driver got into the cab, but the engine obstinately refused to start. (The three in the back of the lorry didn't know Jazdik's plan and thought their attempt had misfired.) Jazdik began turning the crank. Still the engine would not start. Jazdik was tired and he suggested to the driver that they change places. Now Jazdik was in the cab. And the engine immediately let out a roar! The lorry rolled down the slope toward the sentry at the gate. (Jazdik told them later that he had tampered with the throttle while the driver was at the wheel, and quickly turned it on again before he himself took over.) The driver was in no hurry to jump in; he thought that Jazdik would stop the lorry. Instead it passed through the "gate" at speed.

Two shouts of "Halt!" The lorry went on. Sentries opened fire— shooting into the air at first, because it looked very much like a mistake. Perhaps some shots were aimed at the lorry—the runaways couldn't tell; they were lying flat. Around a bend. Once behind the hill they were safe from bullets. The three in the back kept their heads down. It was bumpy, they were traveling fast. Then—suddenly—they came to a stop and Jazdik cried out in despair: he had taken the wrong turn and they were pulled up short by the gates of a mine, with its own camp area and its own watchtowers.

More shooting. Guards ran toward them. The escapers tumbled out onto the ground face downward and covered their heads with their hands. Convoy guards kick, aiming particularly at the head, the ears, the temples, and, from above, at the spine.

The wholesome universal rule "Don't kick a man when he's down" did not apply in Stalin's *katorga!* If a man was down, that's just what they did—kicked him. And if he was on his feet, they shot him.

But the inquiry revealed that *there had been no breakout!* Yes! The lads said in unison that they'd been dozing in the back when the lorry started moving, then there was shooting and it was too late for them to jump off in case they were shot. And Jazdik? He was inexperienced, couldn't handle the lorry. But he'd steered for the mine next door, not for the steppe.

So they got off with a beating.

On May 9, 1950, the fifth anniversary of victory in the Fatherland War, naval veteran Georgi Tenno entered a cell in the celebrated Kengir Prison. It was a select company in the Kengir jail, brought together from various camps. In every cell there were experienced escapers, hand-picked champions. Tenno had found his committed escapers at last!

They were destined never, never to remain long in one place! The committed escapers, like Flying Dutchmen, were driven ever onward by their troubled destiny. If they didn't run away, they were transferred. This whole band of men in a hurry was switched, in handcuffs, to Ekibastuz camp jail.

In something like a month there had been three attempts to escape from Ekibastuz—and still Tenno was not on the run! He was pining away. A jealous longing to outdo them gnawed at him. From the sidelines, you see all the mistakes more clearly and always think that you could do better.

Zhdanok was small, swarthy, very agile. When he caught fire he was very energetic, he put everything he had into his work, into an impulse, a fight, an escape. Of course, he lacked discipline, but Tenno had plenty of that.

Everything pointed to the limekilns as the best place for their escape. One day at the limekilns they damaged the electric cable of a cement mixer. An electrician was called in from outside. While Tenno helped him with his repairs, Zhdanok stole some wire cutters from his pocket.

While they were at the limekilns the would-be escapers made themselves two knives: they chiseled strips of metal from shovels, sharpened them at the blacksmith's shop, tempered them, and cast tin handles for them in clay molds. Tenno's was a "Turkish" knife; it would be a handy weapon to use, and what was more important, the flashing curve of its blade was terrifying. Their intention was to frighten people, not kill them. Wire cutters and knives they carried to the living area held to their ankles by the legs of their underpants, and stowed them away in the foundations of the hut.

Their escape plan hinged on the Culture and Education Section. While the weapons were being made and transferred, Tenno chose a suitable moment to announce that he and Zhdanok would like to take part in a camp concert. Sure enough, Tenno and Zhdanok were given permission to leave the punishment wing after it was locked for the night, and while the camp area as a whole was still alive and in motion for another two hours. They roamed the still unknown camp, noting how and when the guard was changed on the watchtowers, and which were the most convenient spots to crawl under the boundary fence. In the Culture and Education Section itself Tenno carefully read the Pavlodar provincial newspaper, trying to memorize the names of districts, state farms, collective farms, farm chairmen, Party secretaries, shock workers of all kinds. Next he announced that he would put on a sketch, for which he must get hold of his ordinary clothes from the clothing store and borrow a briefcase. (A runaway with a briefcase—that was something out of the ordinary! It would help him to look important.) Permission was given.

The sketch required so much rehearsing that the time left till lights out in the main camp area was too short. So there was one night, and later on another, when Tenno and Zhdanok did not return to the punishment wing at all, but spent the night in the hut which housed the Culture and Education Section, to accustom their own warders to their absence. (Escapers must have at least one night's head start!)

What would be the most propitious moment for escape? Evening roll call. When the lines formed outside the huts, the warders were all busy checking in prisoners, while the prisoners had eyes only for the doors, longing to get to their beds; no one was watching the rest of the camp area. The days were getting shorter, and they must hit on one when roll call would come after sundown, in the twilight, but *before* the dogs were stationed around the boundary fence. They must not let slip those five or ten uniquely precious minutes, because there would be no crawling out once the dogs were there. They chose Sunday, September 17. It would help that Sunday was a nonworking day, so that they could recruit their strength by evening, and take time over the final preparations.

The last night before escape! You can't expect much sleep. You think and think. . . . Shall I be alive this time tomorrow? Possibly not. And if I stay here in the camp? To die the lingering death of a goner by a cesspit? . . . No, you mustn't even begin to accept the idea that you are a prisoner.

The question is this: Are you prepared to die? You are? Then you are also prepared to escape.

A sunny Sunday. To rehearse their sketch, both of them were let out of the punishment wing for the whole day.

The runaways were very short of food: in the punishment wing they were on short rations, and hoarding bread would excite suspicion. They banked on seizing a lorry in the settlement and traveling quickly.

However, that Sunday there was also a parcel from home—his mother's blessing on his escape. Glucose tablets, macaroni, oatmeal—these they could carry in the briefcase. They must also get hold of a "katyusha" —an improvised lighter consisting of a wick in a tube—and a steel and flint to light it. This was better than matches for a man on the run.

Sunday was coming to an end. A golden sun was setting. Tenno, tall and leisurely, and Zhdanok, small and vivacious, now draped padded jackets around their shoulders, took the briefcase (by now everyone in the camp was used to their eccentric appearance), and went to the prearranged departure point—on the grass between some huts, not far from the boundary fence and directly opposite a watchtower. The huts screened them from two other watchtowers. There was only this one sentry facing them. They opened out their padded jackets, lay down on them, and played chess, so that the sentry would get used to them.

The sky turned gray. There was the signal for roll call. The prisoners flocked to their huts. In the half-light, the sentry on his watchtower should not be able to make out that two men were still lying on the grass. His watch was nearly over, and he was less alert than he had been. A stale sentry always makes escape easier.

They intended to cut the wire, not in the open, but directly under the tower. The sentry certainly spent more time watching the boundary fence farther away than the ground under his feet.

Their heads were down near the grass, and besides, it was dusk, so they could not see the spot at which they would shortly crawl under. But it had been thoroughly inspected in advance. Immediately beyond the boundary fence a hole had been dug for a post, and it would be possible to hide there a minute. A little farther on there were mounds of slag: and a road running from the guards' hamlet to the settlement.

The plan was to take a lorry as soon as they reached the settlement. Stop one and say to the driver, "Do you want to earn something? We have to bring two cases of vodka up here from old Ekibastuz." What driver would refuse drink? They would bargain with him. "Half a liter all right? A liter? Right, step on it, but not a word to anybody." Then on the highway, sitting with the driver in his cab, they would overpower him, drive him out into the steppe, and leave him there tied up. While they tore off to reach the Irtysh in a single night, abandon the lorry, cross the river in a little boat, and move on toward Omsk.

It got a little darker still. Up in the towers searchlights were switched on. Their beams lit up the boundary fence, but the runaways for the time being were in a shadowy patch. The very time!

Soon the watch would be changed and the dogs would be brought along and posted for the night.

Now lights were switched on in the huts, and they could see the prisoners going in after roll call. Was it nice inside? It would be warm, comfortable. . . . Whereas here you could be riddled with Tommy-gun bullets, and it would be all the more humiliating because you were lying stretched on the ground.

Just so long as they didn't cough or sneeze under the tower.

Guard away, you guard dogs! Your job is to keep us here, ours is to run away!

Chapter 7

■

The White Kitten (Georgi Tenno's Tale)

In this chapter Georgi Tenno picks up from where the preceding chapter left off and tells his own story. Exercising great daring, he and Kolya Zhdanok remained on the loose for the better part of a month, scrounging for food and water and sometimes stealing. Eventually, they were captured and given long sentences. After his release from camp, Tenno died of cancer.

Chapter 8

■

Escapes—Morale and Mechanics

Escapes from Corrective Labor Camps, provided they were not to somewhere like Vienna or across the Bering Strait, were apparently viewed by Gulag's rulers and by Gulag's regulations with resignation. They saw them as only natural, a manifestation of the waste which is unavoidable in any overextended economic enterprise—a phenomenon of the same sort as cattle losses from disease or starvation, the logs that sink instead of floating, the gap in a wall where a half-brick was used instead of a whole one.

It was different in the Special Camps. In accordance with the particular wish of the Father of the Peoples, these camps were equipped with greatly reinforced defenses and with greatly reinforced armament, at the modern motorized infantry level. At the moment of their foundation it was laid down in the instructions for Special Camps that there *could be no escape* from them, because if one of these prisoners escaped it was just as though a major spy had crossed the frontier, and a blot on the political record of the camp administration and of the officers commanding the convoy troops.

But from that very moment 58's to a man started getting, not *tenners* as before, but *quarters*—i.e., the limit allowed by the Criminal Code. This senseless, across-the-board increase in severity carried with it one disadvantage: just as murderers were undeterred from fresh murders (each time their *tenner* was merely slightly updated), so now political prisoners were no longer deterred by the Criminal Code from trying to escape.

And although there were fewer escapes from the Special Camps than from the Corrective Labor Camps, they were rougher, grimmer, more ruthless, more desperate, and therefore more glorious.

Stories told about them can help us to make up our minds whether our people really was so long-suffering, really was so humbly submissive in those years.

Here is just one of them. In September, 1949, two convicts escaped from the First Division of Steplag (Rudnik, Dzhezkazgan)—Grigory Kudla, a tough, steady, level-headed old man, a Ukrainian (but when his dander was up he had the temper of a Zaporozhian Cossack, and even the hardened criminals were afraid of him), and Ivan Dushechkin, a quiet Byelorussian some thirty-five years old. In the pit where they worked they found a prospecting shaft in an old workings, with a grating at its upper end. When they were on night shift they gradually loosened this grating, and at the same time they took into the shaft dried crusts, knives, and a hot-water bottle stolen from the Medical Section. On the night of their escape attempt, once down the pit each of them separately informed the foreman that he felt unwell, couldn't work, and would lie down a bit. At night there were no warders underground; the foreman was the sole representative of authority and he had to bully discreetly or else he might be found with his head smashed in. The escapers filled the hot-water bottle, took their provisions, and went into the prospecting shaft. They forced the grating and crawled out. The exit turned out to be near the watchtowers but outside the camp boundary. They walked off unnoticed.

They lay down in the daytime and walked at night. Not once did they come across water, and after a week Dushechkin no longer felt like standing up. Kudla got him on his feet with the hope that there might be water in the hills ahead. They dragged themselves that far, but the hollows held no water, only mud. Then Dushechkin said, "I can't go on anyway. *Cut my throat* and drink my blood."

You moralists! What was the right thing to do? Kudla, too, could no longer see straight. Dushechkin was going to die—why should Kudla perish, too? But if he found water soon afterward, *how* could he live with the thought of Dushechkin for the rest of his days? I'll go on a bit, Kudla decided, and if in the morning I come back without water I'll put him out of his misery, and we needn't both perish. Kudla staggered to a hillock, saw a cleft in it and—just as in the most improbable of novels—in the cleft there was water! Kudla slithered to it, fell flat on his face, and drank and drank. (Only in the morning had he eyes for the tadpoles and waterweed in it.) He went back to Dushechkin with the hot-water bottle full. "I've brought you some water—yes, water." Dushechkin couldn't believe it, drank, and still didn't believe it (for hours he had been imagining that he was drinking). They dragged themselves as far as the cleft and stayed there drinking.

When they had drunk, hunger set in. But the following night they climbed over a ridge and went down into a valley like the promised land: with a river, grass, bushes, horses, life. When it got dark Kudla crept up to the horses and killed one of them. They drank its blood straight from the wounds. (Partisans of peace! That very year you were loudly in session in Vienna or Stockholm, and sipping cocktails through straws. Did it occur to you that compatriots of the versifier Tikhonov and the journalist Ehrenburg were sucking the blood of dead horses? Did they explain to you in their speeches that that was the meaning of *peace,* Soviet style?)

They roasted the horse's flesh on fires, ate lengthily, and walked on.

Farther on they frequently came across streams and pools. Kudla also caught and killed a ram. By now they had been a *month* on the run! October was nearing its end; it was getting cold. In the first wood they reached they found a dugout and set up house in it. They couldn't bring themselves to leave this land of plenty. That they settled in such surroundings, that their native places did not call to them or promise them a more peaceful life, meant that their escape lacked a goal and was doomed to fail.

At night they would raid the village nearby, filch a pot or break into a pantry for flour, salt, an ax, some crockery. (Inevitably the escaper, like the partisan, soon becomes a thief, preying on the peaceful folk all around him.) Another time they took a cow from the village and slaughtered it in the forest. But then the first snow came, and to avoid leaving tracks they had to sit tight in their dugout. Kudla went out just once for brushwood and the forester immediately opened fire on him. "So you're the thieves, are you! You're the ones who stole the cow." Sure enough, traces of blood were found around the dugout. They were taken to the village and locked up. The people shouted that they should be shot out of hand and no mercy shown to them. But an investigating officer arrived from the district center with the picture sent around to assist the nationwide search, and addressed the villagers. "Well done!" he said. "These aren't thieves you've caught, but dangerous political criminals."

Suddenly there was a complete change of attitude. The owner of the cow, a Chechen as it turned out, brought the prisoners bread, mutton, and even some money, collected by the Chechens. "What a pity," he said. "You should have come and told me who you were and I'd have given you everything you wanted!" (There is no reason to doubt it; that's how the Chechens are.) Kudla burst into tears. After so many years of savagery, he couldn't stand sympathy.

The prisoners were removed to Kustanai and put in the railroad jail, where their captors not only took away the Chechen's offering (and pocketed it), but gave them no food at all. (Didn't Korneychuk tell you about it at the Peace Congress?) Before they were put on the train out of Kustanai, they were made to kneel on the station platform with their hands handcuffed behind their backs. They were kept like that for some time, for the whole world to see.

If it had been on a station platform in Moscow, Leningrad, or Kiev, or any other flourishing city, everybody would have passed by. The people of Kustanai, however, had little to lose. They were all either "sworn enemies," or persons with black marks against them, or simply exiles. They started crowding around the prisoners, and tossing them makhorka, cigarettes, bread. Kudla's wrists were shackled behind his back, so he bent over to pick up a piece of bread with his teeth—but the guard *kicked it out of his mouth.* Kudla rolled over, and again groveled to pick it up—and the guard kicked the bread farther away. (You progressive film makers, perhaps you will remember this scene and this old man?) The people began pressing forward and making a noise. "Let them go! Let them go!" A militia squad appeared. The policemen had the advantage and dispersed the people.

The train pulled in, and the prisoners were loaded for transport to the Kengir jail.

Chapter 9

■

The Kids with Tommy Guns

The camps were guarded by men in long greatcoats with black cuffs. They were guarded by Red Army men. They were guarded by prisoner guards. They were guarded by elderly reservists. Last came the robust youngsters born during the First Five-Year Plan, who had seen no war service when they took their nice new Tommy guns and set about guarding us.

Twice every day, for an hour at a time, we and they shuffled along, tied together in silent and deadly brotherhood: any one of them was at liberty to kill any one of us.

We walked with never a glance at their sheepskin coats and their Tommy guns—what were they to us? They walked watching our dark ranks all the time. It was there in the regulations that they must watch us all the time. Orders were orders. Duty was duty. Any wrong movement, any false step, they must cut short with a bullet.

How did they think of us, in our dark jackets, our gray caps of Stalin fur, our grotesque felt boots that had served three sentences and shed four soles, our crazy quilt of number patches? Decent people would obviously never be treated like that.

Was it surprising that our appearance inspired disgust? It was intended to do just that.

These kids were not allowed to know anything about us; they were allowed only the right to shoot without warning!

If they had just visited us in our huts of an evening, sat on our bunks and heard why this old man, or the other old fellow over there, was inside . . . those towers would have been unmanned and those Tommy guns would never have fired.

But the whole cunning and strength of the system was in the fact that our deadly bond was forged from ignorance. Any sympathy they showed for us was punishable as treason; any wish to speak to us, as a breach of a solemn oath. And what was the point of talking to us when the political instructor would come at fixed intervals to lead a discussion on the political and moral character of the enemies of the people whom they were guarding.

The political instructor will never contradict himself, never make a slip. He will never tell the boys that some people are imprisoned here simply for believing in God, or simply for desiring truth, or simply for love of justice. Or indeed for nothing at all.

Here is one such political lesson, as remembered by a former convoy guard (at Nyroblag). "Lieutenant Samutin was a lanky, narrow-shouldered man, and his head was flat above the temples. He looked like a snake. He was towheaded and almost without eyebrows. We knew that in the past he had shot people with his own hands. Now he was a political instructor, reciting in a monotonous voice: 'The enemies of the people, over whom you stand guard, are the same as the Fascists, filthy scum. We embody the power and the punitive sword of the Motherland and we must be firm. No sentimentality, no pity.' "

That is how they mold the boys who make a point of kicking a runaway's head when he is down. The boys who can boot a piece of bread out of the mouth of a gray-haired old man in handcuffs. Who can look with indifference at a shackled runaway jounced on the splintery boards of a lorry; his face is bloodied, his head is battered, but they look on unmoved. For they are the Motherland's punitive sword, and he, so they say, is an American colonel.

After Stalin's death, then living in eternal banishment, I was a patient in an ordinary *free* Tashkent clinic. Suddenly I heard a patient, a young Uzbek, telling his neighbors about his service *in the army*. His unit had, he said, kept guard on beasts and butchers. It enraged them that they, the convoy guards, had to freeze on top of watchtowers in the winter (in sheepskin coats down to their heels, it's true), while the enemies of the people, once they entered the working area, scattered about the warming-up shacks (even from the watchtower he could have seen that it was not so) and slept there all day.

Here was an interesting opportunity—to look at a Special Camp through the eyes of a convoy guard! I began asking what kind of reptiles they were and whether my Uzbek friend had talked to them personally. And of course he told me that he had learned all this from political officers, that they had even had "cases" read out to them in their political indoctrination sessions. And his malicious misconceptions about prisoners sleeping all day had of course been reinforced in him by the approving nods of officers.

Woe unto you that cause these little ones to stumble. Better for you had you never been born! . . .

He told us of a number of incidents. Once, for instance, one of his comrades was marching in convoy and fancied that somebody *was about to* run out of the column. He pressed his trigger and killed *five* prisoners with a single burst. Since all the other guards later testified that the column had been moving quietly along, this soldier incurred a terrible punishment: for killing five people he was put in detention for fifteen days (in a warm guardhouse, of course).

But which of the Archipelago's inhabitants has no stories of this sort to tell! . . . We knew so many of them in the Corrective Labor Camps. On a work site which had no fence but an invisible boundary line, a shot rang out and a prisoner fell dead.

Why? Because the man has a gun! Because one man has the arbitrary power to kill or not to kill another.

What's more—it pays! The bosses are always on your side. They'll never punish you for killing somebody. On the contrary, they'll commend you, reward you, and the quicker you are on the trigger—bring him down when he's only put half a foot wrong—the more vigilant you are seen to be, the higher your reward! A month's pay. A month's leave.

Yes, but their underlying common humanity must have been weak, or altogether lacking, if it was not proof against an oath and a few political discussion periods. Not every generation, and not every people, is of the stuff from which such boys are fashioned.

This is surely the main problem of the twentieth century: is it permissible merely to carry out orders and commit one's conscience to someone else's keeping? Can a man do without ideas of his own about good and evil, and merely derive them from the printed instructions and verbal orders of his superiors? Oaths! Those solemn pledges pronounced with a tremor in the voice and intended to defend the people against evildoers: see how easily they can be misdirected to the service of evildoers and against the people!

Chapter 10

■

Behind the Wire the Ground Is Burning

No, the surprising thing is not that mutinies and risings did not occur in the camps, but that in spite of everything they *did*.

Like all embarrassing events in our history—which means three-quarters of what really happened—these mutinies have been neatly cut out, and the gap hidden with an invisible join. Those who took part in them have been destroyed, and even remote witnesses frightened into silence; reports of those who suppressed them have been burned or hidden in safes within safes within safes—so that the risings have already become a myth, although some of them happened only fifteen and others only ten years ago. (No wonder some say that there was no Christ, no Buddha, no Mohammed. There you're dealing in thousands of years. . . .)

When it can no longer disturb any living person, historians will be given access to what is left of the documents, archaeologists will do a little digging, heat something in a laboratory, and the dates, locations, contours of these risings, with the names of their leaders, will come to light.

Perhaps we (no, not we ourselves) shall learn at the same time about the legendary rising in 1948 at public works site No. 501, where the Sivaya Maska-Salekhard railway was under construction. It was legendary because everybody in the camps talked about it in whispers, but no one really knew anything. Legendary also because it broke out not in the Special Camp system, where the mood and the grounds for it by now existed, but in a Corrective Labor Camp, where people were isolated from each other by fear of informers and trampled under foot by thieves, where even their right to be "politicals" was spat upon, and where a prison mutiny therefore seemed inconceivable.

According to the rumors, it was all the work of ex-soldiers (recent ex-soldiers!). It could not have been otherwise. Without them the 58's lacked stamina, spirit, and leadership. But these young men (hardly any of them over thirty) were officers and enlisted men from our fighting armies, or their fellows who had been prisoners of war. These young men still retained in 1948 their wartime élan and belief in themselves, and they could not accept the idea that men like themselves, whole battalions of them, should meekly die. Even escape seemed to them a contemptible half-measure, rather like deserting one by one instead of facing the enemy together.

It was all planned and begun in one particular team. An ex-colonel called Voronin or Voronov, a one-eyed man, is said to have been the leader. A first lieutenant of armored troops, Sakurenko, is also mentioned. The team killed their convoy guards. Then they went and freed a second team, and a third. They attacked the convoy guards' hamlet, then the camp from outside, removed the sentries from the towers, and opened up the camp area.

Arming themselves with weapons taken from the guards, the rebels went on to capture the neighboring Camp Division. With their combined forces they decided to advance on Vorkuta! It was only sixty kilometers away. But this was not to be. Parachute troops were dropped to bar their way. Then low-flying fighter planes raked them with machine-gun fire and dispersed them.

They were tried, more of them were shot, and others given twenty-five or ten years. (At the same time, many of those who had not joined in the operation but remained in the camp had their sentences "refreshed.")

The hopelessness of this rising as a military operation is obvious. But would you say that dying quietly by inches was more "hopeful"?

Riddle: What is the quickest thing in the world? Answer: Thought.

It is and it isn't. It can be slow, too—oh, how slow! Only slowly and laboriously do men, people, society, realize what has happened to them. Realize the truth about their position.

In herding the 58's into Special Camps, Stalin was exerting his strength mainly for his own amusement. He already had them as securely confined as they could be, but he thought he would be craftier than ever and improve on his best. He thought he knew how to make it still more frightening. The results were quite the opposite.

The whole system of oppression elaborated in his reign was based on keeping malcontents apart, preventing them from reading each other's eyes and discovering how many of them there were; instilling it into all of them, even into the most dissatisfied, that no one was dissatisfied except for a few doomed individuals, blindly vicious and spiritually bankrupt.

In the Special Camps, however, there were malcontents by the thousands. They knew their numerical strength. And they realized that they were not spiritual paupers, that they had a nobler conception of what life should be than their jailers, than their betrayers, than the theorists who tried to explain why they must rot in camps.

The old camp mentality—you die first, I'll wait a bit; there is no justice, so forget it; that's the way it was, and that's the way it will be —also began to disappear.

A bold thought, a desperate thought, a thought to raise a man up: how could things be changed so that instead of *us* running from *them, they* would run from *us?*

Once the question was put, once a certain number of people had thought of it and put it into words, and a certain number had listened to them, the age of escapes was over. The age of rebellion had begun.

Suddenly—a suicide. In the Disciplinary Barracks, hut No. 2, a man was found hanging. (I am going through the stages of the process as they occurred in Ekibastuz. But note that the stages were just the same in other Special Camps!) The bosses were not greatly upset; they cut him down and wheeled him off to the scrap heap.

A rumor went around the work team. The man was an informer. He hadn't hanged himself. He had been hanged.

As a lesson to the rest.

"Kill the stoolie!" That was it, the vital link! A knife in the heart of the stoolie! Make knives and cut stoolies' throats—that was it!

Now, as I write this chapter, rows of humane books frown down at me from the walls, the tarnished gilt on their well-worn spines glinting reproachfully like stars through cloud. Nothing in the world should be sought through violence! By taking up the sword, the knife, the rifle, we quickly put ourselves on the level of our tormentors and persecutors. And there will be no end to it. . . .

There will be no end. . . . Here, at my desk, in a warm place, I agree completely.

If you ever get twenty-five years for nothing, if you find yourself wearing four number patches on your clothes, holding your hands permanently behind your back, submitting to searches morning and evening, working until you are utterly exhausted, dragged into the cooler whenever someone denounces you, trodden deeper and deeper into the ground—from the hole you're in, the fine words of the great humanists will sound like the chatter of the well-fed and free.

The oppressed at least concluded that evil cannot be cast out by good.

Murders now followed one another in quicker succession than escapes in the best period. They were carried out confidently and anonymously: no one went with a bloodstained knife to give himself up; they saved themselves and their knives for another deed. At their favorite time—when a single warder was unlocking huts one after another, and while nearly all the prisoners were still sleeping—the masked avengers entered a particular section, went up to a particular bunk, and unhesitatingly killed the traitor, who might be awake and howling in terror or might be still asleep. When they had made sure that he was dead, they walked swiftly away.

They wore masks, and their numbers could not be seen—they were either picked off or covered. But if the victim's neighbors should recognize them by their general appearance, so far from hurrying to volunteer information, they would not now give in even under interrogation, even under threat from the godfathers, but would repeat over and over again: "No, no, I don't know anything, I didn't see anything." And this was not simply in recognition of a hoary truth known to all the oppressed: "What you don't know can't hurt you"; it was self-preservation! Because anyone who *gave names* would have been killed next 5 A.M., and the security officer's good will would have been no help to him at all.

And so murder (although as yet there had been fewer than a dozen) became *the rule,* became a normal occurrence. "Anybody been killed today?" prisoners would ask each other when they went to wash or collect their morning rations. In this cruel sport the prisoner's ear heard the subterranean gong of justice.

Out of five thousand men about a dozen were killed, but with every stroke of the knife more and more of the clinging, twining tentacles fell away. A remarkable fresh breeze was blowing! On the surface, we were prisoners living in a camp just as before, but in reality we had become free—free because for the very first time in our lives, we had started saying openly and aloud all that we thought! No one who has not experienced this transition can imagine what it is like!

An invisible balance hung in the air. In one of its scales all the familiar phantoms were heaped: interrogation officers, punches, beatings, sleepless standing, "boxes" (cells too small to sit or lie down in), cold, damp punishment cells, rats, bedbugs, tribunals, second and third sentences. But this could not all happen at once, this was a slow-grinding bone mill, it could not devour all of us at once and process us in a single day. And even when they had been through it—as every one of us had—men still went on existing.

While in the other scale lay nothing but a single knife—but that knife was meant for you, if you gave in! It was meant for you alone, in the breast, and not sometime or other, but at dawn tomorrow, and all the forces of the Cheka-MGB could not save you from it! It was not a long one, but just right for neat insertion between your ribs. It didn't even have a proper handle, just a piece of old insulation tape wound around the blunt end of the blade—but this gave a very good grip, so that the knife would not slip out of the hand!

And this bracing threat weighed heavier!

So that now the bosses were suddenly blind and deaf.

The information machine on which alone the fame of the omnipotent and omniscient *Organs* had been based in decades past had broken down.

Foremen started *escaping* into the Disciplinary Barracks—hiding behind stone walls! Not only they, but bloodsucking work assigners, stoolies on the brink of exposure or, something told them, next on the list, suddenly took fright and *ran for it!*

This was a new period, a heady and spine-tingling period in the life of the Special Camp. It wasn't we who had taken to our heels—they had, ridding us of their presence! A time such as we had never experienced or thought possible on this earth: when a man with an unclean conscience could not go quietly to bed! Retribution was at hand—not in the next world, not before the court of history, but retribution live and palpable, raising a knife over you in the light of dawn. It was like a fairy tale: the ground is soft and warm under the feet of honest men, but under the feet of traitors it prickles and burns.

The masters of our bodies and souls were particularly anxious not to admit that our movement was political in character. In their menacing orders (warders went around the huts reading them out) the new trend was declared to be nothing but *gangsterism*. This made it all simpler, more comprehensible, somehow cozier. It seemed only yesterday that they had sent us gangsters labeled "politicals." Well, politicals—real politicals for the first time—had now become "gangsters." It was announced, not very confidently, that these gangsters would soon be discovered (so far not one of them had been), and still less confidently, that they would be shot. The orders further appealed to the prisoner mass to *condemn* the gangsters and *struggle* against them!

The prisoners listened and went away chuckling. Seeing the disciplinary officers afraid to call "political" behavior by its name, we were aware of their weakness.

The orders were of no avail. The prisoner masses did not start *condemning* and *struggling* on behalf of their masters. The next measure was to put the whole camp on a punitive routine.

The object of the camp administration was to make things so hard for us that we would betray the murderers out of exasperation. But we braced ourselves to suffer, to hang on a bit: it was worth it! Their other object was to keep the huts closed so that murderers could not come from outside, and so would be easier to find. But another murder took place, and still no one was caught—just as before no one had ever seen anything or knew anything. Then somebody's head was smashed in at work—locked huts are no safeguard against that.

They revoked the punitive regime. Instead they had the bright idea of building the "Great Wall of China." This was a wall two bricks thick and four meters high, to cut across the width of the camp. They were preparing to divide the camp into two parts. We greatly resented that wall—we knew that the bosses had some dirty trick in store—but we had no choice but to build it. Only a little of us was as yet free—our heads and our mouths—but we were still stuck up to our shoulders in the quagmire of slavery.

Two warders came into a hut after work, and casually told a man, "Get ready and come with us."

The prisoner looked around at the other lads and said, "I'm not going."

In fact, this simple everyday situation—*a snatch, an arrest*—which we had never resisted, which we were used to accepting fatalistically, held another possibility: that of saying, "I'm not going!" Our liberated heads understood that now.

The warders pounced on him. "What do you mean, not going?"

"I'm just not going," the zek answered, firmly. "I'm all right where I am."

There were shouts from all around:

"Where's he supposed to go? . . . What's he got to go for? . . . We won't let you take him! . . . We won't let you! . . . Go away!"

And the wolves understood that we were not the sheep we used to be. That if they wanted to grab one of us now they would have to use trickery, or do it at the guardhouse, or send a whole detail to take one prisoner. With a crowd around, they would never take him.

Purged of human filth, delivered from spies and eavesdroppers, we looked about and saw, wide-eyed, that . . . we were thousands! that we were . . . *politicals!* that we could *resist!*

We had chosen well; the chain would snap if we tugged at this link —the stoolies, the talebearers and traitors! Our own kind had made our lives impossible. As on some ancient sacrificial altar, their blood had been shed that we might be freed from the curse that hung over us.

The revolution was gathering strength. The wind that seemed to have subsided had sprung up again in a hurricane to fill our eager lungs.

Chapter 11

■

Tearing at the Chains

The middle ground had now collapsed, the ditch which ran between us and our custodians was now a deep moat, and we stood on opposite slopes, taking the measure of the situation.

"Stood" is of course a manner of speaking. *We* went to work daily, we were never late for work line-up, we never let each other down, there were no shirkers, we chalked up a good day's work—you might think that our masters could be pleased with us.

All the same, we and they were thinking hard about the next stage. Things could not remain as they were: we could not be satisfied with what had happened, nor could they. Someone had to strike a blow.

But what should be our aim? We could now say out loud, without looking nervously around, whatever we liked—all those things which had seethed inside us (and freedom of speech, even if it was only there in the camp, even though it had come so late, was a delight!). But could we hope to spread that freedom beyond the camp, or carry it out with us? Of course not. We, from where we were, could not demand that the country should change completely, nor that it should give up the camps: they would have rained bombs on us.

As to whether we should fight for an eight-hour working day—there was no unanimity among us. . . . We were so unused to freedom that we seemed to have lost all appetite for it.

Ways and means were also discussed. How should we present our demands? What action should we take? Clearly, we could do nothing with bare hands against modern arms, and therefore the course for us to take was not armed rebellion, but a strike.

But the blood in our veins was still slavish, still servile. . . . The word "strike" sounded so terrifying to our ears that we sought firmer ground: by refusing food when we refused to work it seemed to us that our moral right to strike would be reinforced. We felt that we had some sort of right to go on a hunger strike—but to strike in the ordinary sense?

So that by voluntarily undertaking an unnecessary hunger strike, we voluntarily agreed to undermine the physical strength which we needed for the struggle. (Fortunately, no other camp seems to have repeated Ekibastuz's mistake.)

We went over and over the details of the proposed work stoppage-hunger strike. It was talked over in various places, in one group and another; it seemed inevitable and desirable, yet, because it was so novel, somehow impossible.

But our custodians were less likely to lose by acting than by failing to act—and they got their blow in first.

After that, events took on a momentum of their own.

At peace and at ease on our familiar bunks, in our familiar sections, we greeted the new year, 1952. Then on Sunday, January 6, the Orthodox Christmas Eve, when the Ukrainians were getting ready to observe the holiday in style—they would make kutya, fast till the first star appeared, and then sing carols—the doors were locked after morning inspection and not opened again.

So that was it! A regrouping! Guards were posted by the break in the Chinese Wall. It would be bricked up next day. We were taken past the guardhouse and herded, hundreds of us, with sacks and mattresses, like refugees from a burning village, around the boundary fence, past another guardhouse, into the other camp area. Passing those who were being driven in the opposite direction.

In one half of the camp (Camp Division No. 2) only the Ukrainian nationalists, some 2,000 of them, were left. In the half to which we had been driven, and which was to be Camp Division No. 1, there were some 3,500 men belonging to all the other ethnic groups—Russians, Estonians, Lithuanians, Latvians, Tatars, Caucasians, Georgians, Armenians, Jews, Poles, Moldavians, Germans, and a variegated sprinkling of other nationalities. The Ukrainians, the Banderists, the most dangerous rebels, had been moved farther away from it. What did this mean?

We soon learned what. Reliable rumors went around the camp (from the working prisoners who took the gruel to the BUR) that the stoolies in the "safe deposit" had grown cheeky. Suspects, picked up here and there, two or three at a time, had been put in with the stoolies, who were torturing them in their common cell, choking them, beating them, trying to make them "sing" and to name names. *"Who's doing the slashing?"* This made the whole scheme as clear as daylight. They were using torture! Not the dog pack themselves—they probably had no authorization for it, and might run into trouble, so they had entrusted the stoolies with the job: find your murderers yourselves!

So many deep historians have written so many clever books—and still they have not learned how to predict those mysterious conflagrations of the human spirit, to detect the mysterious springs of a social explosion, nor even to explain them in retrospect.

Sometimes you can stuff bundle after bundle of burning tow under the logs, and they will not take. Yet up above, a solitary little spark flies out of the chimney and the whole village is reduced to ashes.

Our three thousand had no plans made, were quite unprepared, but one evening on their return from work the prisoners in a hut next to the BUR began dismantling their bunks, seized the long bars and crosspieces, ran through the gloom to batter down the stout fence around the camp jail. They had neither axes nor crowbars. . . .

There was a hammering noise—they worked like a team of good carpenters, levering the planks away as soon as they gave—and the grating protests of 12-centimeter nails could be heard all over the camp.

The warders hustled over to the BUR, to the half-dark wall where all the activity was—and raced like scalded cats to the staff barracks. Somebody rushed after a warder with a stick. Then, to provide full musical accompaniment, somebody started breaking windows in the staff barracks with stones or a stick. The staff's windowpanes shattered with a merry menacing crash.

What the lads had in mind was not to raise a rebellion, nor even to capture the BUR (no easy matter), but merely to pour petrol into the stoolies' cell, and toss in something burning—meaning: Watch your step, we'll show you yet! But machine-gun fire from the towers suddenly rattled across the camp, and they never did start their blaze.

The warders and Chief Disciplinary Officer Machekhovsky had fled from the camp and informed Division. Division gave telephonic instructions for the corner towers to open fire from machine guns—on three thousand unarmed people who knew nothing of what had happened. (Our team, for instance, was in the mess hall, and we were completely mystified when we heard all the shooting outside.)

Firing at random in the darkness, the machine-gunners blasted away. They pierced the flimsy walls of the huts, and, as always happens, wounded not those who had stormed the BUR, but others, who had no part in it. In hut No. 9 a harmless old man, nearing the end of a ten-year sentence, was killed in his bed. He was due to be released in a month's time.

The besiegers left the prison yard and quickly dispersed to their barracks (where they had to put their bunks together again so as to cover their traces). Many others took the shooting as a warning to stay inside their huts. Yet others, on the contrary, poured out excitedly and scurried about the camp, trying to understand what it was all about.

By then there were no warders left in the camp area. The staff barracks was empty of officers, and terrible jagged holes yawned in its windows. The towers were silent. The curious, and the seekers after truth, roamed the camp.

Suddenly the gates of our Camp Division were flung wide and a platoon of convoy troops marched in with Tommy guns at the ready, firing short bursts at random. They fanned out in all directions, and behind them came the enraged warders, with lengths of iron pipe, clubs, or anything else they had been able to lay their hands on.

A lethal crush developed at the entrance to our hut: the prisoners were so anxious to shove their way in that no one could enter. (Not that the thin boards of the hut walls gave any protection against bullets, but once inside, a man ceased to be a mutineer.) I was one of those by the steps. I remember very well my state of mind: a nauseated indifference to my fate; a momentary indifference whether I survived or not. Why have you fastened your hooks on us, curse you! Why must we go on paying you till the day we die for the crime of being born into this unhappy world? Why must we sit forever in your jails? The prison sickness which is at once nausea and peace of mind flooded my being. Even my constant fear for the as-yet-unrecorded poem and the play I carried within me was in abeyance. In full view of the death which was wheeling toward us in military greatcoats, I made no effort at all to push through the door. This was the true convict mentality; this was what they had brought us to.

Our pursuers did not break into the huts. They locked us in. They hunted down and beat those who had not been quick enough to run inside.

The work stoppage–hunger strike had not been carefully prepared, it was not even a coherent concept, and it began impulsively, with no directing center, no signal system.

Those prisoners in other camps who took over the food stores and then stayed away from work of course behaved more sensibly. But our action, if not very clever, was impressive: three thousand men simultaneously swore off both food and work.

Next morning not a single team sent its man to the bread-cutting room. Not a single team went to the mess hall, where broth and mush awaited them. The warders just could not understand it: twice, three times, four times they came into the huts to summon us with brisk commands, then to drive us out with threats, then to ask us nicely— no farther than to the mess hall, to collect our bread, with never a word for the present about work line-up.

But nobody went. They all lay on their bunks, fully dressed, wearing their shoes, and silent. Only we, the foremen (I had become a foreman in that hot year), felt called upon to answer, since the warders kept addressing themselves to us. We lay on our bunks like the rest and muttered from our pillows: "It's no good, boss. . . ."

This unanimous quiet defiance of a power which never forgave, this obstinate, painfully protracted insubordination, was somehow more frightening than running and yelling as the bullets fly.

In the end they stopped coaxing us and locked up the huts.

In the days that followed, no one left the huts except the orderlies —to carry the latrine buckets out and bring drinking water and coal in. Only bed cases in the Medical Section were by general agreement allowed not to fast.

The bosses could no longer see us, no longer peer into our souls. A gulf had opened between the overseers and the slaves!

None of those who took part will ever forget those three days in our lives. We could not see our comrades in other huts, nor the corpses lying there unburied. Nonetheless, the bonds which united us, at opposite ends of the deserted camp, were of steel.

This was a hunger strike called not by well-fed people with reserves of subcutaneous fat, but by gaunt, emaciated men, who had felt the whip of hunger daily for years on end, who had achieved with difficulty some sort of physical equilibrium, and who suffered acute distress if they were deprived of a single 100-gram ration. Even the goners starved with the rest, although a three-day fast might tip them into irreversible and fatal decline. The food which we had refused, and which we had always thought so beggarly, was a mirage of plenty in the feverish dreams of famished men.

This was a hunger strike called by men schooled for decades in the law of the jungle: "You die first and I'll die later." Now they were reborn, they struggled out of their stinking swamp, they consented to die today, all of them together, rather than to go on living in the same way tomorrow.

In the huts roommates began to treat each other with a sort of ceremonious affection. Whatever scraps of food anyone—this meant mainly those who received parcels—had left were pooled, placed on a piece of rag spread out like a tablecloth, and then, by joint decision of the whole room, some eatables were shared out and others put aside for the next day.

What the bosses would do no one could predict. We thought that perhaps they would start firing on the huts again from the towers. The last thing we expected was any concession. We had never in our lives wrested anything from them, and our strike had the bitter tang of hopelessness.

But there was a sort of satisfaction in this feeling of hopelessness. We had taken a futile, a desperate step, it could only end badly—and that was good. Our bellies were empty, our hearts were in our boots— but some higher need was being satisfied. During those long hungry days, evenings, nights, three thousand men brooded over their three thousand sentences, their families, their lack of families, all that had befallen and would yet befall them, and although the hearts in thousands of breasts could not beat together—and there were those who felt only regret, only despair—yet most of them kept time: Things are as they should be! We'll keep it up to spite you! Things are bad! So much the better!

This, too, is a phenomenon which has never been adequately studied: we do not know the law that governs sudden surges of mass emotion, in defiance of all reason. I felt this soaring emotion myself. I had only one more year of my sentence to serve. I might have been expected to feel nothing but dismay and vexation that I was dirtying my hands on a broil from which I should hardly escape without a new sentence. And yet I had no regrets. Damn and blast the lot of you, I'll serve my time all over again if you like!

Next day we saw from our windows a group of officers making their way from hut to hut. A detail of warders opened the door, went along the corridors, looked into the rooms, and called us (not in the old way, as though we were cattle, but gently): "Foremen! You're wanted at the entrance!"

A debate began among us. It was the teams, not their foremen, who had to decide. Men went from room to room to talk it over.

A decision was reached at last in some invisible quarter. We foremen, six or seven of us, went out to the entrance, where the officers were patiently awaiting us. We lowered our eyes because not one of us could now look at our bosses sycophantically, and rebellious looks would have been foolish.

From both corridors, however, a crowd of zeks pressed into the entranceway, and hiding behind those in front, the back rows could speak freely, call out our demands and our answers.

Officially, the officers with blue-edged epaulets (some we knew, others we had never seen before) saw and addressed only the foremen. Their manner was restrained. They did not try to intimidate us, but their tone was still intended to remind us that we were inferior. It would, so they said, be in our own interest to end the strike and the hunger strike. If we did, we would receive not only today's rations but —something unheard of in Gulag!—yesterday's, too. (They were so used to the idea that hungry men can always be bought!)

There were shouts from the corridor:

"Whoever's to blame for the shooting must be brought to justice!"

"Take the locks off the doors!"

"Off with the numbers!"

The bosses left, and the hut was locked up again.

Although hunger had begun to get many of us down—our heads were heavy and our thoughts lacked clarity—in our hut not a single voice was raised in favor of surrender.

For the second night, the third morning, and the third day hunger clawed at our guts.

But when on the third morning the Chekists, in still greater force, again summoned the foremen to the entrance, and once again we stood there—sullen, unreachable, hangdog—our general resolve was not to give way! We were carried along by inertial force.

The bosses only gave us new strength. The newly arrived brass hat had this to say:

"The administration of the Peschany camp *requests the prisoners to take their food*. The administration will receive any complaints. It will examine them and eliminate the causes of *conflict* between the administration and the prisoners."

Had our ears deceived us? They were *requesting us to take food!* And not so much as a word about work. We had stormed the camp jail, broken windows and lamps, chased warders with knives, and it now turned out that far from being a mutiny, this was a *conflict between* (!) . . . between equals . . . between the administration and the prisoners!

It had taken only two days and two nights of united action—and look how our serfmasters had changed their tone! Never in our lives, not only as prisoners, but as free men, as members of trade unions, had we heard our bosses speak with such unction!

Nonetheless, we started silently dispersing—no one could take a decision *there*. Nor could anyone there promise a decision.

The hut was locked again.

From outside it looked to the bosses as dumb and unyielding as ever. But inside, the sections were the scene of stormy debate. The temptation was too great! Soft speech had affected the undemanding zeks more than any threat would. Voices were heard urging surrender. What more, indeed, could we hope to achieve! . . .

We were tired! We were hungry! The mysterious force which had fused our emotions and borne us aloft was losing height and with tremulous wings bringing us down to earth again.

Yet mouths clamped tight for decades, mouths which had been silent for a lifetime, were now opened.

Give way now? That would mean accepting someone's word of honor. Whose word exactly? That of our jailers, the camp dog pack. In all the time that prisons had existed, in all the time the camps had been there—had they ever once kept their word?

The sediment of ancient sufferings and wrongs and insults was stirred up anew. For the first time ever we had taken the right road—were we to give in so soon? For the first time we had felt what it was like to be human—only to give in so quickly? A keen, a bracing breeze of mischief blew around us. We would go on! We would go on! They'd sing a different tune before we finished! They would give way!

Once more the emotions of two hundred men were fused in a single passion; the wings of the eagle beat the air—he sailed aloft!

We lay down to conserve our strength, trying to move as little as possible and not to talk unnecessarily. Our thoughts were quite enough to occupy us.

Suddenly, in the late afternoon of the third day, when the western sky was clearing and the setting sun could be seen, our observers shouted in anger and dismay:

"Hut nine! . . . Nine has surrendered! . . . Nine's going to the mess hall!"

We all jumped up. Prisoners from the other side of the corridor ran into our room. Through the bars, from the upper and lower bed platforms, some of us on all fours, some looking over other people's shoulders, we watched, transfixed, that sad procession.

Two hundred and fifty pathetic little figures, darker than ever against the sunset, cowed and crestfallen, were trailing slantwise across the camp. On they went, each of them glimpsed briefly in the rays of the setting sun, a dawdling, endless chain, as though those behind regretted that the foremost had set out, and were loath to follow. Some, feebler than the rest, were led by the arm or the hand, and so uncertain were their steps that they looked like blind men with their guides. Many, too, held mess tins or mugs in their hands, and this mean prisonware, carried in expectation of a supper too copious to gulp down onto constricted stomachs, these tins and cups held out like begging bowls, were more degrading and slavish and pitiable than anything else about them.

I felt myself weeping. I glanced at my companions as I wiped away my tears, and saw theirs.

Hut No. 9 had spoken, and decided for us all. It was there that the dead had been lying around for four days, since Tuesday evening.

They went into the mess hall, and it was as though they had decided to forgive the murderers in return for their bread ration and some mush.

We went away from the windows without a word.

It was then that I learned the meaning of Polish pride, and understood their recklessly brave rebellions. The Polish engineer Jerzy Wegierski, whom I have mentioned before, was now in our team. He was serving his ninth and last year. Even when he was a work assigner no one had ever heard him raise his voice. He was always quiet, polite, and gentle.

But now—his face was distorted with rage, scorn, and suffering, as he tore his eyes away from that procession of beggars, and cried in an angry, steely voice:

"Foreman! Don't wake me for supper! I shan't be going!"

He clambered up onto the top bunk, turned his face to the wall— and didn't get up again. That night we went to eat—but he wouldn't get up! He never received parcels, he was quite alone, he was always short of food—but he wouldn't get up. In his mind's eye the steam from a bowl of mush could not veil the ideal of freedom.

If we had all been so proud and so strong, what tyrant could have held out against us?

■

But what had happened had not gone for nothing, and our comrades had not fallen in vain. The atmosphere in the camp would never be as oppressive as before. Meanness was back on its throne, but very precariously. Politics were freely discussed in the huts. No work assigner or foreman would dare kick a zek or take a swing at him. Because every-body knew now how easy it is to make knives and how easily they sink home between the ribs.

Our little island had experienced an earthquake—and ceased to belong to the Archipelago.

This was how Ekibastuz felt. It is doubtful whether Karaganda felt the same. And certain that Moscow did not. The Special Camp system was beginning to collapse in one place after another, but our Father and Teacher had no inkling of it—it was not, of course, reported to him.

Meanwhile the germ of freedom was spreading. Like a legend in chains, our movement entered still servile Kengir, to awaken it, too.

There were other disturbances besides ours, besides those in the Special Camps, but the whole bloody past has been so carefully cleaned up and painted and polished that it is impossible for me now to establish even a bare list of disorders in the camps.

Evidently, the Stalinist camp system, particularly in the Special Camps, was nearing a crisis at the beginning of the fifties. Even in the Almighty One's lifetime the natives were beginning to tear at their chains.

There is no knowing how things would have gone if he had lived. As it was—for reasons which had nothing to do with the laws of economics or society—the sluggish and impure blood suddenly stopped flowing in the senile veins of that undersized and pockmarked *individual.*

However, the tyrant did not die in vain. Something hidden from view slipped and shifted—and suddenly, with a tinny clatter like an empty bucket falling, yet another *individual* came tumbling headlong from the very top of the ladder into the muckiest of bogs.

And now everyone—the vanguard, the rear, even the most wretched natives of Gulag—realized that a new age had arrived.

To us on the Archipelago, Beria's fall was like a thunderclap: he was the Supreme Patron, the Viceroy of the Archipelago! MVD officers were perplexed, embarrassed, dismayed.

"It's all over now," Colonel Chechev said with quivering lips. (But he was mistaken.)

Chapter 12

■

The Forty Days of Kengir

For the Special Camps there was another side to Beria's fall: by raising their hopes it confused, distracted, and disarmed the *katorzhane*. Hopes of speedy change burgeoned. Their anger cooled.

In that fateful year, 1953, the fall of Beria made it urgent for the security ministry to prove its devotion and its usefulness in some signal way. But how?

The mutinies which the security men had hitherto considered a menace now shone like a beacon of salvation. Let's have more disturbances and disorders, so that *measures will have to be taken.* Then staffs, and salaries, will not be reduced.

In less than a year the guards at Kengir opened fire several times on innocent men; and it cannot have been unintentional.

They shot Lida, the young girl from the mortar-mixing gang who hung her stockings out to dry near the boundary fence.

They winged the old Chinaman—nobody in Kengir remembered his name, and he spoke hardly any Russian, but everybody knew the waddling figure with a pipe between his teeth and the face of an elderly goblin. A guard called him to a watchtower, tossed a packet of makhorka near the boundary fence, and when the Chinaman reached for it, shot and wounded him.

Then there was the famous case of the column returning to camp from the ore-dressing plant and being fired on with dumdum bullets, which wounded sixteen men.

This the zeks did not take quietly—it was the Ekibastuz story over again. Kengir Camp Division No. 3 did not turn out for work three days running (but did take food), demanding punishment of the culprits.

A commission arrived and persuaded them that the culprits would be prosecuted. They went back to work.

But in February, 1954, another prisoner was shot at the woodworking plant—"the Evangelist," as all Kengir remembered him (Aleksandr Sisoyev, I think his name was). This man had served nine years and nine months of his *tenner*. His job was fluxing arc-welding rods and he did this work in a little shed which stood near the boundary fence. He went out to relieve himself near the shed—and while he was at it was shot from a watchtower. Guards quickly ran over from the guardhouse and started dragging the dead man into the boundary zone, to make it look as though he had trespassed on it. This was too much for the zeks, who grabbed picks and shovels and drove the murderers away from the murdered man.

The woodworking plant was in an uproar. The prisoners said that they would carry the dead man into camp on their shoulders. The camp officers would not permit it. "Why did you kill him?" shouted the prisoners. The bosses had their explanation ready: the dead man himself was to blame—he had started it by throwing stones at the tower. (Can they have had time to read his identity card; did they know that he had three months more to go and was an Evangelical Christian? . . .)

In the evening after supper, what they did was this. The light would suddenly go out in a section, and someone invisible said from the doorway: "Brothers! How long shall we go on building and taking our wages in bullets? Nobody goes to work tomorrow!" The same thing happened in section after section, hut after hut.

A note was thrown over the wall to the Second Camp Division. In this division, which was multinational, the majority had *tenners* and many were coming to the end of their time—but they joined in just the same.

In the morning the men's Camp Divisions, 2 and 3, did not report for work.

This bad habit—striking without refusing the state's bread and slops—was becoming more and more popular with prisoners.

They held out like this for two days. But the strike was mastered. . . .

For the second time in Kengir, a ripening abscess was lanced before it could burst.

But then the bosses went too far. They reached for the biggest stick they could use on the 58's—for the thieves!

The bosses now renounced the whole principle of the Special Camps, acknowledged that if they segregated political prisoners they had no means of making themselves *understood,* and brought into the mutinous No. 3 Camp Division 650 men, most of them thieves, some of them petty offenders (including many minors). "A *healthy batch* is joining us!" the bosses spitefully warned the 58's. "Now you won't dare breathe."

The bosses understood well enough how the restorers of order would begin: by stealing, by preying on others, and so setting every man against his fellows.

But here again we see how unpredictable is the course of human emotions and of social movements! Injecting in Kengir No. 3 a mammoth dose of tested ptomaine, the bosses obtained not a pacified camp but the biggest mutiny in the history of the Gulag Archipelago.

■

Events followed their inevitable course. It was *impossible* for the politicals not to offer the thieves a choice between war and alliance. It was *impossible* for the thieves to refuse an alliance. And it was *impossible* for the alliance, once concluded, to remain inactive.

The obvious first objective was to capture the service yard, in which all the camp's food stores were also situated. They began the operation in the afternoon of a nonworking day (Sunday, May 16, 1954). . . .

All these quite undisguised operations took a certain time, during which the warders managed to get themselves organized and obtain instructions. . . .

The service yard was now firmly held by the punitive forces, and machine-gunners were posted there. But the Second Camp Division erected a barricade facing the service yard gate. The Second and Third Camp Divisions had been joined together by a hole in the wall, and there were no longer any warders, any MVD authority, in them.

How can we say what feelings wrung the hearts of those eight thousand men, who for so long and until yesterday had been slaves with no sense of fellowship, and now had united and freed themselves, not fully perhaps, but at least within the rectangle of those walls, and under the gaze of those quadrupled guards? So long suppressed, the brotherhood of man had broken through at last!

Proclamations appeared in the mess hall: "Arm yourselves as best you can, and attack the soldiers first!" The most passionate among them hastily scrawled their slogans on scraps of newspaper: "Bash the Chekists, boys!" "Death to the stoolies, the Cheka's stooges!" Here, there, everywhere you turned there were meetings and orators. Everybody had suggestions of his own. What demands shall we put forward? What is it we want? Put the murderers on trial!—goes without saying. What else? . . . No locking huts; take the numbers off! But beyond that? . . . Beyond that came the most frightening thing—the real reason why they had *started it all,* what they really wanted. We want freedom, of course, just freedom—but who can give it to us? The judges who condemned us in Moscow. As long as our complaints are against Steplag or Karaganda, they will go on talking to us. But if we start complaining against Moscow . . . we'll all be buried in this steppe.

Well, then—what do we want? To break holes in the walls? To run off into the wilderness? . . .

Those hours of freedom! Immense chains had fallen from our arms and shoulders! No; whatever happened, there could be no regrets! That one day made it all worthwhile!

Late on Monday, a delegation from command HQ arrived in the seething camp. The delegation was quite well disposed. Our side learned that generals had flown in from Moscow. They found the prisoners' demands *fully justified!* (We simply gasped: justified? We aren't rebels, then? No, no, they're *quite* justified!) "Those responsible for the shooting will be made to answer for it!" "But why did they beat up women?" "Beat up women?" The delegation was shocked. "That can't be true." Anya Mikhalevich brought in a succession of battered women for them to see. The commission was deeply moved: "We'll look into it, never fear!" "Beasts!" Lyuba Bershadskaya shouts at the general. There were other shouts: "No locks on huts!" "We won't lock them any more." "Take the numbers off!" "Certainly we'll take them off." "The holes in the wall between camp areas must remain!" They were getting bolder. "We must be allowed to mix with each other." "All right, mix as much as you like." "Let the holes remain." Right, brothers, what else do we want? We've won, we've won! We raised hell for just one day, enjoyed ourselves, let off steam—and we won! Although some among us shake their heads and say, "It's a trick, it's all a trick," we believe it!

We believe because that's our easiest way out of the situation. . . .

* * *

All that the downtrodden can do is go on hoping. After every disappointment they must find fresh reason for hope.

So on Tuesday, May 18, all the Kengir Camp Divisions went out to work, reconciling themselves to thoughts of their dead.

That morning the whole affair could still have ended quietly. But the exalted generals assembled in Kengir would have considered such an outcome a defeat for themselves. They could not seriously admit that prisoners were in the right!

When the columns of prisoners returned to camp in the evening after giving a day's work to the state, they were hurried in to supper before they knew what was happening, so that they could be locked up quickly. On orders from the general, the jailers had to play for time that first evening—that evening of blatant dishonesty after yesterday's promises.

But before nightfall the long-drawn whistles heard on Sunday shrilled through the camp again—the Second and Third Camp Divisions were calling to each other like hooligans on a spree. The warders took fright, and fled from the camp grounds without finishing their duties.

The camp was in the hands of the zeks, but they were divided. The towers opened fire with machine guns on anyone who approached the inside walls. They killed several and wounded several. Once again zeks broke all the lamps with slingshots, but the towers lit up the camp with flares. . . .

They battered at the barbed wire, and the new fence posts, with long tables, but it was impossible, under fire, either to break through the barrier or to climb over it—so they had to burrow under. As always, there were no shovels, except those for use in case of fire, inside the camp. Kitchen knives and mess tins were put into service.

That night—May 18–19—they burrowed under all the walls and again united all the divisions and the service yard. The towers had stopped shooting now, and there were plenty of tools in the service yard. Under cover of night they broke down the boundary fences, knocked holes in the walls, and widened the passages, so that they would not become traps.

That same night they broke through the wall around the Fourth Camp Division—the prison area—too. The warders guarding the jails fled. The prisoners wrecked the interrogation offices. Among those released from the jail were those who on the morrow would take command of the rising: former Red Army Colonel Kapiton Kuznetsov and former First Lieutenant Gleb Sluchenkov.

Mutinous zeks! These eight thousand men had not so much raised a rebellion as *escaped to freedom,* though not for long! Eight thousand men, from being slaves, had suddenly become free, and now was their chance to . . . live! Faces usually grim softened into kind smiles. Women looked at men, and men took them by the hand. Some who had corresponded by ingenious secret ways, without even seeing each other, met at last! Lithuanian girls whose weddings had been solemnized by priests on the other side of the wall now saw their lawful wedded husbands for the first time—the Lord had sent down to earth the marriages made in heaven! For the first time in their lives, no one tried to prevent the sectarians and believers from meeting for prayer. Foreigners, scattered about the Camp Divisions, now found each other and talked about this strange Asiatic revolution in their own languages. The camp's food supply was in the hands of the prisoners. No one drove them out to work line-up and an eleven-hour working day.

The morning of May 19 dawned over a feverishly sleepless camp which had torn off its number patches. Many took their street clothes from the storerooms and put them on. Some of the lads crammed fur hats on their heads; shortly there would be embroidered shirts, and on the Central Asians bright-colored robes and turbans. The gray-black camp would be a blaze of color.

Orderlies went around the huts summoning us to the big mess hall to elect a commission for negotiations with the authorities and for self-government. For all they knew, they were electing it just for a few hours, but it was destined to become the government of Kengir camp for forty days.

■

The days ran on. And the generals were regretfully forced to conclude that the camp was not disintegrating of its own accord, and that there was no excuse to send troops in to the rescue.

The camp *stood fast* and the negotiations changed their character. Golden-epauleted personages, in various combinations, continued coming into the camp to argue and persuade. They were all allowed in, but they had to pick up white flags, and they had to undergo a body search. In return, the rebel staff *guaranteed* their personal safety! . . .

They showed the generals around, wherever it was allowed (not, of course, around the *secret* sector of the service yard), let them talk to prisoners, and called big meetings in the Camp Divisions for their benefit. Their epaulets flashing, the bosses took their seats in the presidium as of old, as though nothing were amiss.

The discussions sometimes took the form of direct negotiations on the loftiest diplomatic model. Sometime in June a long mess table was placed in the women's camp, and the golden epaulets seated themselves on a bench to one side of it, while the Tommy-gunners allowed in with them as a bodyguard stood at their backs. Across the table sat the members of the Commission, and they, too, had a bodyguard—which stood there, looking very serious, armed with sabers, pikes, and sling-shots. In the background crowds of prisoners gathered to listen to the powwow and shout comments. (Refreshments for the guests were not forgotten!)

The rebels had agreed on their demands (or requests) in the first two days, and now repeated them over and over again:

- Punish the Evangelist's murderer.
- Punish all those responsible for the murders on Sunday night in the service yard.
- Punish those who beat up the women.
- Bring back those comrades who had been illegally sent to closed prisons for striking.
- No more number patches, window bars, or locks on hut doors.
- Inner walls between Camp Divisions not to be rebuilt.
- An eight-hour day, as for free workers.
- An increase in payment for work (here there was no question of equality with free workers).
- Unrestricted correspondence with relatives, periodic visits.
- Review of cases.

Although there was nothing unconstitutional in any of these de-mands, nothing that threatened the foundations of the state (indeed, many of them were requests for a return to the old position), it was impossible for the bosses to accept even the least of them, because these bald skulls under service caps and supported by close-clipped fat necks had forgotten how to admit a mistake or a fault. Truth was unrecogniz-able and repulsive to them if it manifested itself not in secret instruc-tions from higher authority but on the lips of common people.

Still, the obduracy of the eight thousand under siege was a blot on the reputation of the generals, it might ruin their careers, and so they made promises. They promised that nearly all the demands would be satisfied—only, they said, they could hardly leave the women's camp open, that was against the rules (forgetting that in the Corrective Labor Camps it had been that way for twenty years), but they could consider arranging, should they say, *meeting days*. To the demand that the

Commission of Inquiry should start its work inside the camp, the generals unexpectedly agreed. (But Sluchenkov guessed their purpose, and refused to hear of it: while making their statements, the stoolies would expose everything that was happening in the camp.) Review of cases? Well, of course, cases would be re-examined, but prisoners would *have to be patient.* There was one thing that couldn't wait at all—the prisoners must get back to work! to work! to work!

But the zeks knew that trick by now: dividing them up into columns, forcing them to the ground at gunpoint, arresting the ringleaders.

No, they answered across the table, and from the platform. No! shouted voices from the crowd. The administration of Steplag have behaved like provocateurs! We do not trust the Steplag authorities! We don't trust the MVD!

"Don't trust *even* the MVD?" The vice-minister was thrown into a sweat by this treasonable talk. "And who can have inspired in you such hatred for the MVD?"

A riddle, if ever there was one.

There were weeks when the whole war became a war of propaganda. The outside radio was never silent: through several loudspeakers set up at intervals around the camp it interlarded appeals to the prisoners with information and misinformation, and with a couple of trite and boring records that frayed everybody's nerves.

> Through the meadow goes a maiden,
> She whose braided hair I love.

(Still, to be thought worthy even of that not very high honor—having records played to them—they had to rebel. Even rubbish like that wasn't played for men on their knees.) These records also served, in the spirit of the times, as a *jamming device*—drowning the broadcasts from the camps intended for the escort troops.

On the outside radio they sometimes tried to blacken the whole movement, asserting that it had been started with the sole aim of rape and plunder. At other times they tried telling filthy stories about members of the Commission. Then the appeals would begin again. Work! Work! Why should the Motherland keep you for nothing? By not going to work you are doing enormous damage to the state! (This was supposed to pierce the hearts of men doomed to eternal *katorga!*) Whole trainloads of coal are standing in the siding, there's nobody to unload it! (Let them stand there—the zeks laughed—you'll give way all the sooner!)

The Technical Department, however, gave as good as it got. Two portable film projectors were found in the service yard. Their amplifiers were used for loudspeakers, less powerful, of course, than those of the other side. (The fact that the camp had electricity and radio greatly surprised and troubled the bosses. They were afraid that the rebels might rig up a transmitter and start broadcasting news about their rising to foreign countries.)

The camp soon had its own announcers. Programs included the latest news, and news features (there was also a daily wall newspaper, with cartoons). "Crocodile Tears" was the name of a program ridiculing the anxiety of the MVD men about the fate of women whom they themselves had previously beaten up.

But there was not enough power to put on programs for the only potential sympathizers to be found in Kengir—the free inhabitants of the settlement, many of them exiles. It was they whom the settlement authorities were trying to fool, not by radio but with rumors that bloodthirsty gangsters and insatiable prostitutes were ruling the roost inside the camp; that over there innocent people were being tortured and burned alive in furnaces.

How could the prisoners call out through the walls, to the workers one, or two, or three kilometers away: "Brothers! We want only justice! They were murdering us for no crime of ours, they were treating us worse than dogs! Here are our demands"?

The thoughts of the Technical Department, since they had no chance to outstrip modern science, moved backward instead to the science of past ages. Using cigarette paper, they pasted together an enormous air balloon. A bundle of leaflets was attached to the balloon, and slung underneath it was a brazier containing glowing coals, which sent a current of warm air into the dome of the balloon through an opening in its base. To the huge delight of the assembled crowd (if prisoners ever do feel happy they are like children), the marvelous aeronautical structure rose and was airborne. But alas! The speed of the wind was greater than the speed of its ascent, and as it was flying over the boundary fence the brazier caught on the barbed wire. The balloon, denied its current of warm air, fell and burned to ashes, together with the leaflets.

After this failure they started inflating balloons with smoke. With a following wind they flew quite well, exhibiting inscriptions in large letters to the settlement:

"Save the women and old men from being beaten!"

"We demand to see a member of the Presidium."

The guards started shooting at these balloons.

Then some Chechen prisoners came to the Technical Department and offered to make kites. (They are experts.) They succeeded in sticking some kites together and paying out the string until they were over the settlement. There was a percussive device on the frame of each kite. When the kite was in a convenient position, the device scattered a bundle of leaflets, also attached to the kite. The kite fliers sat on the roof of a hut waiting to see what would happen next. If the leaflets fell close to the camp, warders ran to collect them; if they fell farther away, motorcyclists and horsemen dashed after them. Whatever happened, they tried to prevent the free citizens from reading an independent version of the truth. (The leaflets ended by requesting any citizen of Kengir who found one to deliver it to the Central Committee.)

The kites were also shot at, but holing was less damaging to them than to the balloons. The enemy soon discovered that sending up counter-kites to tangle strings with them was cheaper than keeping a crowd of warders on the run.

A war of kites in the second half of the twentieth century! And all to silence a word of truth.

In the meantime the Technical Department was getting its notorious "secret" weapon ready. Let me describe it. Aluminum corner brackets for cattle troughs, produced in the workshops and awaiting dispatch, were packed with a mixture of sulfur scraped from matches and a little calcium carbide. When the sulfur was lit and the brackets thrown, they hissed and burst into little pieces.

But neither these star-crossed geniuses nor the field staff in the bathhouse were to choose the hour, place, and form of the decisive battle. Some two weeks after the beginning of the revolt, on one of those dark nights without a glimmer of light anywhere, thuds were heard at several places around the camp wall. This time it was not escaping prisoners or rebels battering it down; the wall was being demolished by the convoy troops themselves!

In the morning it turned out that the enemy without had made about a dozen breaches in the wall in addition to those already there and the barricaded gateway. (Machine-gun posts had been set up on the other side of the gaps, to prevent the zeks from pouring through them.) This was of course the preliminary for an assault through the breaches, and the camp was a seething anthill as it prepared to defend itself. The rebel staff decided to pull down the inner walls and the mud-brick outhouses and to erect a second circular wall of their own, specially reinforced with stacks of brick where it faced the gaps, to give protection against machine-gun bullets.

How things had changed! The troops were demolishing the boundary wall, the prisoners were rebuilding it, and the thieves were helping with a clear conscience, not feeling that they were contravening their *code.*

Additional defense posts now had to be established opposite the gaps, and every platoon assigned to a gap, which it must run to defend should the alarm be raised at night.

The zeks quite seriously prepared to advance against machine guns with pikes.

There was one attack in the daytime. Tommy-gunners were moved up to one of the gaps, opposite the balcony of the Steplag Administration Building, which was packed with important personages holding cameras or even movie cameras. The soldiers were in no hurry. They merely advanced just far enough into the breach for the alarm to be given, whereupon the rebel platoons responsible for the defense of the breach rushed out to man the barricade—brandishing their pikes and holding stones and mud bricks—and then, from the balcony, movie cameras whirred and pocket cameras clicked (taking care to keep the Tommy-gunners out of the picture). Disciplinary officers, prosecutors, Party officials, and all the rest of them—Party members to a man, of course—laughed at the bizarre spectacle of the impassioned savages with pikes. Well-fed and shameless, these grand personages mocked their starved and cheated fellow citizens from the balcony, and found it *all very funny.*

Then warders, too, stole up to the gaps and tried to slip nooses with hooks over the prisoners, as though they were hunting wild animals or the abominable snowman, hoping to drag out a *talker.*

But what they mainly counted on now were deserters, rebels with cold feet. The radio blared away. Come to your senses! Those who come over will not be tried for mutiny!

The Commission's response, over the camp radio, was this: Anybody who wants to run away can go right ahead, through the main gate if he likes; we are holding no one back!

In all those weeks only about a dozen men fled from the camp.

Why? Surely the rest did not believe in victory. Were they not appalled by the thought of the punishment ahead? They were. Did they not want to save themselves for their families' sake? They did! They were torn, and thousands of them perhaps had secretly considered this possibility. But the social temperature on this plot of land had risen so high that if souls were not transmuted, they were purged of dross, and the sordid laws saying that "we only live once," that being determines consciousness, and that every man's a coward when his neck is at stake, ceased to apply for that short time in that circumscribed place. The laws of survival and of reason told people that they must all surrender together or flee individually, but they did not surrender and they did not flee! They rose to that spiritual plane from which executioners are told: "The devil take you for his own! Torture us! Savage us!"

And the operation, so beautifully planned, to make the prisoners scatter like rats through the gaps in the wall till only the most stubborn were left, who would then be crushed—this operation collapsed because its inventors had the mentality of rats themselves.

No one supported the island of Kengir. It was impossible by now to take off into the wilderness: the garrison was being steadily reinforced. The whole camp had been encircled with a double barbed-wire fence outside the walls. There was only one rosy spot on the horizon: the lord and master (they were expecting Malenkov) was coming to dispense justice. But it was too tiny a spot, and too rosy.

They could not hope for pardon. All they could do was live out their last few days of freedom, and submit to Steplag's vengeance.

There are always hearts which cannot stand the strain. Some were already morally crushed, and were in an agony of suspense for the crushing proper to begin. Some quietly calculated that they were not really involved, and need not be if they went on being careful. Some were newly married (what is more, with a proper religious ceremony —a Western Ukrainian girl, for instance, will not marry without one, and thanks to Gulag's thoughtfulness, there were priests of all religions there). For these newlyweds the bitter and the sweet succeeded each other with a rapidity which ordinary people never experience in their slow lives. They observed each day as their last, and retribution delayed was a gift from heaven each morning.

The believers . . . prayed, and leaving the outcome of the Kengir revolt in God's hands, were as always the calmest of people. Services for all religions were held in the mess hall according to a fixed timetable. The Jehovah's Witnesses felt free to observe their rules strictly and refused to build fortifications or stand guard. They sat for hours on end with their heads together, saying nothing. (They were made to wash the dishes.) A prophet, genuine or sham, went around the camp putting crosses on bunks and foretelling the end of the world.

Some knew that they were fatally compromised and that the few days before the troops arrived were all that was left of life. The theme of all their thoughts and actions must be how to hold out longer. These people were not the unhappiest. (The unhappiest were those who were not involved and who prayed for the end.)

But when all these people gathered at meetings to decide whether to surrender or to hold on, they found themselves again in that heated climate where their personal opinions dissolved, and ceased to exist even for themselves. Or else they feared ridicule even more than the death that awaited them.

And when they voted for or against holding out, the majority were *for.*

Why did it drag on so long? What can the bosses have been waiting for? For the food to run out? They knew it would last a long time. Were they considering opinion in the settlement? They had no need to. Were they carefully working out their plan of repression? They could have been quicker about it. Were they having to seek approval for the operation *up top?* How high up? There is no knowing on what date and at what level the decision was taken.

On several occasions the main gate of the service yard suddenly opened—perhaps to test the readiness of the defenders? The duty picket sounded the alarm, and the platoons poured out to meet the enemy. But no one entered the camp grounds.

In the middle of June several tractors appeared in the settlement. They were working, shifting something perhaps, around the boundary fence. They began working even at night. The unfriendly roar made the night seem blacker.

Then suddenly the skeptics were put to shame! And the defeatists! And all who had said that there would be no mercy, and that there was no point in begging. The orthodox alone could feel triumphant. On June 22 the outside radio announced that the prisoners' demands had been accepted! A member of the Presidium of the Central Committee was on his way!

The rosy spot turned into a rosy sun, a rosy sky! It is, then, possible to get through to them! There *is,* then, justice in our country! They will give a little, and we will give a little. If it comes to it, we can walk about with number patches, and the bars on the windows needn't bother us, we aren't thinking of climbing out. You say they're tricking us again? Well, they aren't asking us to report for work *beforehand!*

Just as the touch of a stick will draw off the charge from an electroscope so that the agitated gold leaf sinks gratefully to rest, so did the radio announcement reduce the brooding tension of that last week.

Even the loathsome tractors, after working for a while on the evening of June 24, stopped their noise.

Prisoners could sleep peacefully on the fortieth night of the revolt. *He* would probably arrive tomorrow; perhaps he had come already. . . .

In the early dawn of Friday, June 25, parachutes carrying flares opened out in the sky, more flares soared from the watchtowers, and the observers on the rooftops were picked off by snipers' bullets before they could let out a squeak! Then cannon fire was heard! Airplanes skimmed the camp, spreading panic. Tanks, the famous T-34's, had taken up position under cover of the tractor noise and now moved on the gaps from all sides. (One of them, however, fell into a ditch.) Some of the tanks dragged concatenations of barbed wire on trestles so that they could divide up the camp grounds immediately. Behind others ran helmeted assault troops with Tommy guns. (Both Tommy-gunners and tank crews had been given vodka first. However *special* the troops may be, it is easier to destroy unarmed and sleeping people with drink inside you.) Operators with walkie-talkies came in with the advancing troops. The generals went up into the towers with the snipers, and from there, in the daylight shed by the flares (and the light from a tower set on fire by the zeks with their incendiary bombs), gave their orders: "Take hut number so-and-so! . . ."

The camp woke up—frightened out of its wits. Some stayed where they were in their huts, lying on the floor as their one chance of survival, and because resistance seemed senseless. Others tried to make them get up and join in the resistance. Yet others ran right into the line of fire, either to fight or to seek a quicker death.

The Third Camp Division fought—the division which had started it all. They hurled stones at the Tommy-gunners and warders, and probably sulfur bombs at the tanks. . . . Nobody thought of the powdered glass. One hut counterattacked twice, with shouts of "Hurrah."

The tanks crushed everyone in their way. (Alla Presman, from Kiev, was run over—the tracks passed over her abdomen.) Tanks rode up onto the porches of huts and crushed people there. The tanks grazed the sides of huts and crushed those who were clinging to them to escape the caterpillar tracks. Semyon Rak and his girl threw themselves under a tank clasped in each other's arms and ended it that way. Tanks nosed into the thin board walls of the huts and even fired blank shells into them. Faina Epstein remembers the corner of a hut collapsing, as if in a nightmare, and a tank passing obliquely over the wreckage and over living bodies; women tried to jump and fling themselves out of the way: behind the tank came a lorry, and the half-naked women were tossed onto it.

The cannon shots were blank, but the Tommy guns were shooting live rounds, and the bayonets were cold steel. Women tried to shield men with their own bodies—and they, too, were bayoneted! Security Officer Belyaev shot two dozen people with his own hand that morning; when the battle was over he was seen putting knives into the hands of corpses for the photographer to take pictures of dead *gangsters*. Suprun, a member of the Commission, and a grandmother, died from a wound in her lung. Some prisoners hid in the latrines, and were riddled with bullets there.

As groups of prisoners were taken, they were marched through the gaps onto the steppe and between files of Kengir convoy troops outside. They were searched and made to lie flat on their faces with their arms stretched straight out. As they lay there thus crucified, MVD fliers and warders walked among them to identify and pull out those whom they had spotted earlier from the air or from the watchtowers. (So busy were they with all this that no one had leisure to open *Pravda* that day. It had a special theme—a day in the life of our Motherland: the successes of steelworkers; more and more crops harvested by machine. The historian surveying our country as it was *that day* will have an easy task.)

The victorious generals descended from the towers and went off to breakfast. Without knowing any of them, I feel confident that their appetite that June morning left nothing to be desired and that they drank deeply. An alcoholic hum would not in the least disturb the ideological harmony in their heads. And what they had for hearts was something installed with a screwdriver.

The number of those killed or wounded was about six hundred, according to the stories, but according to figures given by the Kengir Division's Production Planning Section, which became known some months later, it was more than *seven hundred*.

All day on June 25, the prisoners lay face down on the steppe in the sun (for days on end the heat had been unmerciful), while in the camp there was endless searching and breaking open and shaking out.

The members of the Commission and other suspects were locked up in the camp jail. More than a thousand people were selected for dispatch either to closed prisons or to Kolyma (as always, these lists were drawn up partly by guesswork, so that many who had not been involved at all found their way into them).

May this picture of the pacification bring peace to the souls of those on whom the last chapters have grated.

On June 26, the prisoners were made to spend the whole day taking down the barricades and bricking in the gaps.

On June 27, they were marched out to work. Those trains in the sidings would wait no longer for working hands!

The tanks which had crushed Kengir traveled under their own power to Rudnik and crawled around for the zeks to see. And draw their conclusions. . . .

PART VI

Exile

■

Chapter 1

■

Exile in the First Years of Freedom

Humanity probably invented exile first and prison later. Expulsion from the tribe was of course exile. We were quick to realize how difficult it is for a man to exist, divorced from his own place, his familiar environment.

In the Russian Empire as elsewhere they were not slow to discover exile. It was given legal sanction under Tsar Aleksei Mikhailovich by the Code of Laws of 1648. But even earlier, at the end of the sixteenth century, people were exiled without legal sanction. Our great spaces gave their blessing—Siberia was ours already. Altogether in the course of the nineteenth century *half a million* were exiled, and at the end of the century those in exile at any one time numbered 300,000. Exile was so common in Russia precisely because there weren't enough jails for long-term prisoners.

A feature of the exile system in the Tsarist nineteenth century which was taken for granted and seemed natural to everyone, but to us now seems surprising, was that it concerned itself only with individuals: whether he was dealt with by the courts or administratively, each man was sentenced separately and not as a member of some group.

The conditions of life in exile, the degree of harshness, changed from one decade to the next, and different generations of exiles have left us a variety of evidence. Transported prisoners traveling from transit prison to transit prison had a hard time of it; but we learn from P. F. Yakubovich and from Lev Tolstoi that politicals were transported in quite tolerable conditions. . . .

This mild treatment of exiles was not confined to socially distinguished or famous people. In the twentieth century, too, it was enjoyed by many revolutionaries and frondeurs—by the Bolsheviks in particular, who were not thought dangerous. Stalin, with four escapes behind him, was exiled for a fifth time . . . all the way to Vologda.

Yet exile even under these conditions, lenient as it seems to us, exile with no danger of starving to death, was sometimes taken hard by the exile himself. Many revolutionaries recall how painful they found the move from prison, where they were assured of bread, warmth, shelter, and leisure for their "universities" and their party wrangles, to a place of exile, where you were all by yourself among strangers, and had to use your own ingenuity to find food and shelter. Where there was no need to worry about these things it was, so we are informed (by F. Kon, for instance), still worse: "the horrors of idleness . . . The most dreadful thing of all is that people are condemned to inactivity." Indeed, some of them even abandoned study for moneymaking, for trade, and some simply despaired and took to drink.

Here we see that the threat of exile—of mere displacement, of being set down with your feet tied—has a somber power of its own, the power which even ancient potentates understood, and which Ovid long ago experienced.

Emptiness. Helplessness. A life that is no life at all. . . .

■

The revolution had scarcely taken its first steps on legs still infirm, it was still in its infancy, when it realized that exile was indispensable. Let me quote verbatim Marshall Turkhachevsky about the year 1921 in the province of Tambov. "It was decided to organize large-scale deportation of the families of bandits [i.e., "partisans"—A.S.]. Extensive *concentration camps* were organized, in which these families could be confined while waiting."

Only the fact that it was so convenient to shoot people on the spot rather than carry them elsewhere, guard them and feed them on the way, resettle them and go on guarding them—only this delayed the introduction of a regular exile system until the end of War Communism. Soon after that, in 1922, a permanent Exile Commission was set up under the Commissariat of the Interior to deal with "socially dangerous persons." Thus even in the early twenties exile was a familiar and smoothly operating institution.

Because people are so complacently gullible, the regime's intentions dawned on them slowly: the regime was simply not strong enough yet to eradicate all the unwanted at once. So for the time being they were uprooted not from life itself, but from the memory of their fellows.

There was, however, in the exile system one residual snag: the parasitical attitude of the exiles, who thought the state had an obligation to feed them. The Tsarist government did not dare to try compelling the exiles to increase the national product. And professional revolutionaries considered it beneath them to work.

Lenin received (and did not refuse) his 12 rubles a month, and prices in Siberia were one-half or one-third those in Russia, so that the state's maintenance allowance for exiles was in fact overgenerous. It enabled Lenin to spend three whole years comfortably studying the theory of revolution, not worrying at all about the source of his livelihood.

From 1929 they started elaborating a system of exile in conjunction with forced labor. One worry the exiles were free from was how to cope with senseless idleness. . . . Their one concern now was: how to avoid dying of hunger.

Exile was a temporary pen to hold sheep marked for slaughter. Exiles in the first Soviet decades were not meant to settle but to await the summons—elsewhere. There were clever people—"former" people, and also some simple peasants—who already realized in the twenties all that lay before them. And when they reached the end of their first three-year term they stayed exactly where they were—in Archangel, for instance—just in case. Sometimes this helped them not to be caught under the nit comb again.

This was what exile had become in our time. . . .

You will see that we had heavier burdens than Ovid's homesickness to bear.

Chapter 2

■

The Peasant Plague

This chapter will deal with a small matter. Fifteen million souls. Fifteen million lives.

They weren't educated people, of course. They couldn't play the violin.

In the First World War we lost in all three million killed. In the Second we lost twenty million (so Khrushchev said; according to Stalin it was only seven million. Was Nikita being too generous? Or couldn't Iosif keep track of his capital?). All those odes! All those obelisks and eternal flames! Those novels and poems! For a quarter of a century all Soviet literature has been drunk on that blood!

But about the silent, treacherous Plague which starved fifteen million of our peasants to death, choosing its victims carefully and destroying the backbone and mainstay of the Russian people—about that Plague there are no books.

Our country as well as our European neighbors keep silent about the six million people who were subsequently starved to death during the famine artificially brought about by the Bolsheviks. In the prosperous Poltava region, corpses were left lying around in the villages, on the roads and in the fields. It was impossible to walk through the groves near the stations: the stench of rotting corpses, among them those of babies, would make one faint. The situation was perhaps most horrible in the Kuban region. In many places in Byelorussia special crews had to be brought in from other regions to collect the corpses; those on the spot who were still alive were too few to bury the dead.

No bugles bid our hearts beat faster for them. Not even the traditional three stones mark the crossroads where they went in creaking carts to their doom. Our finest humanists, so sensitive to today's injustices, in those years only nodded approvingly: Quite right, too! Just what they deserve!

It was all kept very dark, every stain carefully scratched out, every whisper swiftly choked.

Where did it all start? With the dogma that the peasantry is *petit bourgeois?* (And who in the eyes of these people is not petit bourgeois? In their wonderfully clear-cut scheme, apart from factory workers [not the skilled workers, though] and big-shot businessmen, all the rest, the whole people—peasants, office workers, actors, airmen, professors, students, doctors—are nothing but the "petite bourgeoisie.") Or did it start with a criminal scheme in high places to rob some and terrorize the rest?

From the last letters which Korolenko wrote to Gorky in 1921, just before the former died and the latter emigrated, we learn that this villainous assault on the peasantry had begun even then, and was taking almost the same form as in 1930.

But as yet their strength did not equal their impudence, and they backed down.

The devastating peasant Plague began, as far as we can judge, in 1929—the compilation of murder lists, the confiscations, the deportations. But only at the beginning of 1930 (after rehearsals were complete, and necessary adjustments made) was the public allowed to learn what was happening—in the decision of the Central Committee of the Party dated January 5. (The Party is "justified in shifting from a policy of restricting the exploiting tendencies of the kulaks to a policy of liquidating the kulaks as a class.")

The savage law of the Civil War (Ten for every one! A hundred for every one!) was reinforced—to my mind an un-Russian law: where will you find anything like it in Russian history? For every activist (which usually meant big-mouthed loafer: A. Y. Olenyev is not the only one to recall that thieves and drunkards were in charge of "dekulakization")—for every *activist* killed in self-defense, hundreds of the most industrious, enterprising, and level-headed peasants, those who should keep the Russian nation on an even keel, were eliminated.

Yells of indignation! What's that? What do you say? What about the *bloodsuckers?* Those who squeezed their neighbors dry? "Take your loan—and pay me back with your hide"?

I suppose that bloodsuckers were a small part of the whole number (but were all the bloodsuckers there among them?). And were they bloodsuckers born? we may ask. Bloodsuckers through and through? Or was it just that all wealth—and all power—corrupts human beings? If only the "cleansing" of mankind, or of a social estate, were so simple! But if they had "cleansed" the peasantry of heartless bloodsuckers with their fine-toothed iron comb, cheerfully sacrificing fifteen millions for the purpose—whence all those vicious, fat-bellied rednecks who preside over collectivized villages (and District Party Committees) today? Those pitiless oppressors of lonely old women and all defenseless people? How was the root of this predatory weed missed during dekulakization? Surely, heaven help us, they can't have sprung from the *activists?* . . .

The principle underlying dekulakization can also be clearly seen in the fate of the children. Take Shurka Dmitriyev, from the village of Masleno (Selishchenskie Kazarmy, near the Volkhov). He was thirteen when his father, Fyodor, died in 1925, and the only son in a family of girls. Who was to manage his father's holding? Shurka took it on. The girls and his mother accepted him as head of the family. A working peasant and an adult now, he exchanged bows with other adults in the street. He was a worthy successor to his hard-working father, and when 1929 came his bins were full of grain. Obviously a kulak! The whole family was driven out!

Adamova-Sliozberg has a moving story about meeting a girl called Motya, who was jailed in 1936 for leaving her place of banishment without permission to go to her native village, Svetlovidovo near Tarussa, *two thousand kilometers on foot!* Sportsmen are given medals for that sort of thing. She had been exiled with her parents in 1929 when she was a little schoolgirl, and deprived of schooling forever. Her teacher's pet name for her was "Motya, our little Edison": the child was not only an excellent pupil, but had an inventive turn of mind, had rigged up a sort of turbine worked by a stream, and invented other things for the school. After seven years she felt an urge to look just once more at the log walls of her unattainable school—and for that "little Edison" went to prison and then to a camp.

Did any child suffer such a fate in the nineteenth century?

Every *miller* was automatically a candidate for dekulakization—and what were millers and blacksmiths but the Russian village's best technicians?

Let us look at one village blacksmith. In fact, we'll start with his father, as Personnel Departments like to do. His father, Gordei Vasilyevich, served for twenty-five years in the Warsaw garrison, and earned enough silver to make a tin button: this soldier with twenty-five years' service was denied a plot of land. He had married a soldier's daughter while he was in the garrison, and after his discharge he went to his wife's native place, the village of Barsuki in the Krasnensky district. The village got him tipsy, and he paid off its tax arrears with half of his savings. With the other half he leased a mill from a landowner, but quickly lost the rest of his money in this venture. He spent his long old age as a herdsman and watchman. He had six daughters, all of whom he gave in marriage to poor men, and an only son, Trifon (their family name was Tvardovsky). The boy was sent away to serve in a haberdasher's shop, but fled back to Barsuki and found employment with the Molchanovs, who had the forge. After a year as an unpaid laborer, and four years as an apprentice, he became a smith himself, built a wooden house in the village of Zagorye, and married. Seven children were born (among them Aleksandr, the poet), and no one is likely to get rich from a forge. The oldest son, Konstantin, helped his father. If they smelted and hammered from one dawn to the next they could make five excellent steel axes, but the smiths of Roslavl, with their presses and their hired workmen, undercut their price. In 1929 their forge was still wood-built, they had only one horse, sometimes they had a cow and a calf, sometimes neither cow nor calf, and besides all this they had eight apple trees—you can see what bloodsuckers they were. . . . The Peasant Land Bank used to sell mortgaged estates on deferred payments. Trifon Tvardovsky had taken eleven desyatins of wasteland, all overgrown with bushes, and the year of the Plague found them still sweating and straining to clear it: they had brought five desyatins into cultivation, and the rest they abandoned to the bushes. The collectivizers marked them down for dekulakization—there were only fifteen households in the village and somebody had to be found. They assessed the income from the forge at a fantastic figure, imposed a tax beyond the family's means, and when it was not paid on time: Get ready to move, you damned kulaks, you!

If a man had a brick house in a row of log cabins, or two stories in a row of one-story houses—there was your kulak: Get ready, you bastard, you've got sixty minutes! There aren't supposed to be any brick houses in the Russian village, there aren't supposed to be two-story houses! Back to the cave! You don't need a chimney for your fire! This is our great plan for transforming the country: history has never seen the like of it.

But we still have not reached the innermost secret. The better off were sometimes left where they were, provided they joined the kolkhoz quickly, while the obstinate poor peasant who failed to apply was deported.

This is very important, the most important thing. The point of it all was not to dekulakize, but to force the peasants into the kolkhoz. Without frightening them to death there was no way of taking back the land which the Revolution had given them, and planting them on that same land as serfs.

It was a second Civil War—this time against the peasants. It was indeed the Great Turning Point, or as the phrase had it, the Great Break. Only we are never told what it was that broke.

It was the backbone of Russia.

■

We can find described in books, or even see in films, barns and pits in the ground, full of grain hoarded by bloodsuckers. What they won't show us is the handful of belongings earned in a lifetime of toil: the livestock, the utensils—things as close to the owner as her own skin—which a weeping peasant woman is ordered to leave forever.

What they will not show us are the little bundles with which the family are allowed onto the state's cart. We shall not learn that in the Tvardovsky house, when the evil moment came there was neither suet nor bread; their neighbor Kuzma saved them: he had several children and was far from rich himself, but brought them food for the journey.

The journey itself, the peasant's *Via Crucis,* is something which our socialist realists do not describe at all. Get them aboard, pack them off—and that's the end of the story. Episode concluded. Three asterisks, please.

They were loaded onto carts . . . if they were lucky enough to be taken in the warm months, but it might be onto sledges in a cruel frost, with children of all ages, babes in arms as well. In February, 1931, when hard frosts were interrupted only by blizzards, the strings of carts rolled endlessly through the village of Kochenevo (Novosibirsk oblast), flanked by convoy troops, emerging from the snowbound steppe and vanishing into the snowbound steppe again. Even going into a peasant hut for a warm-up required special permission from the convoy, which was given only for a few minutes, so as not to hold up the cart train. They all shuffled into the Narym marshes—and in those insatiable quagmires they all remained. Many of the children had already died a wretched death on the cruel journey.

This was the nub of the plan: the peasant's seed must perish together with the adults. Since Herod was no more, only the Vanguard Doctrine has shown us how to destroy utterly—down to the very babes. Hitler was a mere disciple, but he had all the luck: his murder camps have made him famous, whereas no one has any interest in ours at all.

It is hard to believe in such cruelty: on a winter evening out in the taiga they were told: You've arrived! Can human beings really behave like this? Well, they're moved by day so they arrive at nightfall—that's all there is to it. Hundreds and hundreds of thousands were carried into the wilds and dumped down like this, old men, women, children, and all.

As the Plague approached in 1929, all the churches in Archangel were closed: they were due to be closed anyway, but the very real need for somewhere to put the dekulakized hurried things along. Great streams of deported peasants poured through Archangel, and for a time the whole town became one big transit prison. Many-tiered sleeping platforms were put up in the churches, but there was no heat. Consignment after consignment of human cattle was unloaded at the station, and with dogs barking around them, the bast-shod went sullenly to church and a bed of planks. (S., then a boy, would never forget one peasant walking along with a shaft bow around his neck: he had been hurried away before he could decide what would be most useful. Another man carried a gramophone with a horn. Cameramen—there's work for you in this! . . .) In the Church of the Presentation, an eight-tiered bed platform which was not fastened to the wall collapsed in the night and several families were crushed. Their cries brought troops rushing to the church.

This was how they lived in that plague-stricken winter. They could not wash. Their bodies were covered with festering sores. Spotted fever developed. People were dying. Strict orders were given to the people of Archangel not to help the *special resettlers* (as the deported peasants were now called)! Dying peasants roamed the town, but no one could take a single one of them into his home, feed him, or carry tea out to him: the militia seized local inhabitants who tried to do so and took away their passports. A starving man would stagger along the street, stumble, fall—and die. But even the dead could not be picked up (besides the militia, plainclothesmen went around on the lookout for acts of kindness). At the same time market gardeners and livestock breeders from areas near big towns were also being expelled, whole villages at a time (once again—what about the theory that they were supposed to arrest exploiters only?), and the residents of Archangel themselves dreaded deportation. They were afraid even to stop and look down at a dead body.

The plight of these peasants differed from that of all previous and subsequent Soviet exiles in that they were banished not to a center of population, a place made habitable, but to the haunt of wild beasts, into the wilderness, to man's primitive condition. No, worse: even in their primeval state our forebears at least chose places near water for their settlements. For as long as mankind has existed no one has ever made his home elsewhere. But for the *special settlements* the Cheka (not the peasants themselves—they had no right of choice) chose places on stony hillsides. Three or four kilometers off there might be convenient water meadows—but no, according to instructions no one was supposed to settle there. So the hayfields were dozens of kilometers away from the settlement, and the hay had to be brought in by boat. Sometimes settlers were bluntly *forbidden to sow grain crops.* (What they should grow was also determined by the Cheka!) Yet another thing we town folk do not understand—what it means to have lived from time immemorial with animals. A peasant's life is nothing without animals —and here he was condemned for many years never to hear neighing or lowing or bleating; never to saddle, never to milk, never to fill a trough.

On the river Chulym in Siberia, the special settlement of Kuban Cossacks was encircled with barbed wire and towers were put up, as though it were a prison camp.

Everything necessary seemed to have been done to ensure that these odious work fiends should die off quickly and rid our country of themselves and of bread. Indeed, many such special settlements died off to a man. Where they once stood, chance wayfarers are gradually burning what is left of the huts, and kicking the skulls out of sight.

No Genghis Khan ever destroyed so many peasants as our glorious *Organs*, under the leadership of the Party.

Take, for instance, the Vasyugan tragedy. In 1930, 10,000 families (60,000–70,000 people, as families then went) passed through Tomsk and from there were driven farther, at first on foot, down the Tom although it was winter, then along the Ob, then upstream along the Vasyugan—still over the ice. (The inhabitants of villages on the route were ordered out afterward to pick up the bodies of adults and children.) In the upper reaches of the Vasyugan and the Tara they were marooned on patches of firm ground in the marshes. *No food or tools were left for them.* The roads were impassable, and there was no way through to the world outside, except for two brushwood paths, one toward Tobolsk and one toward the Ob. Machine-gunners manned barriers on both paths and let no one through from the death camp. They started dying like flies. Desperate people came out to the barriers begging to be let through, and were shot on the spot.

They died off—every one of them.

■

And yet—exiles survived! Under their conditions it seems incredible—but live they did.

True, when during the war there was a shortage of reckless Russian fighting power at the front, they turned among others to the "kulaks": they must surely be Russians first and kulaks second! They were invited to leave the special settlements and the camps for the front to defend their sacred fatherland.

And—they went. . . .

Not all of them, however. N. Kh——v, a "kulak's" son—whose early years I used for Tyurin, but whose subsequent biography I could not bring myself to recount—was given the chance, denied to Trotskyite and Communist prisoners, however much they yearned to go, of defending his fatherland. Without a moment's hesitation, Kh——v snapped back at the head of the Prisoner Registration and Distribution Section: "It's your fatherland—you defend it, you dung-eaters! *The proletariat has no fatherland!*"

Marx's exact words, I believe.

The things that could have been done with such people if they had been allowed to live and develop freely!!!

The Old Believers—eternally persecuted, eternal exiles—they are the ones who three centuries earlier divined the ruthlessness at the heart of Authority! In 1950 a plane was flying over the vast basin of the Podkamennaya Tunguska. The training of airmen had improved greatly since the war, and the zealous aviator spotted something that no one before him had seen in twenty years: an unknown dwelling place in the taiga. He worked out its position. He reported it. It was far out in the wilds, but to the MVD all things are possible, and half a year later they had struggled through to it. What they had found were the Yaruyevo Old Believers. When the great and longed-for Plague began —I mean collectivization—they had fled from this blessing into the depths of the taiga, a whole village of them. And they lived there without ever poking their noses out, allowing only their headman to go to Yaruyevo for salt, metal fishing and hunting gear, and bits of iron for tools. Everything else they made themselves, and in lieu of money the headman no doubt came provided with pelts. When he had completed his business he would slink away from the marketplace like a hunted criminal. In this way the Yaruyevo Old Believers had won themselves twenty years of life! Twenty years of life as free human beings among the wild beasts, instead of twenty years of kolkhoz misery. They were all wearing homespun garments and homemade knee boots, and they were all exceptionally sturdy.

Well, these despicable deserters from the kolkhoz front were now all arrested, and the charge pinned on them was . . . guess what? Links with the international bourgeoisie? Sabotage? No, Articles 58-10, on Anti-Soviet Agitation (!?!?), and 58-11, on hostile organizations. (Many of them landed later on in the Dzhezkazgan group of Steplag, which is how I know about them.)

In 1946 some other Old Believers were stormed in a forgotten monastery somewhere in the backwoods by our valiant troops, dislodged (with the help of mortars, and the skills acquired in the Fatherland War), and floated on rafts down the Yenisei. Prisoners still, and still indomitable—the same under Stalin as they had been under Peter! —they jumped from the rafts into the waters of the Yenisei, where our Tommy-gunners finished them off.

Warriors of the Soviet Army! Tirelessly consolidate your combat training!

Chapter 3

■

The Ranks of Exile Thicken

Only the peasants were deported so ferociously, to such desolate places, with such frankly murderous intent: no one had been exiled in this way before, and no one would be in the future. Yet in another sense and in its own steady way, the world of exiles grew denser and darker from year to year: more were banished, they were settled more thickly, the rules became more severe.

We could offer the following rough time scheme. In the twenties, exile was a sort of preparatory stage, a way station before imprisonment in a camp. For very few did it all end with exile; nearly all were later raked into the camps.

From the mid-thirties and especially from Beria's time, perhaps because the world of exile became so populous (think how many Leningrad alone contributed!), it acquired a completely independent significance as a totally satisfactory form of restriction and isolation. In the war and postwar years, the exile system steadily grew in capacity and importance together with the camps. It required no expenditure on the construction of huts and boundary fences, on guards and warders, and there was room in its capacious embrace for big batches, especially those including women and children. (At all major transit prisons cells were kept permanently available for women and children, and they were never empty.) Exile made possible a speedy, reliable, and irreversible cleansing of any important region in the "mainland." The exile system established itself so firmly that from 1948 it acquired yet another function of importance to the state—that of rubbish dump or drainage pool, where the waste products of the Archipelago were tipped so that they would never make their way back to the mainland. In spring, 1948, this instruction was passed down to the camps: at the end of their sentences 58's, with minor exceptions, were to be *released into exile.* In other words, they were not to be thoughtlessly unleashed on a country which did not belong to them, but each individual was to be delivered under escort from the camp guardhouse to the commandant's office in an exile colony, from fish trap to fish trap. Since the exile system embraced only certain strictly defined areas, these together constituted yet another separate (though interlocking) country between the U.S.S.R. and the Archipelago—a sort of purgatory in reverse, from which a man could cross to the Archipelago, but not to the mainland.

The years 1944–1945 brought to the exile colonies unusually heavy reinforcements from the "liberated" (occupied) territories, and 1947–1949 yet others from the Western republics. All these streams together, even without the exiled peasants, exceeded many times over the figure of 500,000 exiles which was all that Tsarist Russia, the prison house of nations, could muster in the whole course of the nineteenth century.

For what crimes was a citizen of our country in the thirties and forties punishable by exile or banishment?

The commonest crimes can easily be indicated:

1. Belonging to a criminal nationality (for this see the next chapter).
2. A previous term of imprisonment in the camps.
3. Residence in a criminal environment (seditious Leningrad, or areas in which there was a partisan movement, such as the Western Ukraine or the Baltic States).

And then many of the tributaries enumerated at the very beginning of this book branched out to feed the exile system as well as the camps, continually casting up some of their burden on the shores of exile.

We cannot go into the different types and cases of exile, because all our knowledge of it derives from casual stories or letters. If A. M. Ar——v had not written his letter, the reader would not have the following story. In 1943 news came to a village around Vyatka that one of its kolkhoz peasants, Kozhurin, a private in the infantry, had either been sent to a punitive unit or shot outright. His wife, who had six children (the oldest was ten years, the youngest six months old, and two sisters of hers, spinsters nearing fifty, also lived with her), was immediately visited by the *executants* (you already know the word, reader—

it is a euphemism for "executioners"). They gave the family no time to sell anything (their house, cow, sheep, hay, wood, were all abandoned to the pilferers), threw all nine of them with their smaller possessions onto a sledge, and took them sixty kilometers in a hard frost to the town of Vyatka (Kirov). Why they did not freeze on the way God only knows. They were kept for six weeks at the Kirov Transit Prison, then sent to a small pottery near Ukhta. The spinster sisters ate from rubbish heaps, both went mad and both died. The mother and children stayed alive only thanks to the help (the politically ignorant, unpatriotic, in fact anti-Soviet help) of the local population. The sons all served in the army when they grew up and are said to have "completed their military and political training with distinction." Their mother returned to her native village in 1960—and found not a single log, not a single brick from the stove, where her house had been.

A little cameo like this can surely be threaded on the necklace of our Great Fatherland Victory? But nobody will touch it—it isn't *typical.*

To what necklace will you add, to what category of exiles will you assign soldiers disabled in the Fatherland War, and exiled because of it? We know almost nothing about them. They were exiled to a certain northern island—exiled *because* they had consented to be mutilated in war for the glory of the Fatherland and *in order to* improve the health of a nation, which had by now won such victories in all forms of athletics and ball games. These luckless war heroes are held there on their unknown island, naturally without the right to correspond with the mainland (a very few letters break through, and this is how we know about it), and naturally on meager rations, because they cannot work hard enough to warrant generosity.

I believe they are still living out their days there.

Chapter 4

■

Nations in Exile

The business of banishment was immeasurably improved and speeded up when they drove the first *special settlers* into exile. In the year of the Great Break they designated the dekulakized as "special settlers"—and this made for much greater flexibility and efficiency; it left no grounds for appeal since it was not only kulaks who were dekulakized. Call them "special settlers," and no one can wriggle free.

Then the Great Father gave orders that this word be applied to banished nations.

Even *He* was slow to realize the value of his discovery. His first experiment was very cautious. In 1937 some tens of thousands of those suspicious Koreans were swiftly and quietly transferred from the Far East to Kazakhstan. So swiftly that they spent the first winter in mud-brick houses without windows (where would all that glass have come from!). And so quietly that nobody except the neighboring Kazakhs learned of this resettlement, no one who counted let slip a word about it, no foreign correspondent uttered a squeak.

He liked it. He remembered it. And in 1940 the same method was applied on the outskirts of Leningrad, cradle of the Revolution. But this time the banished were not taken at night and at bayonet point. Instead, it was called a "triumphal send-off" to the (newly conquered) Karelo-Finnish Republic. At high noon, with red flags flapping and brass bands braying, the Leningrad Finns and Estonians were dispatched to settle their new native soil.

These were mere trial runs. Only in July, 1941, did the time come to test the method at full power: the autonomous and of course traitor-ous republic of the Volga Germans (with its twin capitals, Engels and

Marxstadt) had to be expunged and its population hurled somewhere well to the East in a matter of days. Here for the first time the dynamic method of exiling whole peoples was applied in all its purity, and how much easier, how much more rewarding it proved to use a single criterion—that of nationality—rather than all those individual interrogations, and decrees each naming a single person. As for the Germans seized in other parts of Russia (and every last one was gathered in), local NKVD officers had no need of higher education to determine whether a man was an enemy or not. If the name's German—grab him.

The system had been proved and perfected, and henceforward would fasten its pitiless talons on any nation pointed out to it, designated and doomed as treacherous—and more adroitly every time: the Chechens; the Ingush; the Karachai; the Balkars; the Kalmyks; the Kurds; the Crimean Tatars; and finally, the Caucasian Greeks. What made the system particularly effective was that the decision taken by the Father of the Peoples was made known to a particular people not in the form of verbose legal proceedings, but by means of a military operation carried out by modern motorized infantry. Armed divisions enter the doomed people's locality by night and occupy key positions. The criminal nation wakes up and sees every settlement ringed with machine guns and automatic rifles. And they are given twelve hours to get ready whatever each of them can carry in his hands. Then each of them is made to sit cross-legged in the back of a lorry, like a prisoner (old women, mothers with babies at the breast: sit down, all of you; you heard the order!), and the lorries travel under escort to the railway station. From there prison trains take them to a new place.

Neatness and uniformity! That is the advantage of exiling whole nations at once! No special cases! No exceptions, no individual protests! They all go quietly, because . . . they're all in it together. All ages and both sexes go, and that still leaves something to be said. Those still in the womb go, too, and are exiled unborn, by the same decree. Yes, children not yet conceived go into exile, for it is their lot to be conceived under the high hand of the same decree; and from the very day of their birth, whatever that obsolete and tiresome Article 35 of the Criminal Code may say ("Sentence of exile cannot be passed on persons under 16 years of age"), from the moment they thrust their heads out into the light they will be special settlers, exiles in perpetuity. Their coming of age, their sixteenth birthdays, will be marked only by the first of their regular outings to report at the MVD post.

All that the exiles have left behind them—their houses, wide open and still warm, their belongings lying in disorder, the home put together

and improved by ten or even twenty generations—passes without differ-
entiation to the agents of the punitive organs, then some of it to the
state, some to neighbors belonging to more fortunate nations, and
nobody will write to complain about the loss of a cow, a piece of
furniture, or some crockery.

The only crack in the principle of uniformity was made by mixed
marriages (not for nothing has our socialist state always been against
them). When the Germans, and later the Greeks, were exiled, spouses
belonging to other nationalities were not sent with them. But this
caused a great deal of confusion, and left foci of infection in places
supposedly sterilized. (Like those old Greek women who came home
to their children to die.)

Where were the exiled nations sent? Kazakhstan was much favored—and there, together with the ordinary exiles, they formed more
than half the republic's population, so that it could aptly be called Ka-
zek-stan. But Central Asia, Siberia (where very many Kalmyks perished along the Yenisei), the Northern Urals, and the Northern European areas of the U.S.S.R. all received their fair share.

For every nation exiled, an epic will someday be written—on its
separation from its native land, and its destruction in Siberia. Only the
nations themselves can voice their feelings about all they have lived
through: we have no words to speak for them, and we must not get
under their feet.

The tedium of it all! Nothing but the same thing over and over
again. At the beginning of this Part VI we appeared to be discussing
something new: not the camps, but the exile system. And this chapter
made a fresh start: our theme was no longer the administrative exiles,
but the special settlers.

Yet we are back where we started.

Must we—and if so, how often must we repeat ourselves again and
again and again—tell the story of other, and different, exile colonies?
In other places? At other periods? Peopled by other exiled nations?

And if so, which? . . .

Chapter 5

■

End of Sentence

In eight years of prison and prison camp I had never heard anyone who had experienced exile say a good word about it. But from his first days in jail under investigation and in transit, simply because the six flat stone surfaces of a cell press in on him too closely, the dream of exile burns like a secret light in the prisoner's mind, a flickering iridescent mirage, and the wasted breasts of prisoners on their dark bunks heave in sighs of longing: "If only they would sentence me to exile!"

I did not escape the common lot; far from it—the dream of exile had me more powerfully than most in its grip. I even sent a naïve appeal to the Supreme Soviet: for commutation of my eight years in the camps to exile for life, in however remote and wild a place. The elephant did not even sneeze in reply. (I had not yet realized that lifelong exile would always be waiting for me, but that it would come *after*, not *instead of*, the camp.)

In 1952 a dozen prisoners were "released" from the 3,000-strong "Russian" Camp Division at Ekibastuz. It looked very strange at the time: 58's, let out through the gates! Ekibastuz had been in existence for three years by then, and not a single man had been released, nor had anyone reached the end of his sentence. Evidently, for the few who had lived to see the day, the first wartime *tenners* had just ended.

We impatiently awaited letters from them. A few came, directly or indirectly. And we learned that nearly all of them had been taken from the camp to places of exile, although their sentences had not included exile. But this surprised no one. It was clear to our jailers and to us that justice, length of sentence, formal documentation, had nothing to do with it; the point was that once we had been declared enemies, the state would ever after assert the right of the stronger and trample us, crush us, squash us, until the day we died. And we were so used to it, it had become so much part of us, that no other state of affairs would have seemed normal either to the regime or to us.

In Stalin's last years it was not the fate of the exiles that caused alarm, but that of the nominally liberated, those who to all appearances were now safely beyond the gates, and unguarded, those from whom the tutelary gray wing of the MVD had apparently been withdrawn. Exile, which the powers that be obtusely regarded as an additional punishment, was a prolongation of the prisoner's irresponsible existence, the fatalistic routine in which he feels so secure. Exile relieved us of the need to choose a place of residence for ourselves, and so from troublesome uncertainties and errors. No place would have been right, except that to which they had sent us. This was the one and only place in the whole Soviet Union where no one could reproach us as intruders.

As we left the camp under guard we were still careful to respect the final prison superstitions: on no account must you look back at your last prison (or else you will return), and you must do the right thing with your spoon. (What was the right thing, though? Some said take it with you, or you would return for it; others said fling it at the prison, or else the prison would pursue you. I had molded my spoon myself in the foundry, and I took it with me.)

They took us to the jail—and the jail admitted us without the usual body search and bath. The accursed walls were losing some of their harshness! In the morning the block superintendent unlocked the door and said almost in a whisper, "Come out and bring all your belongings."

The devil was unclenching his claws. . . .

We stepped out into the arms of a red spring morning. The dawn light was warming the brick walls of the jail. A lorry was waiting for us in the middle of the yard, with two zeks who were joining our party already sitting in the back. This was the time to breathe deeply, to look around, to steep ourselves in the uniqueness of that moment—but we simply could not waste the chance to strike up an acquaintance. . . .

In the morning they sent up a lorry, and the same escort, after a night out of barracks, came to fetch us. Sixty kilometers farther into the steppe. We got stuck in muddy hollows, and jumped down from the lorry (something we could not have done as zeks) to heave and push it out of the mire, to get the eventful journey over and arrive in perpetual exile more quickly. The escort troops stood in a half-circle and kept guard over us.

The steppe sped by, kilometer after kilometer. To right and to left, as far as the eye could see, there was nothing but harsh gray inedible grass, and only very occasionally a wretched Kazakh village framed with trees. At length the tops of a few poplars (Kok-Terek means "green poplar") appeared ahead of us, over the curve of the steppe.

We had arrived! The lorry sped between Chechen and Kazakh adobe huts, raising a cloud of dust and drawing a pack of indignant dogs in pursuit. Amiable donkeys with little carts made way for us, and from one yard a camel turned slowly and contemptuously to look at us. There were people, too, but we had eyes only for the women—those unfamiliar, forgotten creatures: look at that pretty dark girl in the doorway, shading her eyes with her hand to watch our lorry pass; look at those three walking together in flowery red dresses. Not one of them Russian. "This is all right—we shall find wives for ourselves yet!"

Directly across the street from the MGB stood an amazing building, one story, yet quite high; four Doric columns solemnly upheld a false portico, at the foot of the columns were two steps faced with smooth stone, and over all this—there was a blackened straw roof. My heart could not help beating faster. It was a school! A ten-year school. Stop pounding, be quiet, you nuisance. That building is nothing to you.

Crossing the main street to the magic gate goes a girl with waved hair, neatly dressed in a little wasp-waisted jacket. Surely she is walking on air? She is a *teacher!* She is too young to have graduated from an institute; she must have attended a seven-year school and then a teachers' training college. How I envy her! What a gulf there is between her and a common laborer like me. We belong to different estates, and I would never dare to walk arm in arm with her.

The Commandants were easygoing and allowed us exiles to spend the night not in a locked room but out in the yard, on hay.

A night under the open sky! We had forgotten what it was like. . . . There had always been locks, and bars, always walls and ceilings.

It was only the third of March, but there was not the slightest chill in the night air; it was still almost summery, as it had been in the daytime. Again and again the braying of donkeys rose over the sprawling town of Kok-Terek, long-drawn-out and passionate, telling the she-asses of their love, of the ungovernable strength flooding their bodies. Some of the braying was probably the she-asses answering.

I cannot sleep! I walk and walk and walk in the moonlight. The donkeys sing their song. The camels sing. Every fiber in me sings: I am free! I am free!

In the end I lie down beside my comrades, on some hay under the open-sided shelter. Two steps away from us, horses stand at their mangers peacefully champing hay all night long. Surely there could be no sweeter, no more friendly sound on this our first night of freedom.

Champ away, you mild, inoffensive creatures!

Next day we were allowed to move into private lodgings. I found myself a henhouse to suit my pocket, with a single bleary window and such a low roof that even where it was highest, in the middle, I could not stand upright. "Give me a low-roofed cottage," I once wrote in prison, dreaming of exile. It was not very pleasant, all the same, not being able to raise my head. Still, it was a little house of my own! The floor was earthen. I put my padded camp vest on it, and there was my bed! I had no oil lamp as yet—I had nothing!! an exile must select and buy every single thing he needs, as though he has just landed on this earth—but I did not feel the want of it. All those years, in our cells and our huts, the state's electricity had seared our souls, and now darkness was bliss. Even darkness can be an element of freedom!

What more could I desire? . . .

But the morning of March 6 surpassed anything that I could have wished for! Chadova, my elderly landlady, an exile from Novgorod, whispered, because she dared not say it out loud: "Go and listen to the radio. I'm afraid to repeat what I've just heard."

Something told me to do as she said: I went over to the central square. A crowd of perhaps two hundred people—a lot for Kok-Terek —huddled around the post under the loudspeaker and the sullen sky. Before I could make out what the announcer was saying (he spoke with a histrionic catch in his voice), understanding dawned on me.

This was the moment my friends and I had looked forward to even in our student days. The moment for which every zek in Gulag (except the orthodox Communists) had prayed! He's dead, the Asiatic dictator is dead! The villain has curled up and died!

I could have howled with joy there by the loudspeaker; I could even have danced a wild jig! But alas, the rivers of history flow slowly. My face, trained to meet all occasions, assumed a frown of mournful attention. For the present I must pretend, go on pretending as before.

All the same, my exile had begun with magnificent auguries!

Chapter 6

■

The Good Life in Exile

This autobiographical chapter records Solzhenitsyn's experiences during his time of internal exile in Kazakhstan: his school teaching, his writing, his thoughts at that time.

Chapter 7

■

Zeks at Liberty

We have had a chapter in this book on "Arrest." Do we need one now called "Release"?

Of those on whom the thunderbolt of arrest at one time or another fell (I shall speak only of 58's), I doubt whether a fifth, I should like to think that an eighth lived to experience this "release."

And anyway, release is surely something everybody understands. It has been described so often in world literature, shown in so many films: unlock my dungeon—out into the sunshine—the crowd goes wild —open-armed relatives.

But there is a curse on those "released" under the joyless sky of the Archipelago, and as they move into freedom the clouds will grow darker.

Only in its long-windedness, its leisureliness, its otiose flourishes (what need has the law to hurry now?), does release differ from the lightning stroke of arrest. In all other respects, release is arrest all over again, the same sort of punishing transition from state to state, shattering your breast, the structure of your life and your ideas, and promising nothing in return.

Because in this country, whenever someone is released, somewhere an arrest must follow.

The space between two arrests—that is what release meant throughout the forty pre-Khrushchev years.

A life belt thrown between two islands—splash your way from camp to camp! . . . The walk from one camp boundary to the next— that's what is meant by release.

You will not be given a residence permit in a town, even a small one, nor will you ever get a decent job. In the camp at least you received your rations, but here you do not.

Moreover, your freedom of movement is illusory. . . .

Not "released," but "deprived of exile" would be the best description of these unfortunates. Denied the blessings of an exile decreed by fate, they cannot force themselves to go into the Krasnoyarsk taiga or the Kazakh desert, where there are so many of their own kind, so many *ex*es, all around. No, they plunge deep into the tormented world of *freedom*, where everyone recoils from them, and where they are marked men, candidates for a new spell inside.

It's a vicious circle: no job without a residence permit, no residence permit unless you have a job. And without a job you have no bread card either. Former zeks did not know the rule that the MVD is required to find them work. And those who did know were afraid to apply in case they were *put back inside*. . . .

You may be free, but your troubles are only beginning.

But on the Kolyma there was really not much choice: they *hung on to* people. The discharged zek immediately signed a *voluntary* undertaking to go on working for Dalstroi. (Permission to leave for the mainland was even harder to obtain than your discharge.)

■

Just as a common illness develops differently in different people, the effects of freedom upon us varied greatly.

Its physical effects, to begin with . . . Some had overstrained themselves in the fight to end their time in the camps alive. They had endured it all like men of steel, consuming for ten whole years a fraction of what the body requires; working and slaving; breaking stones half-naked in freezing weather—and never catching cold. But once their sentence was served, once the inhuman pressure from outside was lifted, the tension inside them also slackened. Such people are destroyed by a sudden drop in pressure. The giant Chulpenyov, who had never caught cold in seven years as a lumberjack, contracted a variety of illnesses once he was freed.

There used to be a saying: The hard times brace you, and the soft times drive you to drink. Sometimes a man's teeth would all fall out in a year. Sometimes he would grow old overnight. Another man's strength would give out as soon as he got home, and he would die burned out.

Yet there were others who took heart when they were released. For them, it was time to grow younger and spread their wings. It comes as a sudden revelation: life after all is so *easy* when you're free! There, on the Archipelago, the force of gravity is quite different, your legs are as heavy as an elephant's, but here they move as nimbly as a sparrow's. All the problems which tease and torment men who have always been free we solve with a single click of the tongue. We have our own cheerful standards: "Things have been worse!" Things used to be worse —so now everything is quite easy. We never get tired of repeating it: Things have been worse! Things have been worse!

But the pattern of a man's future may be even more firmly drawn by the emotional crisis which he undergoes at the moment of release. This crisis can take very different forms. Only on the threshold of the guardhouse do you begin to feel that what you are leaving behind you is both your prison and your homeland. This was your spiritual birth-place, and a secret part of your soul will remain here forever—while your feet trudge on into the dumb and unwelcoming expanse of *freedom*.

The camps bring out a man's character—but so does release! This is how Vera Alekseyevna Korneyeva, whom we have met before in our story, took leave of a Special Camp in 1951. "The five-meter gates closed behind me, and although I could hardly believe it myself, I was weeping as I walked out to freedom. Weeping for what? . . . I felt as though I had torn my heart away from what was dearest and most precious to it, from my comrades in misfortune. The gates closed—and it was all finished. I should never see those people again, never receive any news from them. *It was as though I had passed on to the next world. . . .*"

To the next world! . . . Release as a form of death. Perhaps we had not been released? Perhaps we had died, to begin a completely new life beyond the grave? A somewhat ghostly existence, in which we cautiously felt the objects about us, trying to identify them.

It was as though our freedom was stolen, not authentic. Those who felt like this seized their scrap of stolen freedom and ran with it to some lonely place. "While in the camp almost all my closest comrades thought, as I did, that if ever God allowed us to leave the camp alive, we would not live in towns, or even in villages, but somewhere in the depths of the forest. We would find work as foresters, rangers, or failing that, as herdsmen, and stay as far away as we could from people, politics, and all the snares and delusions of the world." (V. V. Pospelov) For some time after he was discharged Avenir Borisov shunned other people and took refuge in the countryside. "I felt like hugging and kissing every birch tree, every poplar. The rustle of fallen leaves (I was released in autumn) was like music to me, and tears came to my eyes. I didn't give a damn that I only got 500 grams of bread—I could listen to the silence for hours on end, and read books, too. Any sort of work seemed easy and simple now that I was free, the days flew by like hours, my thirst for life was unquenchable. If there is any happiness in the world at all, it is certainly that which comes to any zek in the first year of his life as a free man!"

It is a long time before people like this want to *own* anything: they remember that property is easily lost, vanishes into thin air. They have an almost superstitious aversion to new things, go on wearing the same old clothes, sitting on the same old broken chairs. One friend of mine had furniture so rickety that there was nothing you could safely sit or lean on. They made a joke of it. "This is the way to live—between camps." (His wife had also been inside.)

But people vary. And many crossed the line to freedom with quite different feelings (especially in the days when the Cheka-KGB seemed to be closing its eyes a little). Hurrah! I'm free! One thing I solemnly swear: Never to land inside again! Now I'm going all out to make up for what I've missed.

For two centuries Europe has been prating about equality—but how very different we all are! How unlike are the furrows life leaves on our souls. We can forget nothing in eleven years—or forget everything the day after. . . .

Each year on the anniversary of my arrest I organize myself a "zek's day": in the morning I cut off 650 grams of bread, put two lumps of sugar in a cup and pour hot water on them. For lunch I ask them to make me some broth and a ladleful of thin mush. And how quickly I get back to my old form: by the end of the day I am already picking up crumbs to put in my mouth, and licking the bowl. The old sensations start up vividly.

I had also brought out with me, and still keep, my number patches. Am I the only one? In some homes they will be shown to you like holy relics.

Associations of former zeks gather once a year, varying the place from time to time, to drink and reminisce. "And strangely enough," says V. P. Golitsyn, "the pictures of the past conjured up are by no means all dark and harrowing; we have many warm and pleasant memories."

Another normal human characteristic. And not the worst.

"My identification number in camp began with *yery*," V. L. Ginzburg rapturously informs me. "And the passport they issued to me was in the 'Zk' series!"

You read it—and feel a warm glow. No, honestly—however many letters you receive, those from zeks stand out unmistakably. Such extraordinary toughness they show! Such clarity of purpose combined with such vigor and determination! In our day, if you get a letter completely free from self-pity, genuinely optimistic—it can only be from a former zek. They are used to the worst the world can do, and nothing can depress them.

I am proud to belong to this mighty race! We were not a race, but they made us one! They forged bonds between us, which we, in our timid and uncertain twilight, where every man is afraid of every other, could never have forged for ourselves. The orthodox and the stoolies automatically removed themselves from our midst when we were freed. We need no explicit agreement to support each other. We no longer need to test each other. We meet, look into each other's eyes, exchange a couple of words—and what need for further explanation? We are ready to help each other out. Our kind has friends everywhere. And there are millions of us!

■

Freedom has something else in store for former convicts—reunion with family and friends. Reunion of fathers with sons. Of husbands with wives. And it is not often that good comes out of these reunions. In the ten or fifteen years lived apart from us, how could our sons grow in harmony with us: sometimes they are simply strangers, sometimes they are enemies. Nor are women who wait faithfully for their husbands often rewarded: they have lived so long apart, long enough for a person to change completely, so that only his name is the same. His experience and hers are too different—and it is no longer possible for them to come together again.

This is a subject which others can make into films and novels, but there is no room for it in this book.

PART VII

Stalin Is No More

■

"Neither repented they of their murders . . ."

<small>REVELATION 9:21</small>

Chapter 1

■

Looking Back on It All

We never, of course, lost hope that our story *would* be told: since sooner or later the truth is told about all that has happened in history. But in our imagining this would come in the rather distant future—after most of us were dead. And in a completely changed situation. I thought of myself as the chronicler of the Archipelago, I wrote and wrote, but I, too, had little hope of seeing it in print in my lifetime.

History is forever springing surprises even on the most perspicacious of us. We could not foresee what it would be like: how for no visible compelling reason the earth would shudder and give, how the gates of the abyss would briefly, grudgingly part so that two or three birds of truth would fly out before they slammed to, to stay shut for a long time to come.

So many of my predecessors had not been able to finish writing, or to preserve what they had written, or to crawl or scramble to safety —but I had this good fortune: to thrust the first handful of truth through the open jaws of the iron gates before they slammed shut again.

Like matter enveloped by antimatter, it exploded instantaneously!

Its explosion touched off in turn an explosion of letters—that was to be expected. But also an explosion of newspaper articles—written with gritted teeth, with ill-concealed hatred and resentment: an explosion of official praise that left a sour taste in my mouth.

When former zeks heard this fanfare from all the newspapers in unison, learned that some sort of story about the camps had come out and that the journalists were slavering over it, their unanimous conclusion was: "More lying nonsense! Nothing's safe from those crafty liars!" That our newspapers, with their habitual immoderation, might suddenly start falling over each other to praise the truth was something no one could possibly imagine! Some of them were reluctant to risk soiling their hands on my story [of Ivan Denisovich].

But when they started reading it, a single groan broke from all those thousands—a groan of joy and of pain. Letters poured in.

I treasure those letters. Only too rarely do our fellow countrymen have a chance to speak their mind on matters of public concern—and former prisoners still more rarely. Their faith had proved false, their hopes had been cheated so often—yet now they believed that the era of truth was really beginning, that at last it was possible to speak and write boldly!

And they were disappointed, of course, for the hundredth time. . . .

Truth, it seems, is always bashful, easily reduced to silence by the too blatant encroachment of falsehood.

The prolonged absence of any free exchange of information within a country opens up a gulf of incomprehension between whole groups of the population, between millions and millions.

We simply *cease to be a single people,* for we speak, indeed, different languages.

■

Nonetheless, a breakthrough had been made! Oh, it was stout, the wall of lies, it looked so secure, looked built to last forever—but a breach yawned, and news broke through. Only yesterday we had had no camps, no Archipelago, and today there they were for the whole people, the whole world to see—prison camps! Camps, what is more, of the Fascist type!

When Khrushchev, wiping the tear from his eye, gave permission for the publication of *Ivan Denisovich,* he was quite sure that it was about Stalin's camps, and that he had none of his own.

I myself was taken by surprise when I received a stream of letters —from *present-day* zeks. On crumpled scraps of paper, in a blurred pencil scrawl, in stray envelopes often addressed and posted by free employees, in other words, *on the sly,* today's Archipelago sent me its criticisms, and sometimes its angry protests.

These letters, too, were a single many-throated cry. But a cry that said: *"What about us!!??"*

And the zeks set up a howl: What do you mean, never happen again? We're *here inside now,* and our conditions are just the same!

"Nothing has changed since Ivan Denisovich's time"—the message was the same in letters from many different places.

"Any zek who reads your book will feel bitterness and disgust because everything is just as it was."

"What has changed, if all the laws providing for twenty-five years' imprisonment issued under Stalin are still in force?"

After reading all these letters, I who had been thinking myself a hero saw that I hadn't a leg to stand on: in ten years I had lost my vital link with the Archipelago.

For *them,* for *today*'s zeks, my book is no book, my truth is no truth unless there is a continuation, unless I go on to speak of them, too. Truth must be told—and things must change!

■

In a declaration by the Soviet government dated December, 1964, we read: "The perpetrators of monstrous crimes must never and in no circumstances escape just retribution. . . . The crimes of the Fascist murderers, who aimed at the destruction of whole peoples, have no precedent in history."

This was to prevent the Federal German Republic from introducing a statute of limitations for war criminals after twenty years had elapsed.

But they show no desire to face judgment *themselves,* although they, too, "aimed at the destruction of whole peoples."

But in the U.S.S.R. no one would have to *answer.* No one would be *looked into.*

While in the records office they carry out a leisurely inspection and destroy all unwanted documents: lists of people shot, orders committing prisoners to solitary confinement or the Disciplinary Barracks, files on investigations in the camps, denunciations from stoolies, superfluous information about practical workers and convoy guards.

Chapter 2

■

Rulers Change, the Archipelago Remains

The Special Camps must have been among the best-loved brain-children of Stalin's old age. After so many experiments in punishment and re-education, this ripe perfection was finally born: a compact, faceless organization of numbers, not people, psychologically divorced from the Motherland that bore it, having an entrance but no exit, devouring only enemies and producing only industrial goods and corpses. It is difficult to imagine the paternal pain which the Visionary Architect would have felt if he had witnessed in turn the bankruptcy of this great system of his. While he yet lived it was shaken, it was giving off sparks, it was covered with cracks—but probably caution prevailed and these things were not reported to him. If the Great Coryphaeus had lived a year or eighteen months longer, it would have been impossible to conceal these explosions from him, and his weary senile brain would have been burdened with a new decision: either to abandon his pet scheme and mix the camps again, or, on the contrary, to crown it by systematically shooting all the index-lettered thousands.

But, amid weeping and wailing, the Thinker died, and soon afterward his frozen hand brought crashing down his still rosy-cheeked, still hale and vigorous comrade in arms—the Minister of those extraordinarily extensive, intricate, and irresolvable Internal Affairs.

The fall of the Archipelago's Boss tragically accelerated the breakdown of the Special Camps. Number patches—the supreme discovery of twentieth-century prison-camp science—were hurriedly ripped off, thrown away, and forgotten! This alone was enough to rob the Special

Camps of their austere uniformity. It hardly mattered, when the bars had also been removed from hut windows and locks from hut doors, so that the Special Camps had lost the pleasant jail-like peculiarities which distinguished them from Corrective Labor Camps. They also lifted restrictions on correspondence, which more than anything had made prisoners in Special Camps really feel buried alive. They even allowed visits—dreadful thought! Visits! . . . The tide of liberalism swept on so irresistibly over the erstwhile Special Camps that prisoners were allowed to choose their own hair styles. Instead of credit accounts, instead of Special Camp coupons, the natives were allowed to handle ordinary Soviet currency and settle their bills with cash like people outside.

Carelessly, recklessly they demolished the system which had fed them—the system which they had spent decades weaving and binding and lashing together.

And were those hardened criminals at all mollified by this pampering? They were not! On the contrary! They showed their depravity and ingratitude by adopting the profoundly inappropriate, offensive, and nonsensical word "Beria-ites"—and now whenever something upset them they would yell this insult at conscientious convoy guards, long-suffering warders, and their solicitous guardians, the camp chiefs. Not only did the word pain the tenderhearted practical workers, it could even be dangerous so soon after Beria's fall, because someone might make it the starting point of an accusation.

For this reason the head of one of the Kengir Camp Divisions was compelled to deliver the following address from the platform: "Men!" (In those few short years, from 1954 to 1956, they found it possible to call the prisoners "men.") "You hurt the feelings of the supervisory staff and the convoy troops by shouting 'Beria-ites' at them! Please stop it." To which the diminutive V. G. Vlasov replied: "Your feelings have been hurt in the last few months. But I've heard nothing but 'Fascist' from your guards for eighteen years. Do you think we have no feelings?" And so the major promised to cut out the abusive word "Fascist." A fair trade.

After all these pernicious and destructive reforms we may consider the separate history of the Special Camps concluded in 1954, and need no longer distinguish them from Corrective Labor Camps.

Throughout the topsy-turvy Archipelago easier times set in from 1954 and lasted till 1956—an era of unprecedented indulgences. . . .

■

Historians attracted to the ten-year reign of Nikita Khrushchev—when certain physical laws to which we had grown accustomed suddenly seemed to stop operating, when objects miraculously began defying the forces in the electromagnetic field, defying the pull of gravity—will inevitably be astounded to see how many opportunities were briefly concentrated in those hands, and how playfully, how frivolously they were used before they were nonchalantly tossed aside. Endowed with greater power than anyone in our history except Stalin, a power which though impaired was still enormous, he used it like Krylov's Mishka in the forest clearing, rolling his log first this way, then that, and all to no purpose. He was given the chance to draw the lines of freedom three times, five times more firmly, and he failed to understand his duty, abandoned it as though it were a game—for space, for maize, for rockets in Cuba, for Berlin ultimatums, for persecution of the church, for the splitting of Oblast Party Committees, for the battle with abstract art.

He never carried anything through to its conclusion—least of all the fight for freedom! Stir him up against the intelligentsia? Nothing could have been simpler. Use his hands, the hands that wrecked Stalin's camps, to reinforce the camps now? That was easily achieved! And just think *when!*

In 1956, the year of the Twentieth Congress, the first orders limiting relaxation of the camp regime were promulgated! They were extended in 1957—the year when Khrushchev achieved undivided power.

But the caste of practical workers was still not satisfied. Scenting victory, they went over to the offensive. We can't go on like this! The camp system is the main prop of the Soviet regime and it is collapsing!

Yielding to this pressure, without examining anything closely, without pausing to reflect that crime had not increased in those last five years (or that if it had, the causes must be sought in the political system), without considering how these new measures could be squared with his faith in the triumphal advance of Communism, or attempting to study the matter in detail, or even to look at it with his own eyes—this Tsar who had spent "all his life on the road" light-heartedly signed the order for nails to knock the scaffold together again quickly, in its old shape and as sturdy as ever.

And all this happened in the very year—1961—when Nikita made his last, expiring effort to tug the cart of freedom up into the clouds. It was in 1961—the year of the Twenty-second Congress—that a decree was promulgated on the death penalty in the camps for "terrorist acts against reformed prisoners [in other words, stoolies] and against supervisory staff" (something which had never happened), and the plenum of the Supreme Court confirmed (in June, 1961) *regulations for four disciplinary categories in camps*—Khrushchev's camps now, not Stalin's.

When he climbed onto the Congress platform for another attack on Stalin's tyranny, Nikita had only just allowed the screws of his very own system to be turned no less tight. And he sincerely believed that all this could be fitted together and made consistent!

The camps today are as approved by the Party before the Twenty-second Congress. Six years later they are just as they were then.

They differ from Stalin's camps not in regime, but in the composition of their population: there are no longer millions and millions of 58's. But there are still millions inside, and just as before, many of them are helpless victims of perverted justice: swept in simply to keep the system operating and well fed.

Rulers change, the Archipelago remains.

It remains because *that particular* political regime could not survive without it. If it disbanded the Archipelago, it would cease to exist itself.

■

Every story must have an end. It must be broken off somewhere. To the best of our modest and inadequate ability we have followed the history of the Archipelago from the crimson volleys which greeted its birth to the pink mists of rehabilitation. In the glorious period of leniency and disarray on the eve of Khrushchev's measures to make the camps harsher again, on the eve of a new Criminal Code, let us consider our story ended. Other historians will appear—historians who to their sorrow know the Khrushchev and post-Khrushchev camps better than we do.

Two have in fact appeared already: S. Karavansky and Anatoly Marchenko. And they will float to the surface in great numbers.

Chapter 3

■

The Law Today

The reader has seen throughout this book that from the very beginning of the Stalin age there have been no *politicals* in our country. The crowds, the millions driven past while you watched, all those millions of 58's, were merely common *criminals.*

Besides, merry, mouthy Nikita Sergeyevich took so many bows from so many platforms: *Politicals?* Not a one!! We just don't have them!

And as grief grew forgetful, as distance softened craggy contours, as fat formed under the skin—we almost believed it! Even former zeks did. Millions of zeks were released for all to see—so perhaps there really were no politicals left? We had returned, others joined us, our friends and families were back. The gaps in our little world of urban intellectuals seemed to be filled, the ring closed. You could sleep undisturbed, and no one would have been taken from the house when you awoke. Friends would telephone—no one was missing. Not that we altogether believed it—but for practical purposes we accepted that there were no longer any politicals in jail.

And Nikita was there, glued to his platform. "There can be no return to deeds and occurrences such as these, either in the Party or in the country generally" (May 22, 1959—that was before Novocherkassk). "Now everyone in our country can breathe freely . . . with no need to worry about the present or the future" (March 8, 1963, *after* Novocherkassk).

Novocherkassk! A town of fateful significance in Russia's history. As though the Civil War had not left scars enough, it thrust itself beneath the saber yet again.

Novocherkassk! A whole town rebels—and every trace is licked clean and hidden. Even under Khrushchev the fog of universal ignorance remained so thick that no one abroad got to know about Novocherkassk, there were no Western broadcasts to inform us of it, and even local rumor was stamped out before it could spread, so that the majority of our fellow citizens do not know what event is associated with the name Novocherkassk and the date June 2, 1962.

Let me then put down here all that I have been able to gather.

We can say without exaggeration that this was a turning point in the modern history of Russia. If we leave out the Ivanovo weavers at the beginning of the thirties (theirs was a large-scale strike, but it ended without violence), the flare-up at Novocherkassk was the first time the people had spoken out in forty-one years (since Kronstadt and Tambov): unorganized, leaderless, unpremeditated, it was a cry from the soul of a people who could no longer live as they had lived.

On Friday, June 1, one of those carefully considered enactments of which Khrushchev was so fond was published throughout the Union —raising the prices of meat and butter. On that very same day, as demanded by another and quite separate economic plan, piece rates at the huge Electric Locomotive Works in Novocherkassk (NEVZ) were lowered, in some cases by 30 percent. That morning the workers in two shops (the forge and the foundry), usually obedient creatures of habit, geared to their jobs, could not force themselves to work—so hot had things become for them. Their loud, excited discussions developed into a spontaneous mass meeting. An everyday event in the West, an extraordinary one for us.

By noon the strike had spread throughout the enormous locomotive works. (Runners were sent to other factories, where the workers wavered but did not come out in support.) The Moscow-Rostov railway line runs close to the works. Either to make sure that the news would reach Moscow more quickly, or to prevent troops and tanks from moving in, a large number of women sat down on the tracks to hold up trains, whereupon the men began pulling up the rails and building barriers. Strike action of such boldness is unusual in the history of the Russian workers' movement. Slogans appeared on the works building: "Down with Khrushchev!" "Use Khrushchev for sausage meat!"

While all this was happening, troops and police began converging on the works. Tanks took up position on the bridge over the Tuzlov. From evening until the following morning, movement inside the city or across the bridge was completely forbidden. Even during the night the workers' settlement did not quiet down for a moment. Overnight about thirty workers were arrested as "ringleaders" and carried off to the city police station.

On the morning of June 2, some other enterprises in the town struck (but by no means all of them). Another spontaneous mass meeting at NEVZ decided on a protest march into the town to demand the release of the arrested workers. The procession (only about three hundred strong to begin with—you had to be brave!), with women and children in its ranks, carrying portraits of Lenin and peaceful slogans, marched over the bridge past the tanks without obstruction, then uphill into the town. Here their numbers were quickly swelled by curious onlookers, individual workers from other enterprises, and little boys. At several places in the city people stopped lorries and used them as platforms for speech-making. The whole town was seething. The NEVZ demonstrators marched along the main street (Moskovskaya) and some of them began trying to break down the locked doors of the town police station in the belief that their arrested comrades were inside. They were met with pistol shots. All the streets were choked with people and here, on the square, the crowd was densest.

The Party offices were found to be empty—the city authorities had fled to Rostov.

It was about 11 A.M. There were no police to be seen in the town, but there were more and more troops. (A revealing picture: at the first slight shock the civil authorities hid behind the army.) Soldiers had occupied the post office, the radio station, the bank. By this time the whole of Novocherkassk was beleaguered, and every entry and exit barred. Tanks crawled slowly along Moskovskaya Street, following the route the demonstrators had taken toward Party headquarters. Boys started scrambling onto the tanks and obstructing the observation slits. The tanks fired a few blank shells, rattling the windows of shops and houses all along the street. The boys scattered and the tanks crawled on.

And the students? Novocherkassk is of course a town of students! Where were they all? . . . The students of some institutes, including the Polytechnic, and of some technical secondary schools, had been *locked in* their dormitories or in other school buildings from early morning. Their rectors had thought quickly. But we may as well say it: the students for their part showed little civic courage. They were presumably glad of this excuse to do nothing. It would take more than the turn of a key to hold back rebel students in the West today (and took more in Russia in days gone by).

A scuffle broke out inside the Party building, step by step the speakers were dragged back inside and soldiers emerged onto the balcony, more and more of them. A file of riflemen began forcing the crowd back from the small square immediately before the palace toward the railings of the garden. (Several witnesses say unanimously that *these* soldiers were all non-Russians—Caucasians brought in from the other end of the oblast to replace the cordon from the local garrison previously posted there. But not all witnesses agree that the previous cordon had been ordered to open fire, and that the order was not carried out because the captain who received it killed himself in front of his men rather than pass it on. That an officer committed suicide is beyond doubt, but accounts of the circumstances are vague and no one knows the name of this hero of conscience.) It is not known who gave the order, but *these* soldiers raised their rifles and fired a first volley over the heads of the crowd.

The burst fired over the heads of the crowd found the trees in the little garden and the boys who had climbed into them, some of whom fell to the ground. The crowd, it seems, gave a roar, whereupon the soldiers, whether at a command, or because they saw red, or in panic, started firing freely into the crowd, and—yes—with dumdum bullets. (Remember Kengir? The sixteen at the guardhouse?) The crowd fled in panic, jamming the narrow paths around the garden, but the troops *went on firing at their backs as they retreated.* Information from a variety of sources is more or less unanimous that some seventy or eighty people were killed. The soldiers looked around for lorries and buses, commandeered them, loaded them with the dead and the wounded, and dispatched them to the high-walled military hospital. (For a day or two afterward these buses went around with bloodstained seats.)

That day, just as in Kengir, movie cameras took pictures of the rebels on the streets.

The firing ceased, the terror passed, the crowd poured back onto the square, and *was fired upon again.*

All this happened between noon and 1 P.M.

This is what an observant witness saw at 2 P.M.: "There are about eight tanks of different types standing on the square in front of Party headquarters. A cordon of soldiers stands before them. The square is almost deserted, there are only small groups of people, mostly youngsters, standing about and shouting at the soldiers. On the square puddles of blood have formed in the depressions in the pavement. I am not exaggerating; I never suspected till now that there could be so much blood. The benches in the public garden are spattered with blood, there are bloodstains on its sanded paths and on the whitewashed tree trunks in the public garden. The whole square is scored with tank tracks. A red flag, which the demonstrators had been carrying, is propped against the wall of Party headquarters, and a gray cap splashed with red-brown blood has been slung over the top of its pole. Across the façade of the Party building hangs a red banner, there for some time past: 'The People and the Party are one.'

"People go up to the soldiers, to curse them or to appeal to their conscience. 'How could you do it?' 'Who did you think you were shooting at?' 'Your own people you were shooting at!' They make excuses: 'It wasn't us! We've only just been brought in and posted here. We had nothing to do with it.'

"That's how efficient our murderers are (and yet people talk about bureaucratic sluggishness). He knows his business, that General Pliev. . . ."

Toward five or six o'clock the square gradually filled with people again. (They *were* brave, the people of Novocherkassk! The town radio kept appealing to them: "Citizens, do not fall for provocation, go home quietly!" The riflemen still stood there, the blood had not been mopped up, and again they pressed forward.) Shouts from the crowd, more and more people, and another impromptu meeting. They knew by now that six senior members of the Central Committee had flown in (probably arriving before the first shootings?), among them, needless to say, Mikoyan (the expert on Budapest-type situations), Suslov, and Frol Kozlov. A delegation of younger workers from NEVZ was sent to tell them what had happened. A buzz went through the crowd: "Let Mikoyan come down here! Let him see all this blood for himself!" Mikoyan wouldn't come down, thank you. But a reconnaissance heli-

General Pliev

copter flew low over the square around six o'clock. Inspected it. Flew off again.

Shortly afterward the workers' delegation came back. The delegates reported to the crowd that they had seen the Central Committee members and told them about this "bloody Saturday," and that *Kozlov had wept* when he heard about the children falling from the trees at the first volley. (You know Frol Kozlov, the Leningrad Party gang boss, the cruelest of Stalinists? He wept! . . .) The Central Committee members had promised to investigate these events and severely punish those responsible, but for the present everyone must go home to prevent the outbreak of fresh disorders in the town.

The meeting, however, did not disperse! The crowd grew ever denser toward the evening. The desperate courage of Novocherkassk!

Around nine in the evening they tried to drive the people away from the palace with tanks. But as soon as the drivers switched on their engines people clustered around the tanks, blocking the hatches and the observation slits. The tanks stalled. The riflemen stood by and made no effort to help the tank crews.

An hour later tanks and armored personnel carriers appeared from the opposite side of the square, with an escort of Tommy-gunners perched on top of them.

At last, toward midnight, the riflemen began firing tracer bullets into the air and the crowd slowly dispersed.

(What power there is in a popular disturbance! How quickly it changes the whole political situation! The night before there had been a curfew, and people had been frightened anyway, but now the whole town was strolling about and hooting at the soldiers. A people transformed—can it be so near to breaking through the crust of this half-century, into a completely different atmosphere?)

On June 3 the town radio broadcast speeches by Mikoyan and Kozlov. Kozlov did not weep. Nor did they any longer promise to find the culprits. What they now said was that *these events were the result of enemy provocation,* and that *these enemies would be severely punished.* Mikoyan said further that *dumdum bullets had never been adopted as part of the equipment of Soviet troops, and that they must therefore have been used by enemies of the state.*

(But who were these enemies? How had they parachuted into the country? Where were they hiding? Show us just one! We are so used to being treated like fools: "Enemies," they say, and all is explained. In the Middle Ages it was "devils.")

The shops were immediately the richer for butter, sausage, and many other things not seen in those parts for a long time, or anywhere outside the capitals.

The wounded all vanished without trace; not one of them went home. Instead, the *families* of the wounded and the killed (who of course wanted to know what had become of their kin) *were deported to Siberia.* So were many of those involved in the demonstration who had been noticed or photographed. Some participants were dealt with in a series of trials in camera. There were also two "public" trials (with entry by ticket for factory Party officials and for the town apparatchiki). At one of these, nine men were sentenced to be shot and two women to fifteen years' imprisonment.

On the Saturday following "bloody Saturday," the town radio announced that the "workers of the Electric Locomotive Works have solemnly undertaken to fulfill their seven-year plan ahead of time."

At Alexandrovo in 1961, a year before Novocherkassk, the police beat a man to death while he was under arrest and then would not allow his body to be carried past their "precinct" to the cemetery. The crowd was furious and burned down the police station. Arrests followed immediately. (There was a similar incident about the same time in Murom.) What would the appropriate charge now be? Under Stalin, even a tailor who stuck a needle in a newspaper could get Article 58. Now a more sensible view was taken: wrecking a police station should not be regarded as a political act. It was ordinary banditry. *Instructions were handed down* to this effect: "mass disorders" should not be treated as political offenses. (If they are not political, what is?)

So all at once—there were no more *politicals.*

But one stream has never dried up in the U.S.S.R., and still flows. A stream of criminals untouched by the "beneficent wave summoned to life . . ." etc. A stream which flowed uninterruptedly through all those decades—whether "Leninist norms were infringed" or strictly observed—and flowed in Khrushchev's day more furiously than ever.

I mean the believers. Those who resisted the new wave of cruel persecution, the wholesale closing of churches. Monks who were slung out of their monasteries. Stubborn sectarians, especially those who refused to perform military service: there's nothing we can do about it, we're really very sorry, but you're directly aiding imperialism; we let you off lightly nowadays—it's five years first time around.

These are in no sense politicals, they are "religionists," but still they have to be *re-educated.* Believers must be dismissed from their jobs merely for their faith; Komsomols must be sent along to break the windows of believers; believers must be officially compelled to attend antireligious lectures, church doors must be cut down with blow-torches, domes pulled down with hawsers attached to tractors, gatherings of old women broken up with fire hoses. (Is this what you mean by *dialogue,* French comrades?)

As the monks of the Pochayev Monastery were told in the Soviet of Workers' Deputies: *"If we always observe Soviet laws, we shall have to wait a long time for Communism."*

Only in extreme cases, when *educational* methods do not help, is recourse to the *law* necessary.

Here we can dazzle the world with the diamond-pure nobility of our laws today. We no longer try people in closed courts, as under Stalin, we no longer try them in absentia, we try them semi-publicly (that is to say, in the presence of a semi-public).

I hold in my hand a record of the trial of some Baptists at Nikitovka in the Donbas, in January, 1964.

This is how it's done. On the pretense that their identity must be checked, the Baptists who arrived to attend the trial were held in jail for three days (until the trial was over, and to give them a fright). Someone (a free citizen!) who threw flowers to the defendants got ten days. So did a Baptist who kept a record of the trial, and his notes were taken away (but another record survived). A bunch of hand-picked Komsomols were let in before the general public by a side door, so that they could occupy the front rows. While the trial was in progress there were shouts from the spectators: "Pour kerosene over the lot and set fire to them!" The court did nothing to curb this righteous indignation. Typical of its procedures: it admitted the evidence of hostile neighbors and also that of terrorized minors; little girls of nine and eleven were brought before the court (who the hell cares what effect it has on them as long as we get our verdict). Their exercise books with texts from the Scriptures were introduced as exhibits.

One of the defendants, Bazbei, father of *nine* children, was a miner who had never received any support from the Union committee at his pit because he was a Baptist. But they managed to confuse his daughter Nina, a schoolgirl in the eighth grade, and to suborn her with fifty rubles from the Union committee and a promise to place her in an institute later on, so that during the investigation she made fantastic statements against her father: he had tried to poison her with a sour fruit drink; when the believers were hiding in the woods for their prayer meetings (because they were persecuted in the settlement) they had had a radio transmitter—"a tall tree with wire wound all around it." Afterward these lying statements began to prey on Nina's mind, she became mentally ill and was put in the violent ward of an asylum. Nonetheless, she was produced in court in the expectation that she would stick to her evidence. But she repudiated every word of it! "The interrogator dictated what I had to say himself." It made no difference. The shameless judge ignored her latest statements and regarded only her earlier evidence as valid. (Whenever depositions favorable to the prosecution come unstuck, this is the typical and regular dodge used by the courts: they ignore what is brought out in court and base themselves on faked evidence obtained in the preliminary investigation: "Now, what do you mean by that? It says here in your deposition . . . You testified during the investigation . . . What right have you to retract now? That's an offense, too, you know!")

The judge is not at all interested in the substance of the case, in the truth. The Baptists are persecuted because they do not accept preachers sent by an atheist plenipotentiary of the state, but prefer their own. (Under Baptist rules, any brother can preach the Gospel.) There is a directive from the Oblast Party Committee: put them on trial and forcibly take their children from them. And this will be carried out, although with its left hand the Presidium of the Supreme Soviet has just (July 2, 1962) signed the world convention on "the fight against discrimination in the sphere of education." One of its points is that "parents must be allowed to provide for the religious and moral education of their children in accordance with their own convictions." But that is precisely what we cannot allow! Anyone who speaks in court on the substance of the case, anyone who tries to clarify the issue, is invariably interrupted, diverted from his train of thought, deliberately confused by the judge, who conducts the debate on this level: "How can you talk about the end of the world when we are committed to the building of Communism?"

This is from the closing statement made by one young girl, Zhenya Khloponina. "Instead of going to the cinema or to dances, I used to read the Bible and say my prayers—and just for that you are taking my freedom from me. Yes, to be free is a great happiness, but to be free from sin is a greater still. Lenin said that only in Turkey and Russia did such shameful phenomena as religious persecution still exist. I've never been in Turkey and know nothing about it, but how things are in Russia you can see for yourselves." She was cut short.

The sentences: Two of them got five years in the camps, two of them four years, and Bazbei, father of all those children, got three. The defendants accepted their sentences *joyfully,* and said a prayer. The "representatives from enterprises" shouted: "Not long enough! Make it more!" (Throw kerosene over them and put a match to it. . . .)

The long-suffering Baptists took note and kept count: and set up a "Council of Prisoners' Relatives," which began issuing manuscript bulletins about all the persecutions. From these bulletins we learn that from 1961 to June, 1964, 197 Baptists were condemned, 15 of them women. (They are all listed by name. Prisoners' dependents, now left without means of support, have also been counted: 442, of whom 341 are under school age.) The majority get five years of exile, but some get five years in a *strict* regime camp (narrowly escaping the hardened criminals' motley!), with three to five years of exile in addition. B. M. Zdorovets from Olshany in Kharkov oblast got seven years of strict regime for his faith. A seventy-six-year-old, Y. V. Arend, was put inside, as were the whole Lozovoy family (father, mother, and son). Yevgeny M. Sirokhin, a (Group 1) disabled veteran of the Fatherland War, *blind in both eyes,* was condemned in the village of Sokolovo, Zmievski district, Kharkov oblast, to three years in a camp for bringing up his children Lyuba, Nadya, and Raya as Christians, and they were taken away from him by order of the court.

The court trying the Baptist M. I. Brodovsky (at Nikolayev, October 6, 1966) was not too squeamish to use crudely faked documents; when the defendant protested—"This is dishonest of you!"—they barked back at him: "The *law* will crush you, smash you, destroy you!"

The law, my friend. Not one of your acts of "extrajudicial vengeance," as practiced in the years when "norms were still observed."

We recently got to know S. Karavansky's soul-chilling "Petition," which was transmitted from a camp to the outside world. The author had been sentenced to twenty-five years, had served sixteen of them (1944–1960), had been released (evidently under the "two-thirds" rule), had married, had begun a university course—but no! In 1965 they came for him again. Get yourself ready! You still have nine years to go.

Where else is this possible, under what other code of law on earth except ours?

There are quite a few people like this. People who were not affected by the epidemic of releases under Khrushchev, the teammates, cellmates, transit prison acquaintances whom we left behind. We have long ago forgotten them in our new lives, but they still shuffle hopelessly, drearily, numbly about the same little patches of trampled earth, with the same watchtowers and barbed-wire fences all around them. The faces in the papers change, the speeches from platforms change, people fight against the *cult* and then stop fighting—but the twenty-five-year prisoners, Stalin's godchildren, are still inside. . . .

Karavansky cites the blood-freezing prison careers of several such people.

All you freedom-loving "left-wing" thinkers in the West! You left laborites! You progressive American, German, and French students! As far as you are concerned, none of this amounts to much. As far as you are concerned, this whole book of mine is a waste of effort. You may suddenly understand it all someday—but only when you *yourselves* hear "hands behind your backs there!" and step ashore on our Archipelago.

■

Still, there really is no comparison between the numbers of political prisoners now and in Stalin's time; they are no longer counted in millions or in hundreds of thousands.

Is this because the *law* has been reformed?

No, it is just that the ship has changed course (for a time).

We called this chapter "The Law Today." It should rightly be called *"There Is No Law."*

The same treacherous secrecy, the same fog of injustice, still hangs in our air, worse than the smoke of city chimneys.

For half a century and more the enormous state has towered over us, girded with hoops of steel. The hoops are still there. There is no law.

Afterword

Instead of my writing this book alone, the chapters should have been shared among people with special knowledge, and we should then have met in editorial conference and helped each other to put the whole in true perspective.

But the time for this was not yet. Those whom I asked to take on particular chapters would not do so, but instead offered stories, written or oral, for me to use as I pleased. I suggested to Varlam Shalamov that we write the whole book together, but he also declined.

What was really needed was a well-staffed office. To advertise in the newspapers and on the radio ("Please reply!"), to carry on open correspondence, to do what was done with the story of the Brest fortress.

Not only could I not spread myself like this; I had to conceal the project itself, my letters, my materials, to disperse them, to do everything in deepest secrecy. I even had to camouflage the time I spent working on the book with what looked like work on other things.

As soon as I began the book, I thought of abandoning it. I could not make up my mind: should I or should I not be writing such a book by myself? And would I have the stamina for it? But when, in addition to what I had collected, prisoners' letters converged on me from all over the country, I realized that since all this had been given to me, I had a duty.

I must explain that *never once* did this whole book, in all its parts, lie on the same desk at the same time! In September, 1965, when work on the Archipelago was at its most intensive, I suffered a setback: my archive was raided and a novel impounded. At this point the parts of the Archipelago already written, and the materials for the other parts, were scattered, and never reassembled: I could not take the risk, especially when all the names were given correctly. I kept jotting down reminders to myself to check this and remove that, and traveled from place to place with these bits of paper. The jerkiness of the book, its imperfections, are the true mark of our persecuted literature. Take the book for what it is.

I have stopped work on the book not because I regard it as finished, but because I cannot spend any more of my life on it.

Besides begging for indulgence, I want to cry aloud: When the time and the opportunity come, gather together, all you friends who have survived and know the story well, write your own commentaries to go with my book, correct and add to it where necessary (but do not make it too unwieldy, do not duplicate what is there already). Only then will the book be definitive. God bless the work!

I am surprised to have finished it safely, even in this form. I have several times thought they would not let me.

I am finishing it in the year of a double anniversary (and the two anniversaries are connected): it is fifty years since the revolution which created Gulag, and a hundred since the invention of barbed wire (1867).

This second anniversary will no doubt pass unnoticed.

Ryazan—Ukryvishche
April 27, 1958–February 22, 1967

P.P.S.

I was in a hurry when I wrote what you have just read, because I expected that even if I did not perish in the explosion set off by my letter to the Writers' Congress I should lose my freedom to write and access to my manuscripts. But as things turned out, I was not only not arrested as a result of the letter, but found myself on a granite footing. I realized then that I must and could complete and correct this book.

A few friends have now read it. They have helped me to see the serious defects in it. I did not try it out on a wider circle, and if this ever becomes possible, it will be too late for me.

In this last year I have done what I could to improve it. Let no one blame me for its incompleteness; there is no end to the additions which could be made, and every single person who has had the slightest contact with the subject or thought seriously about it will always be able to add something—often something precious. But there are laws of proportion. In size my book has reached the utmost limit. Push in a few more little grains and the whole cliff will come tumbling down.

For sometimes expressing myself badly, for repetition in places and loose construction in others, I ask forgiveness. I was not granted a quiet year after all, and during the last few months the ground has been burning under my feet again, and the desk under my hand. Even while preparing this last version I have *never once* seen the whole book together, never once had it all on my desk at one time.

The full list of those without whom this book could not have been written, revised, or kept safe cannot yet be entrusted to paper. They know who they are. They have my homage.

Rozhdestvo-na-Iste
May, 1968

About the author

About the book

Read on

Insights,
Interviews
& More . . .

Meet Aleksandr I. Solzhenitsyn

© The Nobel Foundation

ALEKSANDR I. SOLZHENITSYN was born in Kislovodsk, Russia, on December 11, 1918. He earned a degree in mathematics and physics from Rostov University and studied literature through a correspondence course from the Moscow Institute of History, Philosophy, and Literature. A captain in the Soviet Army during World War II, he was arrested in 1945 for criticizing Stalin and the Soviet government in private letters. He was sentenced to eight years of incarceration, to be followed by "perpetual" internal exile, but was cleared of all charges in 1957 as part of Nikita Khrushchev's campaign of de-Stalinization. Solzhenitsyn vaulted from unknown schoolteacher to internationally famous writer in 1962

❝ A captain in the Soviet Army during World War II, he was arrested in 1945 for privately criticizing Stalin and the Soviet government in private letters. ❞

with the publication of his novella *One Day in the Life of Ivan Denisovich*, which Khrushchev himself authorized. The writer's increasingly vocal opposition to the regime resulted in another arrest, a charge of treason, and expulsion from the USSR in 1974. For eighteen years of his exile, he and his family lived in Vermont. In 1994 he returned to Russia, thus fulfilling his longstanding prediction. He died at his home in Moscow on August 3, 2008. Solzhenitsyn's major works include the novels *In the First Circle* and *Cancer Ward*, the memoirs *The Oak and the Calf* and *Invisible Allies*, a cycle of historical novels with the series title *The Red Wheel*, and the monumental history of the Soviet prison system *The Gulag Archipelago*, which *Time* magazine named the "Best Nonfiction Work of the Twentieth Century." In 1970 Solzhenitsyn received the Nobel Prize in Literature.

☙

Written in Secret
The Nobel Lecture

While the Nobel Lecture traditionally is delivered at the annual award ceremony in Stockholm, Solzhenitsyn could not risk the trip in December 1970, for fear he would be barred re-entry into the Soviet Union. Instead, he stayed in Moscow and worked on the lecture in secret. The finished manuscript was photographed and transferred onto a film negative; Solzhenitsyn then arranged for it to be smuggled to Sweden via a network of supporters. Transported across the Soviet border inside a portable radio, Solzhenitsyn's lecture finally was presented to the Swedish Academy in 1972. In this lecture Solzhenitsyn introduced to the world the term "Gulag Archipelago." He was already at work on the monumental history of the same name.

1

Just as that puzzled savage who has picked up—a strange cast-up from the ocean?—something unearthed from the sands?—or an obscure object fallen down from the sky?—intricate in curves, it gleams first dully and then with a bright thrust of light. Just as he turns it this way and that, turns it over, trying to discover what to do with it, trying to discover some mundane function within his own grasp, never dreaming of its higher function. So also we, holding

Art in our hands, confidently consider ourselves to be its masters; boldly we direct it, we renew, reform and manifest it; we sell it for money, use it to please those in power; turn to it at one moment for amusement—right down to popular songs and night-clubs, and at another—grabbing the nearest weapon, cork or cudgel—for the passing needs of politics and for narrow-minded social ends. But art is not defiled by our efforts, neither does it thereby depart from its true nature, but on each occasion and in each application it gives to us a part of its secret inner light. But shall we ever grasp the whole of that light? Who will dare to say that he has *defined* Art, enumerated all its facets? Perhaps once upon a time someone understood and told us, but we could not remain satisfied with that for long; we listened, and neglected, and threw it out there and then, hurrying as always to exchange even the very best—if only for something new! And when we are told again the old truth, we shall not even remember that we once possessed it. One artist sees himself as the creator of an independent spiritual world; he hoists onto his shoulders the task of creating this world, of peopling it and of bearing the all-embracing responsibility for it; but he crumples beneath it, for a mortal genius is not capable of bearing such a burden. Just as man in general, having declared himself the center of existence, has not succeeded in creating a balanced spiritual system. And if misfortune overtakes him, he casts the blame upon the age-long disharmony of the world, upon the complexity of today's ruptured soul, or upon the stupidity of the public. Another artist, recognizing a higher power above, gladly works as a humble apprentice beneath God's heaven; then, however, his responsibility for everything that is written or drawn, for the souls which perceive his work, is more exacting than ever. But, in return, it is not he who has created this world, not he who directs it, there is no doubt as to its foundations; the artist has merely to be more keenly aware than others of the harmony of the world, of the beauty and ugliness of the human contribution to it, and to communicate this acutely to his fellow-men. And in misfortune, and even at the depths of existence—in destitution, in prison, in sickness—his sense of ▶

stable harmony never deserts him. But all the irrationality of art, its dazzling turns, its unpredictable discoveries, its shattering influence on human beings—they are too full of magic to be exhausted by this artist's vision of the world, by his artistic conception or by the work of his unworthy fingers. Archeologists have not discovered stages of human existence so early that they were without art. Right back in the early morning twilights of mankind we received it from Hands which we were too slow to discern. And we were too slow to ask: *for what purpose* have we been given this gift? What are we to do with it? And they were mistaken, and will always be mistaken, who prophesy that art will disintegrate, that it will outlive its forms and die. It is we who shall die—art will remain. And shall we comprehend, even on the day of our destruction, all its facets and all its possibilities? Not everything assumes a name. Some things lead beyond words. Art inflames even a frozen, darkened soul to a high spiritual experience. Through art we are sometimes visited—dimly, briefly—by revelations such as cannot be produced by rational thinking. Like that little looking-glass from the fairy-tales: look into it and you will see—not yourself—but for one second, the Inaccessible, whither no man can ride, no man fly. And only the soul gives a groan . . .

2

One day Dostoevsky threw out the enigmatic remark: "Beauty will save the world." What sort of a statement is that? For a long time I considered it mere words. How could that be possible? When in bloodthirsty history did beauty ever save anyone from anything? Ennobled, uplifted, yes—but whom has it saved? There is, however, a certain peculiarity in the essence of beauty, a peculiarity in the status of art: namely, the convincingness of a true work of art is completely irrefutable and it forces even an opposing heart to surrender. It is possible to compose an outwardly smooth and elegant political speech, a headstrong article, a social program, or a philosophical system on the basis of both a mistake and a lie. What is hidden, what distorted, will not immediately become obvious. Then a contradictory speech, article, program, a differently constructed philosophy rallies in opposition—and all just as elegant and smooth, and once again it works. Which is why such things are both trusted and mistrusted. In vain to reiterate what does not reach the heart. But a work of art bears within itself its own verification: conceptions which are devised or stretched do not stand being portrayed in images, they all come crashing down, appear sickly and pale, convince no one. But those works of art which have scooped up the truth and presented it to us as a living force—they take hold of us, compel us, and nobody ever, not even in ages to come, will appear to refute them. So perhaps that ancient trinity of Truth, Goodness and Beauty is not simply an empty, faded formula as we thought in the days of our self-confident, materialistic youth? If the tops of these three trees converge, as the scholars maintained, but the too blatant, too direct stems of Truth and Goodness are crushed, cut down, not allowed through—then perhaps the fantastic, unpredictable, unexpected stems of Beauty will push through and soar *to that very same place*, and in so doing will fulfill the work of all three? In that case Dostoevsky's remark, "Beauty will save the world," was not a careless phrase but a prophecy? After all *he* was granted to see much, a man of fantastic illumination. And in that case art, literature might really be able to help the world today? It is the small insight which, over the years, I have succeeded in gaining into this matter that I shall attempt to lay before you here today.

3

In order to mount this platform from which the Nobel lecture is read, a distant platform offered only once in a lifetime, I have climbed not three or four makeshift steps, but hundreds and even thousands of them; unyielding, precipitous, frozen steps, leading out of the darkness and cold where it was my fate to survive, while others—perhaps with a greater gift and stronger than I—have ▶

perished. Of them, I myself met but a few on the Archipelago of GULAG, shattered into its fractionary multitude of islands; and beneath the millstone of shadowing and mistrust I did not talk to them all, of some I only heard, of others still I only guessed. Those who fell into that abyss already bearing a literary name are at least known, but how many were never recognized, never once mentioned in public? And virtually no one managed to return. A whole national literature remained there, cast into oblivion not only without a grave, but without even underclothes, naked, with a number tagged on to its toe. Russian literature did not cease for a moment, but from the outside it appeared a wasteland! Where a peaceful forest could have grown, there remained, after all the felling, two or three trees overlooked by chance. And as I stand here today, accompanied by the shadows of the fallen, with bowed head allowing others who were worthy before to pass ahead of me to this place, as I stand here, how am I to divine and to express what *they* would have wished to say? This obligation has long weighed upon us, and we have understood it. In the words of Vladimir Solov'ev:

> Even in chains we ourselves must complete
> That circle which the gods have mapped out for us.

Frequently, in painful camp seethings, in a column of prisoners, when chains of lanterns pierced the gloom of the evening frosts, there would well up inside us the words that we should like to cry out to the whole world, if the whole world could hear one of us. Then it seemed so clear: what our successful ambassador would say, and how the world would immediately respond with its comment. Our horizon embraced quite distinctly both physical things and spiritual movements, and it saw no lop-sidedness in the indivisible world. These ideas did not come from books, neither were they imported for the sake of coherence. They were formed in conversations with people now dead, in prison cells and by forest fires, they were tested against *that* life, they grew out of *that* existence. When at last the outer pressure grew a little weaker, my and our horizon broadened and gradually, albeit through a minute chink, we saw and knew "the whole world." And to our amazement the whole world was not at all as we had expected, as we had hoped—that is to say a world living "not by that," a world leading "not there"; a world which could exclaim at the sight of a muddy swamp, "what a delightful little puddle!" and at concrete neck stocks, "what an exquisite necklace!"; but instead a world where some weep disconsolate tears and others dance to a lighthearted musical. How could this happen? Why the yawning gap? Were we insensitive? Was the world insensitive? Or is it due to language differences? Why is it that people are not able to hear each other's every distinct utterance? Words cease to sound and run away like water—without taste, color, smell. Without trace. As I have come to understand this, so through the years has changed and changed again the structure, content and tone of my potential speech. The speech I give today.

And it has little in common with its original plan, conceived on frosty camp evenings.

4

From time immemorial man has been made in such a way that his vision of the world, so long as it has not been instilled under hypnosis, his motivations and scale of values, his actions and intentions are determined by his personal and group experience of life. As the Russian saying goes, "Do not believe your brother, believe your own crooked eye." And that is the most sound basis for an understanding of the world around us and of human conduct in it. And during the long epochs when our world lay spread out in mystery and wilderness, before it became encroached by common lines of communication, before it was transformed into a single, convulsively pulsating lump—men, relying on experience, ruled without mishap within their limited areas, within their communities, within their societies, and finally on their national territories. At that time it was possible for individual human beings to perceive and accept a general ▶

scale of values, to distinguish between what is considered normal, what incredible; what is cruel and what lies beyond the boundaries of wickedness; what is honesty, what deceit. And although the scattered peoples led extremely different lives and their social values were often strikingly at odds, just as their systems of weights and measures did not agree, still these discrepancies surprised only occasional travelers, were reported in journals under the name of wonders, and bore no danger to mankind which was not yet one. But now during the past few decades, imperceptibly, suddenly, mankind has become one—hopefully one and dangerously one—so that the concussions and inflammations of one of its parts are almost instantaneously passed on to others, sometimes lacking in any kind of necessary immunity. Mankind has become one, but not steadfastly one as communities or even nations used to be; not united through years of mutual experience, neither through possession of a single eye, affectionately called crooked, nor yet through a common native language, but, surpassing all barriers, through international broadcasting and print. An avalanche of events descends upon us—in one minute half the world hears of their splash. But the yardstick by which to measure those events and to evaluate them in accordance with the laws of unfamiliar parts of the world—this is not and cannot be conveyed via sound waves and in newspaper columns. For these yardsticks were matured and assimilated over too many years of too specific conditions in individual countries and societies; they cannot be exchanged in mid-air. In the various parts of the world men apply their own hard-earned values to events, and they judge stubbornly, confidently, only according to their own scales of values and never according to any others. And if there are not many such different scales of values in the world, there are at least several; one for evaluating events near at hand, another for events far away; aging societies possess one, young societies another; unsuccessful people one, successful people another. The divergent scales of values scream in discordance, they dazzle and daze us, and in order that it might not be painful we steer clear of all other values, as though from insanity, as though from illusion, and we confidently judge the whole world according to our own home values. Which is why we take for the greater, more painful and less bearable disaster not that which is in fact greater, more painful and less bearable, but that which lies closest to us. Everything which is further away, which does not threaten this very day to invade our threshold—with all its groans, its stifled cries, its destroyed lives, even if it involves millions of victims—this we consider on the whole to be perfectly bearable and of tolerable proportions. In one part of the world, not so long ago, under persecutions not inferior to those of the ancient Romans, hundreds of thousands of silent Christians gave up their lives for their belief in God. In the other hemisphere a certain madman, (and no doubt he is not alone), speeds across the ocean to *deliver* us from religion—with a thrust of steel into the high priest! He has calculated for each and every one of us according to his personal scale of values! That which from a distance, according to one scale of values, appears as enviable and flourishing freedom, at close quarters, and according to other values, is felt to be infuriating constraint calling for buses to be overthrown. That which in one part of the world might represent a dream of incredible prosperity, in another has the exasperating effect of wild exploitation demanding immediate strike. There are different scales of values for natural catastrophes: a flood craving two hundred thousand lives seems less significant than our local accident. There are different scales of values for personal insults: sometimes even an ironic smile or a dismissive gesture is humiliating, while for others cruel beatings are forgiven as an unfortunate joke. There are different scales of values for punishment and wickedness: according to one, a month's arrest, banishment to the country, or an isolation-cell where one is fed on white rolls and milk, shatters the imagination and fills the newspaper columns with rage. While according to another, prison sentences of twenty-five years, isolation-cells where the walls are covered with ice and the prisoners stripped to their ▶

underclothes, lunatic asylums for the sane, and countless
unreasonable people who for some reason will keep running
away, shot on the frontiers—all this is common and accepted.
While the mind is especially at peace concerning that exotic
part of the world about which we know virtually nothing, from
which we do not even receive news of events, but only the trivial,
out-of-date guesses of a few correspondents. Yet we cannot
reproach human vision for this duality, for this dumbfounded
incomprehension of another man's distant grief, man is just
made that way. But for the whole of mankind, compressed into
a single lump, such mutual incomprehension presents the threat
of imminent and violent destruction. One world, one mankind
cannot exist in the face of six, four or even two scales of values:
we shall be torn apart by this disparity of rhythm, this disparity
of vibrations. A man with two hearts is not for this world, neither
shall we be able to live side by side on one Earth.

5

But who will co-ordinate these value scales, and how? Who will
create for mankind one system of interpretation, valid for good
and evil deeds, for the unbearable and the bearable, as they are
differentiated today? Who will make clear to mankind what is
really heavy and intolerable and what only grazes the skin
locally? Who will direct the anger to that which is most
terrible and not to that which is nearer? Who might succeed
in transferring such an understanding beyond the limits of his
own human experience? Who might succeed in impressing upon
a bigoted, stubborn human creature the distant joy and grief of
others, an understanding of dimensions and deceptions which he
himself has never experienced? Propaganda, constraint, scientific
proof—all are useless. But fortunately there does exist such a
means in our world! That means is art. That means is literature.
They can perform a miracle: they can overcome man's detrimental
peculiarity of learning only from personal experience so that the
experience of other people passes him by in vain. From man to

man, as he completes his brief spell on Earth, art transfers the whole weight of an unfamiliar, lifelong experience with all its burdens, its colors, its sap of life; it recreates in the flesh an unknown experience and allows us to possess it as our own. And even more, much more than that; both countries and whole continents repeat each other's mistakes with time lapses which can amount to centuries. Then, one would think, it would all be so obvious! But no; that which some nations have already experienced, considered and rejected, is suddenly discovered by others to be the latest word. And here again, the only substitute for an experience we ourselves have never lived through is art, literature. They possess a wonderful ability: beyond distinctions of language, custom, social structure, they can convey the life experience of one whole nation to another. To an inexperienced nation they can convey a harsh national trial lasting many decades, at best sparing an entire nation from a superfluous, or mistaken, or even disastrous course, thereby curtailing the meanderings of human history. It is this great and noble property of art that I urgently recall to you today from the Nobel tribune. And literature conveys irrefutable condensed experience in yet another invaluable direction; namely, from generation to generation. Thus it becomes the living memory of the nation. Thus it preserves and kindles within itself the flame of her spent history, in a form which is safe from deformation and slander. In this way literature, together with language, protects the soul of the nation. (In recent times it has been fashionable to talk of the leveling of nations, of the disappearance of different races in the melting-pot of contemporary civilization. I do not agree with this opinion, but its discussion remains another question. Here it is merely fitting to say that the disappearance of nations would have impoverished us no less than if all men had become alike, with one personality and one face. Nations are the wealth of mankind, its collective personalities; the very least of them wears its own special colors and bears within itself a special facet of divine intention.) But woe to that nation whose literature is ▶

disturbed by the intervention of power. Because that is not just a violation against "freedom of print," it is the closing down of the heart of the nation, a slashing to pieces of its memory. The nation ceases to be mindful of itself, it is deprived of its spiritual unity, and despite a supposedly common language, compatriots suddenly cease to understand one another. Silent generations grow old and die without ever having talked about themselves, either to each other or to their descendants. When writers such as Achmatova and Zamjatin—interred alive throughout their lives— are condemned to create in silence until they die, never hearing the echo of their written words, then that is not only their personal tragedy, but a sorrow to the whole nation, a danger to the whole nation. In some cases moreover—when as a result of such a silence the whole of history ceases to be understood in its entirety—it is a danger to the whole of mankind.

6

At various times and in various countries there have arisen heated, angry and exquisite debates as to whether art and the artist should be free to live for themselves, or whether they should be for ever mindful of their duty towards society and serve it albeit in an unprejudiced way. For me there is no dilemma, but I shall refrain from raising once again the train of arguments. One of the most brilliant addresses on this subject was actually Albert Camus' Nobel speech, and I would happily subscribe to his conclusions. Indeed, Russian literature has for several decades manifested an inclination not to become too lost in contemplation of itself, not to flutter about too frivolously. I am not ashamed to continue this tradition to the best of my ability. Russian literature has long been familiar with the notions that a writer can do much within his society, and that it is his duty to do so. Let us not violate the *right* of the artist to express exclusively his own experiences and introspections, disregarding everything that happens in the world beyond. Let us not *demand* of the artist, but—reproach, beg, urge and entice him—that we may be allowed to do. After all, only in part does he himself develop his talent; the greater part of it is blown into him at birth as a finished product, and the gift of talent imposes responsibility on his free will. Let us assume that the artist does not *owe* anybody anything: nevertheless, it is painful to see how, by retiring into his self-made worlds or the spaces of his subjective whims, he *can* surrender the real world into the hands of men who are mercenary, if not worthless, if not insane. Our Twentieth Century has proved to be crueler than preceding centuries, and the first fifty years have not erased all its horrors. Our world is rent asunder by those same old cave-age emotions of greed, envy, lack of control, mutual hostility which have picked up in passing respectable pseudonyms like class struggle, racial conflict, struggle of the masses, trade-union disputes. The primeval refusal to accept a compromise has been turned into a theoretical principle and is considered the virtue of orthodoxy. It demands millions of sacrifices in ceaseless civil wars, it drums into our souls that there is no such thing as unchanging, universal concepts of goodness and justice, that they are all fluctuating and inconstant. Therefore the rule—always do what's most profitable to your party. Any professional group no sooner sees a convenient opportunity to *break off a piece*, even if it be unearned, even if it be superfluous, than it breaks it off there and then and no matter if the whole of society comes tumbling down. As seen from the outside, the amplitude of the tossings of western society is approaching that point beyond which the system becomes metastable and must fall. Violence, less and less embarrassed by the limits imposed by centuries of lawfulness, is brazenly and victoriously striding across the whole world, unconcerned that its infertility has been demonstrated and proved many times in history. What is more, it is not simply crude power that triumphs abroad, but its exultant justification. The world is being inundated by the brazen conviction that power can do anything, justice nothing. Dostoevsky's *Devils*—apparently a provincial nightmare fantasy of the last century—are crawling across the whole world in front of our very eyes, infesting countries where they could not ▶

have been dreamed of; and by means of the hijackings, kidnappings, explosions and fires of recent years they are announcing their determination to shake and destroy civilization! And they may well succeed. The young, at an age when they have not yet any experience other than sexual, when they do not yet have years of personal suffering and personal understanding behind them, are jubilantly repeating our depraved Russian blunders of the Nineteenth Century, under the impression that they are discovering something new. They acclaim the latest wretched degradation on the part of the Chinese Red Guards as a joyous example. In shallow lack of understanding of the age-old essence of mankind, in the naive confidence of inexperienced hearts they cry: let us drive away *those* cruel, greedy oppressors, governments, and the new ones (we!), having laid aside grenades and rifles, will be just and understanding. Far from it! . . . But of those who have lived more and understand, those who could oppose these young—many do not dare oppose, they even suck up, anything not to appear "conservative." Another Russian phenomenon of the Nineteenth Century which Dostoevsky called *slavery to progressive quirks*. The spirit of Munich has by no means retreated into the past; it was not merely a brief episode. I even venture to say that the spirit of Munich prevails in the Twentieth Century. The timid civilized world has found nothing with which to oppose the onslaught of a sudden revival of barefaced barbarity, other than concessions and smiles. The spirit of Munich is a sickness of the will of successful people, it is the daily condition of those who have given themselves up to the thirst after prosperity at any price, to material well-being as the chief goal of earthly existence. Such people—and there are many in today's world—elect passivity and retreat, just so as their accustomed life might drag on a bit longer, just so as not to step over the threshold of hardship today—and tomorrow, you'll see, it will all be all right. (But it will never be all right! The price of cowardice will only be evil; we shall reap courage and victory only when we dare to make sacrifices.) And on top of this we are threatened by destruction in the fact that the physically compressed, strained world is not allowed to blend spiritually; the molecules of knowledge and sympathy are not allowed to jump over from one half to the other. This presents a rampant danger: *the suppression of information* between the parts of the planet. Contemporary science knows that suppression of information leads to entropy and total destruction. Suppression of information renders international signatures and agreements illusory; within a muffled zone it costs nothing to reinterpret any agreement, even simpler—to forget it, as though it had never really existed. (Orwell understood this supremely.) A muffled zone is, as it were, populated not by inhabitants of the Earth, but by an expeditionary corps from Mars; the people know nothing intelligent about the rest of the Earth and are prepared to go and trample it down in the holy conviction that they come as "liberators."A quarter of a century ago, in the great hopes of mankind, the United Nations was born. Alas, in an immoral world, this too grew up to be immoral. It is not a United Nations organization but a United Governments organization where all governments stand equal; those which are freely elected, those imposed forcibly, and those which have seized power with weapons. Relying on the mercenary partiality of the majority, the UN jealously guards the freedom of some nations and neglects the freedom of others. As a result of an obedient vote it declined to undertake the investigation of private appeals—the groans, screams and beseechings of humble individual *plain people*— not large enough a catch for such a great organization. The UN made no effort to make the Declaration of Human Rights, its best document in twenty-five years, into an *obligatory* condition of membership confronting the governments. Thus it betrayed those humble people into the will of the governments which they had not chosen. It would seem that the appearance of the contemporary world rests solely in the hands of the scientists; all mankind's technical steps are determined by them. It would seem that it is precisely on the international goodwill of scientists, ▶

and not of politicians, that the direction of the world should depend. All the more so since the example of the few shows how much could be achieved were they all to pull together. But no; scientists have not manifested any clear attempt to become an important, independently active force of mankind. They spend entire congresses in renouncing the sufferings of others; better to stay safely within the precincts of science. That same spirit of Munich has spread above them its enfeebling wings. What then is the place and role of the writer in this cruel, dynamic, split world on the brink of its ten destructions? After all we have nothing to do with letting off rockets, we do not even push the lowliest of hand-carts, we are quite scorned by those who respect only material power. Is it not natural for us too to step back, to lose faith in the steadfastness of goodness, in the indivisibility of truth, and to just impart to the world our bitter, detached observations: how mankind has become hopelessly corrupt, how men have degenerated, and how difficult it is for the few beautiful and refined souls to live amongst them? But we have not even recourse to this flight. Anyone who has once taken up the *word* can never again evade it; a writer is not the detached judge of his compatriots and contemporaries, he is an accomplice to all the evil committed in his native land or by his countrymen. And if the tanks of his fatherland have flooded the asphalt of a foreign capital with blood, then the brown spots have slapped against the face of the writer forever. And if one fatal night they suffocated his sleeping, trusting Friend, then the palms of the writer bear the bruises from that rope. And if his young fellow citizens breezily declare the superiority of depravity over honest work, if they give themselves over to drugs or seize hostages, then their stink mingles with the breath of the writer. Shall we have the temerity to declare that we are not responsible for the sores of the present-day world?

7

However, I am cheered by a vital awareness of *world literature* as of a single huge heart, beating out the cares and troubles of our world, albeit presented and perceived differently in each of its corners. Apart from age-old national literatures there existed, even in past ages, the conception of world literature as an anthology skirting the heights of the national literatures, and as the sum total of mutual literary influences. But there occurred a lapse in time: readers and writers became acquainted with writers of other tongues only after a time lapse, sometimes lasting centuries, so that mutual influences were also delayed and the anthology of national literary heights was revealed only in the eyes of descendants, not of contemporaries. But today, between the writers of one country and the writers and readers of another, there is a reciprocity if not instantaneous then almost so. I experience this with myself. Those of my books which, alas, have not been printed in my own country have soon found a responsive, worldwide audience, despite hurried and often bad translations. Such distinguished western writers as Heinrich Böll have undertaken critical analysis of them. All these last years, when my work and freedom have not come crashing down, when contrary to the laws of gravity they have hung suspended as though on air, as though on *nothing*—on the invisible dumb tension of a sympathetic public membrane; then it was with grateful warmth, and quite unexpectedly for myself, that I learnt of the further support of the international brotherhood of writers. On my fiftieth birthday I was astonished to receive congratulations from well-known western writers. No pressure on me came to pass by unnoticed. During my dangerous weeks of exclusion from the Writers' Union, the *wall of defense* advanced by the world's prominent writers protected me from worse persecutions; and Norwegian writers and artists hospitably prepared a roof for me, in the event of my threatened exile being put into effect. Finally even the advancement of my name for the Nobel Prize was raised not in the country where I live and write, but by François Mauriac and his colleagues. And later still entire national writers' unions have expressed their support for me. Thus I have understood and felt that world literature is no longer an abstract anthology, nor ▶

Written in Secret *(continued)*

a generalization invented by literary historians; it is rather a certain common body and a common spirit, a living heartfelt unity reflecting the growing unity of mankind. State frontiers still turn crimson, heated by electric wire and bursts of machine fire; and various ministries of internal affairs still think that literature too is an "internal affair" falling under their jurisdiction; newspaper headlines still display: "No right to interfere in our internal affairs!" Whereas there are no *internal affairs* left on our crowded Earth! And mankind's sole salvation lies in everyone making everything his business; in the people of the East being vitally concerned with what is thought in the West, the people of the West vitally concerned with what goes on in the East. And literature, as one of the most sensitive, responsive instruments possessed by the human creature, has been one of the first to adopt, to assimilate, to catch hold of this feeling of a growing unity of mankind. And so I turn with confidence to the world literature of today—to hundreds of friends whom I have never met in the flesh and whom I may never see. Friends! Let us try to help if we are worth anything at all! Who from time immemorial has constituted the uniting, not the dividing, strength in your countries, lacerated by discordant parties, movements, castes and groups? There in its essence is the position of writers: expressers of their native language—the chief binding force of the nation, of the very earth its people occupy, and at best of its national spirit. I believe that world literature has it in its power to help mankind, in these its troubled hours, to see itself as it really is, notwithstanding the indoctrinations of prejudiced people and parties. World literature has it in its power to convey condensed experience from one land to another so that we might cease to be split and dazzled, that the different scales of values might be made to agree, and one nation learn correctly and concisely the true history of another with such strength of recognition and painful awareness as it had itself experienced the same, and thus might it be spared from repeating the same cruel mistakes. And perhaps under such conditions we artists will be able to cultivate within ourselves a field of vision to embrace the *whole world*: in the center observing like any other human being that which lies nearby, at the edges we shall begin to draw in that which is happening in the rest of the world. And we shall correlate, and we shall observe world proportions.

And who, if not writers, are to pass judgment—not only on their unsuccessful governments, (in some states this is the easiest way to earn one's bread, the occupation of any man who is not lazy), but also on the people themselves, in their cowardly humiliation or self-satisfied weakness? Who is to pass judgment on the light-weight sprints of youth, and on the young pirates brandishing their knives? We shall be told: what can literature possibly do against the ruthless onslaught of open violence? But let us not forget that violence does not live alone and is not capable of living alone: it is necessarily interwoven with falsehood. Between them lies the most intimate, the deepest of natural bonds. Violence finds its only refuge in falsehood, falsehood its only support in violence. Any man who has once acclaimed violence as his *method* must inexorably choose falsehood as his *principle*. At its birth violence acts openly and even with pride. But no sooner does it become strong, firmly established, than it senses the rarefaction of the air around it and it cannot continue to exist without descending into a fog of lies, clothing them in sweet talk. It does not always, not necessarily, openly throttle the throat, more often it demands from its subjects only an oath of allegiance to falsehood, only complicity in falsehood. And the simple step of a simple courageous man is not to partake in falsehood, not to support false actions! Let *that* enter the world, let it even reign in the world—but not with my help. But writers and artists can achieve more: they can *conquer falsehood*! In the struggle with falsehood art always did win and it always does win! Openly, irrefutably for everyone! Falsehood can hold out against much in this world, but not against art. And no sooner will falsehood be dispersed than the nakedness of violence will be revealed in all its ugliness—and violence, decrepit, will fall. That is why, my ▶

friends, I believe that we are able to help the world in its white-hot hour. Not by making the excuse of possessing no weapons, and not by giving ourselves over to a frivolous life—but by going to war! Proverbs about truth are well-loved in Russian. They give steady and sometimes striking expression to the not inconsiderable harsh national experience: *one word of truth shall outweigh the whole world.* And it is here, on an imaginary fantasy, a breach of the principle of the conservation of mass and energy, that I base both my own activity and my appeal to the writers of the whole world.

Copyright © The Nobel Foundation

More from Aleksandr I. Solzhenitsyn

THE FIRST CIRCLE

The thrilling cold war masterwork, published in full for the first time

Moscow, Christmas Eve, 1949. The Soviet secret police intercept a call made to the American embassy by a Russian diplomat who promises to deliver secrets about the nascent Soviet Atomic Bomb program. On that same day, a brilliant mathematician is locked away inside a Moscow prison that houses the country's brightest minds. He and his fellow prisoners are charged with using their abilities to sleuth out the caller's identity, and they must choose whether to aid Joseph Stalin's repressive state—or refuse and accept transfer to the Siberian Gulag camps . . . and almost certain death. First written between 1955 and 1958, In the First Circle is Solzhenitsyn's fiction masterpiece. In order to pass through Soviet censors, many essential scenes—including nine full chapters—were cut or altered before it was published in a hastily translated English edition in 1968. Now with the help of the author's most trusted translator, Harry T. Willetts, here for the first time is the complete, definitive English edition of Solzhenitsyn's powerful and magnificent classic.

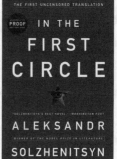

THE FIRST UNCENSORED TRANSLATION

PROOF

IN THE
**FIRST
CIRCLE**

"SOLZHENITSYN'S BEST NOVEL." —WASHINGTON POST

ALEKSANDR
WINNER OF THE NOBEL PRIZE IN LITERATURE
SOLZHENITSYN

"Solzhenitsyn's best novel."

—*Washington Post*

"A classic. . . . Future generations will read it with wonder and awe."

—*New York Times*

"So profound in its vision and its implications that it transcends both its locale and the specificities of its subject matter." —*New Republic*

More from Aleksandr I. Solzhenitsyn
(continued)

THE GULAG ARCHIPELAGO, VOLUME 1

The explosive first volume of *The Gulag Archipelago* details Solzhenitsyn's arrest and interrogation, revealing the vast bureaucracy of secret police that haunted Soviet society.

THE GULAG ARCHIPELAGO, VOLUME 2

"Volume Two is concerned with the daily life and death of the prisoners, among whom Solzhenitsyn spent eight years. . . . [P]assionate and sharply ironic. . . . Both a powerful chronicle of brutal abuses and at the same time a testament to the tensile strength of the human spirit."

—Newsweek

THE GULAG ARCHIPELAGO, VOLUME 3

"[An] enthralling record of camp uprisings, of escapes, of defiance by individuals and groups of victims. . . .
In poignant closing chapters, [Solzhenitsyn] recalls his own resurrection from the house of the dead."

—New Yorker ↝

Don't miss the next book by your favorite author. Sign up now for AuthorTracker by visiting www.AuthorTracker.com.